Artificial Intelligence in Heat Transfer

Artificial Intelligence in Heat Transfer shows how artificial intelligence (AI) tools and techniques, such as artificial neural networks, machine learning algorithms, genetic algorithms, etc., provide practical benefits specific to thermal sciences. It presents case studies involving heat and mass transfer, multi-objective optimization, conjugate heat transfer, nanofluids, thermal radiation, heat transfer through porous media (metal foam), and more.

Drawing on the collective expertise of leading researchers and experts in multiple fields, the book provides an in-depth understanding of the possibilities that emerge when these tools are applied to problems related to thermal sciences. AI is an ever-evolving discipline that has created new and groundbreaking opportunities to advance the mechanical engineering field, particularly in the area of numerical heat transfer. This volume, *Advances in Numerical Heat Transfer*, explores various ways AI is used in heat transfer to solve engineering problems.

This book will serve as an important resource for upper-level undergraduate students, researchers, engineers, and professionals, equipping them with the knowledge and inspiration to push the boundaries of the thermal sciences through AI-driven tools and techniques.

Artificial Intelligence in Heat Transfer

Advances in Numerical Heat Transfer

Volume VI

Edited by J.P. Abraham and J.M. Gorman

CRC Press
Taylor & Francis Group
Boca Raton London New York

CRC Press is an imprint of the
Taylor & Francis Group, an **informa** business

Designed cover image: John Gorman, PhD, 2023

First edition published 2025
by CRC Press
2385 NW Executive Center Drive, Suite 320, Boca Raton FL 33431

and by CRC Press
4 Park Square, Milton Park, Abingdon, Oxon, OX14 4RN

CRC Press is an imprint of Taylor & Francis Group, LLC

ISBN: 978-1-032-68810-7 (hbk)
ISBN: 978-1-032-68811-4 (pbk)
ISBN: 978-1-032-68812-1 (ebk)

DOI: 10.1201/9781032688121

Typeset in Times
by KnowledgeWorks Global Ltd.

Contents

Preface vii
Editors xi
List of Contributors xiii

Chapter 1 Physics-informed Neural Networks for Solving Partial Differential Equations 1

Prakhar Sharma, Michelle Tindall, and Perumal Nithiarasu

Chapter 2 Multi-objective Optimization of Heat Transfer Problems 21

Andrea Fragnito, Marcello Iasiello, Gerardo Maria Mauro, Wilson K. S. Chiu, and Nicola Bianco

Chapter 3 CFD/HT Simulations and DNN Modelling of Conjugate Heat Transfer in Metal Foams 63

Ubade Kemerli, Muhsin Gokhan Gunay, and Yogendra Joshi

Chapter 4 Integrating Artificial Intelligence in Nanofluid Heat Transfer: A Deep Dive into Artificial Intelligence Applications 108

Andaç Batur Çolak

Chapter 5 Developing an Artificial Neural Network Algorithm for Heat and Mass Transfer Assessment in Ternary Hybrid Nanofluid Flow 135

Shilpa B. and Naveen Kumar R.

Chapter 6 Physics-Informed Deep Learning Approaches for Industrial Heat Exchangers 155

Vishal Jadhav, Ritam Majumdar, Anirudh Deodhar, Shirish Karande, Lovekesh Vig, and Venkataramana Runkana

Chapter 7 AI-Based Analysis for Optimizing Radiative Jeffery-Hamel Flow for Cross-Diffusion Effects: A Physics-Informed Machine Learning 187

Muhammad Naeem Aslam, Nadeem Shaukat, and Arshad Riaz

Chapter 8 Machine Learning Process on Double Diffusive Convection in a Parallelogram-Shaped Cavity 222

Bengisen Pekmen and Hakan F. Oztop

Index 245

Preface

Artificial intelligence (AI) is an ever-evolving discipline that has created new and groundbreaking opportunities to advance the mechanical engineering field, particularly in the area of numerical heat transfer. This volume, *Advances in Numerical Heat Transfer*, explores various ways in which AI is used in heat transfer to solve engineering problems. Through a series of insightful chapters, this book aims to show how AI tools and techniques are currently providing practical benefits specific to thermal sciences. Drawing on the collective expertise of leading researchers and experts in multiple fields, we aim to provide an in-depth understanding of the possibilities that emerge when these tools are applied to problems related to heat and mass transfer, multi-objective optimization, conjugate heat transfer, and more.

Chapter 1 explores physics-informed neural networks (PINNs), which have gained popularity across different engineering disciplines as an alternative to conventional numerical techniques for solving partial differential equations (PDEs). PINNs are physics-based deep learning frameworks that seamlessly integrate the measurements and the PDE in a multitask loss function. In forward problems, these measurements are initial (IC) and boundary conditions (BCs), whereas in the inverse problems, they are sparse measurements such as temperature recorded by thermocouples. The scope of PDEs applicable in PINNs could include integer-order PDEs, integrodifferential equations, fractional PDEs, or even stochastic PDEs. This chapter includes a state-of-the-art overview of PINNs for solving PDEs and focuses on solutions to parametric problems, approaches to tackle stiff PDEs, and problems involving complex geometries. The advantages and disadvantages of several PINN frameworks are also discussed.

In Chapter 2, the authors focus on multi-objective optimization principles in complex thermal engineering problems. As heat transfer problems grow in sophistication, numerical tools become vital, especially when including different heat transfer mechanisms like conduction, convection, and radiation, which require customized analysis methods. A detailed system analysis opens the door to possible improvements, i.e., optimization. Choosing between gradient-based and gradient-free techniques, e.g., genetic algorithms (GAs), depends on system size and complexity. Large-scale systems like industrial heat exchangers benefit from GAs, prioritizing thermal performance optimization; they effectively navigate complex design spaces. Conversely, small-scale systems like microfluidic devices favor precision, making gradient-based techniques more suitable.

This chapter serves as a comprehensive guide for readers in defining and selecting an appropriate optimization tool based on modeling and the system/component scale. It begins by emphasizing the significance of a clear objective setting and constraint identification during the preliminary phase. Throughout the discussion, diverse examples from literature, such as heat exchanger systems, illustrate the complex nature of engineering challenges. Once classifications are provided, attention shifts to understanding multi-objective optimization principles, particularly focusing on GAs. At the core of the exploration lies the crucial decision of selecting the most suitable optimization method. Through a systematic evaluation of current optimization techniques and modern approaches, readers are equipped to make informed choices. Emphasis is placed on the benefits and bottlenecks of the approaches presented based on the desired solution. Thus, the chapter culminates in a detailed examination of several optimization examples, guiding readers toward successful resolutions of heat transfer challenges and offering insights and practical advice on methodologies, solvers, and constraints.

Chapter 3 presents a case study of machine learning/artificial intelligence (ML/AI) approaches that characterize single-phase fluid flow and heat transfer characteristics of metal foams. Metal foams were scanned using high-resolution micro-computed tomography. The scanned images were numerically investigated using OpenFOAM for combined heat conduction and convection characteristics. The overall pressure drop and heat transfer results agreed well with experimental data and the empirical correlations in the literature. Longitudinal flow mixing across pores due to the blockage of nodes has been analyzed and depicted for different porosities. The temperature distribution, local heat transfer coefficient, and heat flux on the metal foam fluid interface are characterized to reveal the underlying physical reasons for different metal foams exhibiting distinct thermal characteristics. Additionally, the variation of these parameters with height has been examined for different velocities. The results showed higher local heat transfer coefficients for thinner filaments. However, the temperature difference between the fluid and solid portions is marginal due to the lower effective thermal conductivity for higher porosity (low-density) metal foams.

The study also showed that the effective interfacial area used for heat transfer decreases with inlet velocity and porosity. The local heat transfer coefficients and local heat fluxes on the interface are analyzed in detail. The regions of strong heat transfer are reported. The integrated values from the computational fluid dynamics/heat transfer (CFD/HT) solutions were used to train a deep neural network (DNN) algorithm. Three-dimensional (3D) surface plots were generated to show the agreement of the numerical data and the predictions obtained by using the DNN algorithm. The computational time can be reduced from several hours to a few minutes by using the DNN model. This study can provide guidance in improved channel design and metal foam selection for high-performance heat exchangers.

Chapter 4 uses AI as an important tool to overcome the difficulties of traditional modeling approaches in simulating the complex structure of nanofluids. In this chapter's case study, the usability of AI techniques was investigated to understand nanofluids' thermophysical properties and flow properties, which are very important for various industrial and engineering applications. In the research, various AI methodologies, including artificial neural networks (ANNs) and machine learning algorithms, were examined in predicting basic thermophysical properties such as thermal conductivity, viscosity, and specific heat capacity, as well as heat transfer characteristics in various applications where nanofluids are used. As demonstrated in this chapter, AI-supported models appear to exhibit a higher prediction performance than traditional numerical techniques.

Chapter 5 studies the influence of thermal radiation and thermophoretic particle deposition influence on heat and mass transport of ternary hybrid nanofluid flow in a stretching/shrinking convergent channel. The modeling expressions for the considered mathematical model are addressed with the aid of Galerkin finite element analysis, and obtained results are compared with existing literature under the limiting instances. The results disclosed that an enhancement in thermal radiation term accelerates the warmth of the fluid. As the thermophoretic parameter enhances, the concentration decreases because the thermophoretic force draws additional particles to the surface when there is a large heat difference. An ANN is trained on simulated data using the Levenberg–Marquardt technique to enable it to identify and comprehend the underlying structures. The mass and heat transfer in the convergent channels that are stretching and shrinking are then computed using the trained neural network. This approach offers a workable replacement for traditional parametric research, enabling more precise predictions and comprehension of the behavior of complex systems. Ternary hybrid nanofluid flow has the highest mass and heat transfer rates in the stretched channel. Throughout all examined situations, the ANN algorithm's mean-squared error (MSE) values range from 0 to 0.03972. The heat and mass transfer properties of the ternary hybrid nanofluid are predicted through this algorithm, and also the regression plot, error histograms, performance plot, and learning curves are presented for the predicted model. The research results enhance our comprehension of the properties and behavior of ternary hybrid nanofluid flow in convergent channels, which helps with mass and heat transfer design and optimization.

Chapter 6 uses physics-informed deep learning (PIDL) as a new modeling framework to combine physics-based and data-driven models to overcome limitations and take advantage of both these modeling approaches. Recent work on various PIDL approaches like physics-informed neural networks (PINNs), domain decomposition and transfer learning of PINNs, hypernetwork-based PINNs for adapting the models easily for different boundary conditions, symbolic regression (SR), physics-informed symbolic networks (PISNs), and symbolic regression via pruning based differential program architecture (DPA) for deriving interpretable mathematical models are presented in this chapter. These modeling methodologies are demonstrated for an air preheater (APH), a critical heat exchanger in thermal power plants. A computational fluid dynamics (CFD) model for representing heat transfer in the APH is also presented. The performance of PIDL approaches is compared with numerical simulation results with the CFD model and found to be reasonably satisfactory. More importantly, the inference times of the PIDL models are quite low, indicating their suitability for real-time applications and the design of complex processes that require many simulations.

In Chapter 7, the impact of Joule heating, Soret Dufour factors, and chemical reaction effects on convergent/divergent channel flow using supervised and unsupervised physics-informed machine learning algorithms are investigated. The governing coupled system of partial differential equations (PDEs) is transformed into a nonlinear system of ordinary differential equations (ODEs) through appropriate transformations. Further, the ODEs are transformed into fitness functions based on physics-informed machine learning. The optimal value of the fitness function utilized the meta-heuristic algorithms, especially particle swarm optimization hybrid with neural network algorithms. The obtained results of the proposed algorithm are compared with the NDSolve Mathematica solver as a reference solution for algorithm validation. In addition, 200 independent runs are conducted for the efficiency of the propped approach. Further, the ANN-BLMT technique is used for the analysis of supervised learning. The statistical analysis is undertaken for 50 independent runs to check the convergence of the proposed algorithm. The obtained optimal biases and weights are seen in the range of −10 to 10 using hybrid PSO-NNA. The physics-informed machine learning algorithms are found to be much better than ANN-BLMT supervised learning.

Chapter 8 studies the average Nusselt number (Nu) and the average Sherwood number (Sh) of a double-diffusive natural convection flow of Fe3O4–water nanofluid in a parallelogram-shaped cavity by using neural networks. The data for modeling is obtained from the numerical simulation of the problem in different problem parameter combinations of Rayleigh (Ra), Lewis (Le) numbers, buoyancy ratio parameter (Nr), nanoparticle concentration (φ), and the inclination angle of the parallelogram (θ). The governing dimensionless equations are numerically solved by cubic polyharmonic spline radial basis function collocation method using Gauss–Chebyshev–Lobatto node distribution. Modeling of important problem features enables one to interpret the enhancement in convective heat and mass transfer immediately in some parameter concert instead of performing numerical calculations many times. The collaboration between CFD and ANNs shows a promising way to control heat and mass transfer characteristics quickly and efficiently.

In compiling this collection of chapters, we extend our sincere gratitude to the authors whose expertise and dedication have made this endeavor possible. Their contributions demonstrate the remarkable potential AI has to contribute to the thermal sciences, both in research and practical problem-solving.

As we look into the future, the potential of AI in the field of heat transfer seems limitless. It is our hope that this book will serve as an important resource for students, researchers, engineers, and professionals, equipping them with the knowledge and inspiration to push the boundaries of the thermal sciences through AI-driven tools and techniques.

J.P. Abraham
J.M. Gorman

Editors

J.P. Abraham is a professor of thermal sciences at the University of St. Thomas School of Engineering, Minnesota, and the current editor-in-chief of Numerical Heat Transfer. His areas of research include thermodynamics, heat transfer, fluid flow, numerical simulation, and energy. After gaining his doctorate at the University of Minnesota in 2002, he joined St. Thomas as an adjunct instructor, later becoming a full-time faculty member. He has produced more than 400 publications, books, book chapters, conference presentations, and patents in areas including biological heat transfer and fluid flow, biomedical device design, energy, burn injuries, climate change, fundamental heat transfer and fluid mechanics, and manufacturing processes.

J.M. Gorman is currently an independent consultant in the mechanical engineering field. In addition to teaching as an adjunct professor, he is also a part of several startup companies with biomedical, automotive, and novel heat recovery systems. After receiving his doctorate at the University of Minnesota in 2014, he was a research associate at the University of Minnesota until 2020. His research encompasses all facets of mechanical engineering, and his teaching is focused on modeling and numerical simulation in the thermal sciences. He has published more than 60 papers in archival journals, along with 5 book chapters. He has also been a serial editor of *Advances in Heat Transfer* and the *Advances in Numerical Heat Transfer* book series.

Contributors

Muhammad Naeem Aslam
Department of Physics and Applied Mathematics (DPAM)
Pakistan Institute of Engineering & Applied Sciences
Nilore, Islamabad
Center for Mathematical Sciences (CMS)
Pakistan Institute of Engineering & Applied Sciences
Nilore, Islamabad
Department of Mathematics
Lahore Garrison University
Lahore, Pakistan
Department of Mathematics, Division of Science and Technology
University of Education
Lahore, Pakistan

Shilpa B
Department of Mathematics
Dayananda Sagar College of Engineering
Bangalore, India

Nicola Bianco
Department of Industrial Engineering
Università degli Studi di Napoli Federico II
Napoli, Italy

Wilson K. S. Chiu
School of Mechanical, Aerospace and Manufacturing Engineering
University of Connecticut
Connecticut, USA

Andaç Batur Çolak
Information Technologies Application and Research Center
Istanbul Ticaret University
İstanbul, Türkiye

Anirudh Deodhar
Tata Research Development and Design Centre
TCS Research
Tata Consultancy Services Limited
Pune, India

Andrea Fragnito
Department of Engineering
Università degli Studi del Sannio
Benevento, Italy

Muhsin Gokhan Gunay
Technology Faculty, Mechanical Engineering Department
Isparta University of Applied Sciences
Isparta, Turkey

Marcello Iasiello
Department of Industrial Engineering
Università degli Studi di Napoli Federico II
Napoli, Italy

Vishal Jadhav
Tata Research Development and Design Centre
TCS Research
Tata Consultancy Services Limited
Pune, India

Yogendra Joshi
G.W. Woodruff School of Mechanical Engineering
Georgia Institute of Technology
Atlanta, Georgia, USA

Shirish Karande
Tata Research Development and Design Centre
TCS Research
Tata Consultancy Services Limited
Pune, India

Ubade Kemerli
Technology Faculty, Mechanical Engineering Department
Isparta University of Applied Sciences
Isparta, Turkey

Ritam Majumdar
Tata Research Development and Design Centre
TCS Research
Tata Consultancy Services Limited
Pune, India

Gerardo Maria Mauro
Department of Engineering
Università degli Studi del Sannio
Benevento, Italy

Perumal Nithiarasu
Zienkiewicz Institute for Modelling, Data and AI
Faculty of Science and Engineering
Swansea University
Swansea, United Kingdom

Hakan F. Oztop
Department of Mechanical Engineering, Technology Faculty
Fırat University
Elazig, Turkey
University Centre for Research & Development
Chandigarh University
Mohali, Punjab

Bengisen Pekmen
Department of Mathematics
TED University
Ankara, Turkey

Naveen Kumar R
Department of Mathematics
Amrita School of Engineering, Amrita Vishwa Vidyapeetham
Bengaluru, India

Arshad Riaz
Department of Mathematics
Lahore Garrison University
Lahore, Pakistan
Department of Mathematics, Division of Science and Technology
University of Education
Lahore, Pakistan

Venkataramana Runkana
Tata Research Development and Design Centre
TCS Research
Tata Consultancy Services Limited
Pune, India

Prakhar Sharma
Zienkiewicz Institute for Modelling, Data and AI
Faculty of Science and Engineering
Swansea University
Swansea, United Kingdom

Nadeem Shaukat
Department of Physics and Applied Mathematics (DPAM) and Center for Mathematical Sciences (CMS)
Pakistan Institute of Engineering & Applied Sciences
Nilore, Islamabad

Michelle Tindall
Zienkiewicz Institute for Modelling, Data and AI
Faculty of Science and Engineering
Swansea University
Swansea, United Kingdom
Culham Science Centre, United Kingdom Atomic Energy Authority
Abingdon, United Kingdom

Lovekesh Vig
Tata Research Development and Design Centre
TCS Research
Tata Consultancy Services Limited
Pune, India

1 Physics-informed Neural Networks for Solving Partial Differential Equations

Prakhar Sharma, Michelle Tindall, and Perumal Nithiarasu

1.1 INTRODUCTION

Conventional numerical techniques for solving partial differential equations (PDEs) have been a cornerstone in engineering design for years. However, their implementation poses challenges for a variety of problems. Some of the common issues encountered include: mesh dependency of solutions, computational expense in high-dimensional parametric solutions, stress concentration at sharp corners, challenges in achieving convergence and stability, and difficulties in generating adaptive meshes.

Conventional numerical techniques such as finite elements and finite volumes were primarily developed for forward problems; however, significant challenges were encountered when solving inverse problems. The ill-conditioned nature of inverse problems, the need to integrate noisy experimental data, and the existence of non-unique solutions makes these problems intractable for conventional numerical techniques. This chapter does not cover these challenges associated with inverse problems, instead it focuses on the advancements and applications of physics-informed neural networks (PINNs) in forward problems.

Although early efforts were made to solve differential equations with neural networks, these were constrained by the limitations of smaller-scale neural networks (NNs) and less efficient optimisers [1–4]. The recent advancement in existing algorithms and computing power have led to significant achievements in the field [5–7].

Recently, study discussed in reference [8] introduced the groundbreaking concept of PINNs, a method to seamlessly integrate both data and PDE within a deep learning framework. They demonstrated its capability to effectively solve both forward and inverse problems. Unlike conventional numerical techniques, PINNs are inherently meshless, effectively addressing several aforementioned issues such as mesh dependency and adaptive mesh generation. In addition, by utilising information from previously solved problems, one can accelerate the solution to newer problems through the so called transfer learning method.

The PINNs are particularly noteworthy due to their ability to solve PDEs without the need of additional ground truth, unlike standard training-based approaches. This represents

DOI: 10.1201/9781032688121-1

a significant advantage over traditional NN-based PDE solvers, which typically requires a large number of ground truth, such as simulation results, for model training.

1.2 PHYSICS-DRIVEN VS. DATA-DRIVEN APPROACHES

There exist a plethora of methodologies for tackling physics-based problems, leveraging both conventional numerical solvers and machine learning (ML) models. Before the advent of PINNs, the predominant approaches were either physics-driven or data-driven. As depicted in Figure 1.1, physics-driven techniques such as conventional numerical solvers, are on the left, whereas purely data-driven methods, including traditional supervised ML models, are on the right.

There are two distinct data-driven methods for enforcing the underlying physics: physics-guided NN (PGNN) and physics-encoded NN (PENN). PGNNs follow a classical supervised learning framework, constructing surrogate models from the data derived from experiments and simulations. These models typically necessitate extensive datasets to achieve generalisation [9, 10].

PENN-based models aim to directly encode underlying physical principles within the NN's architecture, offering an advantage when the explicit forms of the differential equations are not well-defined. Among these, two notable approaches stand out: physics-encoded recurrent convolutional NN (PERCNN) [11] and neural ordinary differential equations (NeuralODE) [12].

The PERCNN approach integrates the nonlinear systems directly into the NN by replacing the traditional activation function with a novel element-wise product operation. This modification allows the NN to directly encode the dynamics of physical systems into its computational process, enabling it to handle complex, nonlinear behaviours more effectively than traditional NNs.

In contrast, NeuralODE approach reimagines the structure of NNs to parallel the behaviour of ordinary differential equations (ODEs) that can be solved using Euler's method. By designing the NN's layers to represent discrete steps in solving an ODE, NeuralODE allows for the direct application of numerical methods within the NN. This architecture creates a bridge between numerical analysis and ML.

PINNs belong to the region between physics-driven and hybrid approach. This signifies that PINNs are capable of not only solving PDEs without needing additional ground truth, but they can seamlessly integrate noisy experimental data which is a significant advantage over conventional numerical solvers.

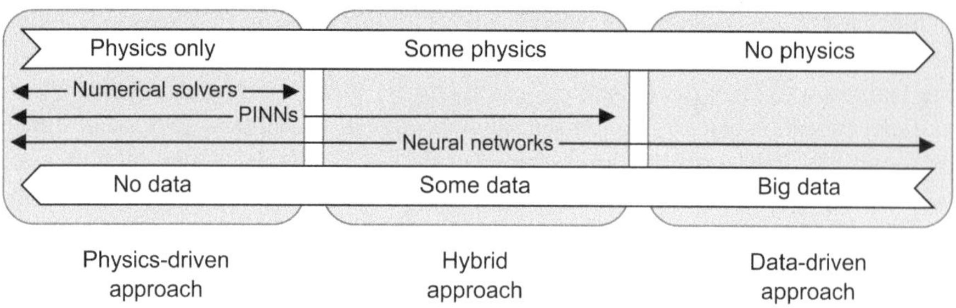

FIGURE 1.1 A broad classification of different approaches for solving physics problems.

TABLE 1.1
Conventional Numerical Techniques vs. PINNs

	Conventional Numerical Techniques	PINNs
Basis function	Piecewise polynomial	Neural network
Solution methodology	Numerical approximation and iterative methods	Optimisation problem
PDE embedding	Discretised equations	Loss function
Geometric representation	Mesh	Point cloud

Abbreviation: PDE, partial differential equation.

1.3 HOW PINNs ARE DIFFERENT FROM CONVENTIONAL NUMERICAL TECHNIQUES?

Conventional numerical techniques such as finite elements method (FEM) and finite volumes method (FVM) were primarily developed for forward problems [13–15]. In these methods, the domain is discretised into a mesh consisting of elements (in FEM) or control volumes (in FVM), with corners or boundaries defined by nodes. In FEM, each element uses a basis function, often described by a piecewise polynomial function, to interpolate the solution within the element based on the nodal values of the field variable. In contrast, FVM divides the domain into control volumes and the solution is directly approximated within these volumes.

The PINNs are NNs that model the forward problem as an optimisation problem. Instead of discretising the PDE, PINNs formulate them as a part of the loss function. Traditional mesh-based discretisation is replaced with a point cloud throughout the domain. The solution is then inferred at these points once the network has been fitted to minimise the loss function. The key distinctions between conventional numerical techniques and the PINNs are summarised in Table 1.1.

1.4 THE BASELINE PINN

Consider a well-posed PDE problem:

$$\begin{aligned} \mathbf{u}_t + \mathcal{N}_x[\mathbf{u}] = 0, \quad & \mathbf{x} \in \Omega, t \in [0,T] \\ \mathbf{u}(\mathbf{x},0) = h(\mathbf{x}), \quad & \mathbf{x} \in \partial\Omega \\ \mathbf{u}(\mathbf{x},t) = g(\mathbf{x},t), \quad & \mathbf{x} \in \partial\Omega, t \in [0,T] \end{aligned} \tag{1.1}$$

where the first equation represents the PDE with a temporal derivative u_t and a spatial derivative operator $\mathcal{N}_x[u]$. Here, $u(x,t)$ is the dependent variable, where x and t denote the independent spatial and temporal variables, respectively. The Ω and $\partial\Omega$ denote the spatial domain and the boundary of the problem. The function $g(x)$ specifies the BCs and $h(x,0)$ denotes the IC at $t = 0$.

The inputs to a PINN for a two-dimensional (2D) time-dependent problem are the spatio-temporal coordinates denoted by x, y, z, and t. Unlike the mesh generation in conventional numerical methods like FEM, where the structure of the grid can significantly influence solution, the sampling of coordinates for a PINN can be conducted in a more arbitrary manner. There are certain techniques to effectively reduce the number of random points needed while still ensuring comprehensive domain coverage as discussed in Section 1.5.1. These random points are then fed into the NN, as illustrated in Figure 1.2. The baseline PINN employs a feed-forward NN (FNN), which is composed of several fully connected layers leading to the predicted output, represented by $\hat{u}$ [16].

Similar to traditional supervised ML techniques, in PINNs, the predicted output $\hat{u}$ plays a crucial role in formulating the constraints which are represented through a loss function [17]. Unlike simpler models, PINNs require multiple loss functions to simultaneously satisfy the PDE, the BCs, and the IC, if the problem is transient. This approach results in a multitask loss function, comprising the total loss ($\mathcal{L}$) and the individual loss terms ($\mathcal{L}_{PDE}$, $\mathcal{L}_{BC}$, $\mathcal{L}_{IC}$), which are defined as follows:

$$\mathcal{L} = \lambda_{PDE}\mathcal{L}_{PDE} + \lambda_{BC}\mathcal{L}_{BC} + \lambda_{IC}\mathcal{L}_{IC} \tag{1.2}$$

$$\begin{aligned}
\mathcal{L}_{PDE} &= \frac{1}{N_r}\sum_{i=1}^{N_r}\left|\hat{u}_t(\mathbf{x}_i,t_i) + \mathcal{N}_x\left[\hat{u}(\mathbf{x}_i,t_i)\right]\right|^2 \\
\mathcal{L}_{BC} &= \frac{1}{N_b}\sum_{i=1}^{N_b}\left|\hat{u}(\mathbf{x}_i,t_i) - g(\mathbf{x}_i,t_i)\right|^2 \\
\mathcal{L}_{IC} &= \frac{1}{N_0}\sum_{i=1}^{N_0}\left|\hat{u}(\mathbf{x}_i,0) - h(\mathbf{x}_i)\right|^2
\end{aligned} \tag{1.3}$$

In these equations, N_r, N_b, and N_0 represent the number of data points sampled to satisfy the PDE, BCs, and IC, respectively, as mentioned in Eqn. 1.2. The coefficients λ_{PDE}, λ_{BC} and λ_{IC} are weighting factors in Eqn. 1.2 that help in achieving better convergence and accuracy in the model. The PDE loss $\mathcal{L}_{PDE}$ is the mean squared error (MSE) of the residual of the PDE. Similarly, the BC loss $\mathcal{L}_{BC}$ and the IC loss $\mathcal{L}_{IC}$ are MSEs of the difference between $\hat{u}(x,y,z,t)$ and the known BC and IC at their respective locations. Figure 1.2 illustrates the overall architecture of a PINN for a three-dimensional (3D) spatio-temporal problem.

In the baseline PINN framework, gradient-based optimisers are employed to minimise the total loss $\mathcal{L}$, by adjusting the weights of the NN during the training process [18]. Among these optimisers, adaptive moment estimation (Adam) and limited memory Broyden–Fletcher–Goldfarb–Shanno (L-BFGS) are frequently utilised due to their efficiency in handling large-scale optimisation problems.

Both Adam and L-BFGS bring distinct advantages to the training process of PINNs. Adam's adaptive learning mechanism can lead to faster convergence, especially in the early stages of training. Whereas, L-BFGS is often preferred in the later

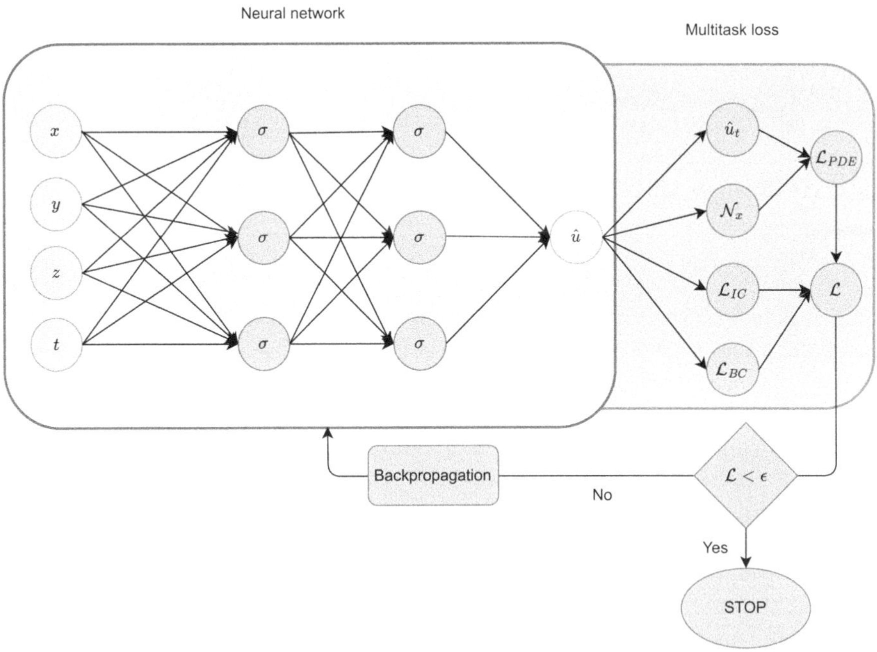

FIGURE 1.2 The architecture of a PINN with two hidden layers for a 3D spatio-temporal forward problem. Here the inputs are x, y, z, and t, σ is the activated neuron and $\hat{u}$ is the predicted output or the solution to the PDE.

stages of training when fine-tuning around minima is required, as it can provide more accurate updates by approximating second-order curvature information.

Given that the problem is well-posed, there exists a unique solution [19]. The enforcement of the loss terms $\mathcal{L}_{PDE}$, $\mathcal{L}_{BC}$, and $\mathcal{L}_{IC}$ within the PINN framework contributes to maintaining the well-posedness of the problem, thereby facilitating the convergence towards a unique solution.

1.4.1 Nature of Solutions in PINNs

The PINNs were originally developed as solvers for PDEs. In conventional applications, once a PDE is solved within the specified domain with prescribed BCs and IC, further inference on new spatio-temporal locations is typically unnecessary. However, by incorporating validation dataset, PINNs can be generalised to interpolate or extrapolate the field variables at new spatio-temporal locations. This predictive capacity aligns with the conventional ML techniques, where the availability of ground truth enables the model to learn and make accurate predictions on new spatio-temporal locations. It is important to note that this generalisation approach deviates from the traditional use-case of PDE solvers, which do not rely on ground truth data.

1.5 DEVELOPMENTS IN THE PINN FRAMEWORKS

The PINNs have rapidly evolved, with significant advancements in each of their core components such as sampling strategy, network architecture, activation function, etc. These enhancements have not only improved the accuracy and efficiency of PINNs but also have contributed to the broader field of ML. This section will briefly discuss these developments, highlighting how they address previous limitations of the PINNs.

1.5.1 Sampling

Sampling plays a crucial role in the training of PINNs, just as it does in other ML techniques. For the PINNs this involves generating a point cloud within the domain of interest which serves as the training data. Various strategies can be employed to sample these points effectively.

A common practice in PINNs is to employ low-discrepancy quasi-random sequences. These sequences are advantageous as they require fewer points than uniformly distributed random points to achieve a comparable level of domain coverage. Essentially, these sequences "spread out" the points in such a way that they are evenly distributed across all the dimensions. Figure 1.3 demonstrates the distribution of 10 points within a unit square for various sampling methods, including random, grid, Latin hypercube sampling (LHS), Sobol, Halton, and Hammersley sequence. The random sampling shows no pattern, which can lead to clustering and gaps. The grid pattern does not randomise the locations, which may not capture the local variations in the solution. In contrast, quasi-random sequences such as the Sobol, Halton, and Hammersley methods provide a more uniform distribution without clustering, which is beneficial for capturing the local variations in the solution [20, 21].

Importance sampling can be seen as substitute of adaptive mesh refinement in PINNs. Rather than using the same sampled points in each training iteration,

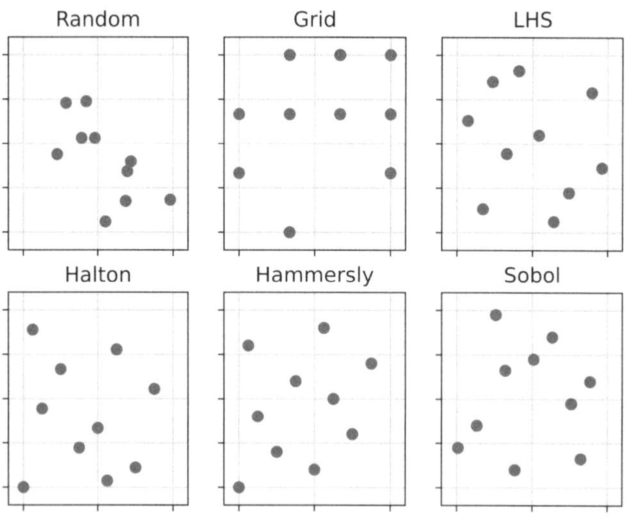

FIGURE 1.3 The comparison of various sampling methods.

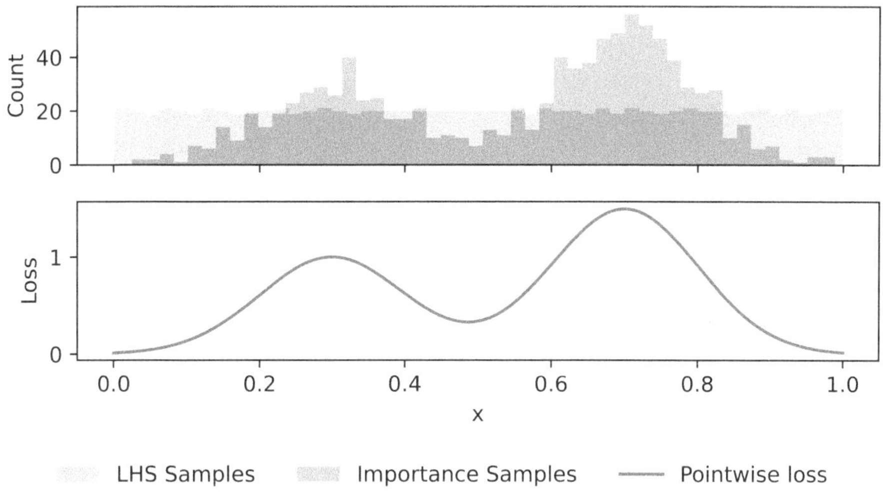

FIGURE 1.4 The comparison of initial and importance sampling strategies in a 1D feature space x. The top histogram shows initial samples obtained via LHS, and the target distribution of samples derived from importance sampling. The bottom plot shows the pointwise total loss $\mathcal{L}$ across the 1D feature space x, highlighting regions of higher loss where more points are sampled.

the points are drawn from a distribution that is proportional to the total loss, $\mathcal{L}$. Consequently, regions with higher pointwise total loss are sampled more densely, thereby focusing areas where the model needs the most improvement [22–24].

Figure 1.4 illustrates the evolution of sampling strategies in a one-dimension (1D) feature space, x, ranging from 0 to 1. Initially, 1000 sample points are distributed uniformly across the feature space using LHS, as depicted by the evenly spaced histogram in blue in the top plot. As the training progresses, samples are drawn from a distribution that aligns with the total loss $\mathcal{L}$. This distribution, represented by the histogram in orange, is concentrated around regions where the pointwise total loss $\mathcal{L}$, shown in the bottom plot, is higher. By dynamically adjusting the sampling density in accordance with the pointwise total loss $\mathcal{L}$, the PINN effectively focuses on learning complex dynamics within the feature space.

1.5.2 Imbalanced Loss Terms

The individual loss terms in a PINN can exhibit significant differences in magnitude, leading to imbalanced contributions to the total loss. For instance, the $\mathcal{L}_{PDE}$, which often includes higher-order derivatives, might be substantially lower than the $\mathcal{L}_{BC}$. This disparity can result in the PINN predominantly learning to satisfy the BCs while ignoring the PDE. Such an imbalance can yield erroneous behaviour, as the problem effectively becomes ill-posed.

A predominant approach to address this issue is the introduction of balancing coefficients for each loss term. These coefficients denoted as λ_{PDE}, λ_{BC}, and λ_{IC} are multiplied with the respective terms in the total loss function (Eqn. 1.2). By adjusting these coefficients, the relative magnitude of respective loss terms can be balanced [25].

Efforts to automatically adjust these coefficients have led to notable developments, such as self-adaptive PINNs [26] and the self-adaptive weight PINN [27], and the implementation of algorithms like learning rate annealing [28] and neural tangent kernel [29].

1.5.3 Activation Function

An activation function, denoted as σ, imparts nonlinearity to a NN enabling it to learn complex input–output relationships. Selection of a suitable activation function can affect the convergence. The choice of an appropriate activation function is crucial for convergence in PINNs, as they require smooth activation functions to compute higher-order derivatives present in PDEs. Thus, activation functions with discontinuities, such as the rectified linear unit (ReLU), exponential linear unit (ELU), or scaled exponential linear units (SELU), should generally be avoided [30].

Mathematically, a NN is a function, where the linear combination of network's weights w and previous layer's input is passed through the activation function σ which serves as the input to the next layer (Eqn. 1.4). Reference [31] proposed the concept of a global adaptive activation function (GAAF), where a trainable parameter A is also passed through the activation function (Eqn. 1.5). This parameter, acting as the slope of the activation function, allows for more sophisticated feature transformations between the hidden layers. Later, reference [32], developed the layer-wise locally adaptive activation functions (L-LAAF), incorporating a distinct trainable parameter, denoted as $A^{(2)}$, in each hidden layer (Eqn. 1.6). This layer-specific adaptability further enhances the NN's capacity to capture complex behaviours. The standard GAAF and L-LAAF structures are given as:

$$\hat{u} = w^{(3)}\sigma\left(w^{(2)}\sigma\left(w^{(1)}X\right)\right) \tag{1.4}$$

$$\hat{u} = w^{(3)}\sigma\left(Aw^{(2)}\sigma\left(Aw^{(1)}X\right)\right) \tag{1.5}$$

and $$\hat{u} = w^{(3)}\sigma\left(A^{(2)}w^{(2)}\sigma\left(A^{(1)}w^{(1)}X\right)\right). \tag{1.6}$$

1.5.4 Spectral Bias in the NN

Spectral bias is a learning bias of NNs towards low-frequency functions. This is a challenge when dealing with high-frequency functions that represent sharp variations, especially in solutions within low-dimensional domains. In Figure 1.5, we present a histogram of the magnitude spectrum derived from a Fourier transform of 2D spatial data exhibiting discontinuities. While the distribution mostly consists of low-frequency components, there are a few high-frequency components attributable to the discontinuities. These high-frequency components pose a challenge for traditional FNN architectures, potentially leading to non-convergence issues during training.

Reference [33] proposed the Fourier NN, an approach that employs input encoding to project data from low-dimensional domains into a higher-dimensional Fourier

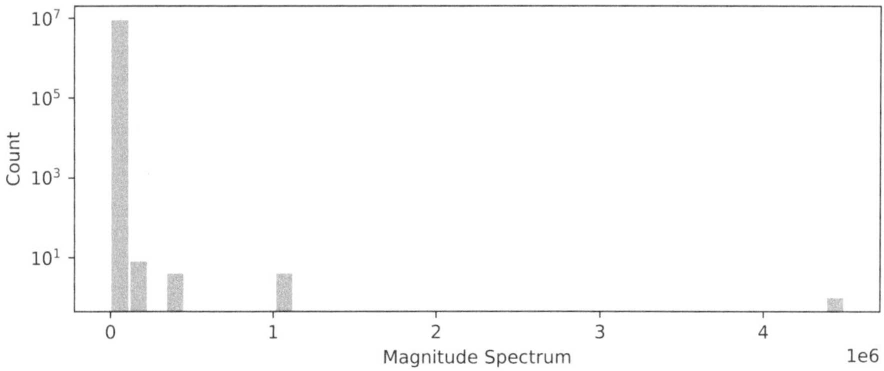

FIGURE 1.5 Histogram of the magnitude spectrum obtained from the Fourier transform of 2D spatial data, indicating a the presence of high-frequency components due to discontinuities.

space using a frequency matrix. Eqn. 1.7 shows the high-dimensional training dataset, mitigating the effects of spectral bias.

$$\begin{bmatrix} \sin(2\pi fX) \\ \cos(2\pi fX) \end{bmatrix}^T X \tag{1.7}$$

where f is trainable frequency matrix and X is the the data in low-dimensional domains. Similar input encodings have been utilised in modified Fourier network [29], sinusoidal representation networks (SiReNs) [34], and the deep Galerkin method (DGM) network [35]. A comprehensive survey by Sharma et al. [36] discusses solutions to discontinuous problems with PINNs, detailing these architectures among others.

1.6 APPLICATIONS OF PINNs TO FORWARD PROBLEMS

In this section, we will briefly discuss the application of PINNs to forward problems. We highlight scenarios where PINNs offer solutions to challenges commonly faced by conventional numerical methods, such as handling parametric problems, complex geometries, and transfer learning.

1.6.1 Parametric Problems

The PINNs can be easily extended to solve the problem over a range of parameters by including them as additional features in the training dataset. These parameters can encompass BCs, IC, coefficients of the PDE, and even the geometry of the domain. Consider a training dataset where X represents the input features $[x, y, t]$ for a 2D time-dependent problem, alongside a range of parameters k_i, where i ranges from 1 to n. To learn the parametric solutions, the PINN's training dataset is constructed by concatenating the X with each instance of k_i as shown in Figure 1.6 [37].

Recently, reference [38] proposed physics-informed deep operator network (PIDeepONets), an operator learning architecture to solve parametric problems.

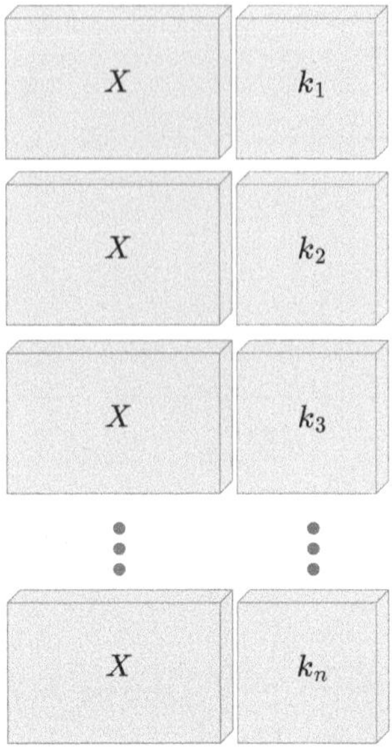

FIGURE 1.6 Schematic representation of the training dataset with varying parameter k.

Similar to PINNs, PIDeepONets only require the PDE, IC, and BCs. While a discussion on PIDeepONets is beyond the scope of this chapter, those interested can refer to a survey in reference [39].

1.6.2 Complex Geometry

Reference [40] proposed a so-called conservative PINN, a space decomposition for the PINNs. This is similar to the concept of elements in FEM, where each element has its own trail function. However, the domain can be decomposed in any arbitrary way without needing any special algorithm as opposed to FEM. Specifically, two additional loss terms were introduced to account for the mismatch in the $\mathcal{L}_{PDE}$ and $\hat{u}$ at the interface of two neighbouring sub-domains. XPINNs further advanced this by handling space-time domain decomposition (DD) for any irregular geometry [41]. The XPINNs were able to handle problems with sharp gradient over complicated geometry at the cost of longer training time. Parallel PINNs addressed this by introducing efficient parallel algorithms [42]. The work in reference [43] developed theoretical insights on the convergence and generalisation properties of PINNs, enabling accurate modelling of discontinuities, like shockwaves, with prior knowledge of their locations.

1.6.3 Transfer Learning

Transfer learning stands out as a key advantage of PINNs when compared to conventional numerical methods. It allows the utilisation of a model trained on one problem, referred to as the base task, to solve similar problem, known as target task.

The base task is generally a simpler problem, which may differ from the target task in terms of the geometry, BC, or PDE. By utilising a PINN trained on the base task, we can approach more complicated target task, leveraging the pretrained model and avoid the lengthy and computationally intensive training from scratch [44, 45].

Figure 1.7 illustrates transfer learning of a pretrained model to solve the target task with a different geometry and BCs denoted by "*". We refer the

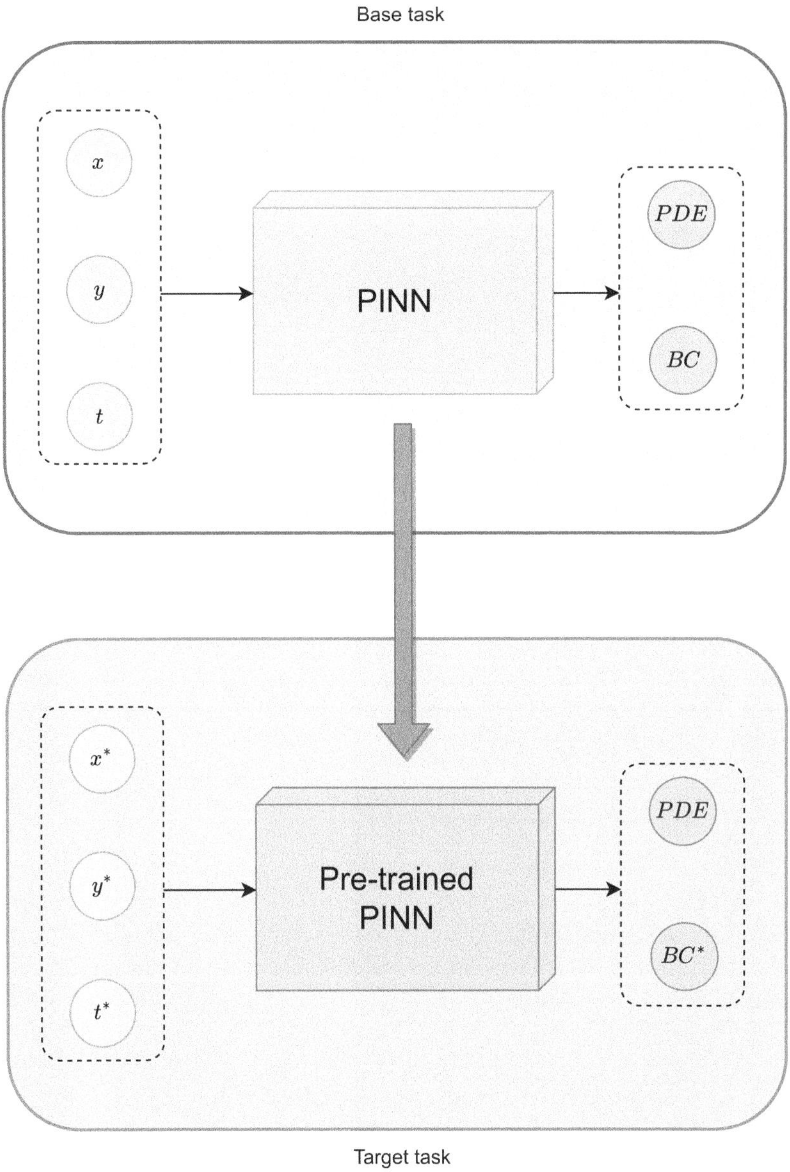

FIGURE 1.7 Transfer learning from a base task to a target task with different geometry and BC, indicated by "*".

reader to comprehensive overview of transfer learning with PINNs presented in reference [46].

1.7 EXAMPLES

We present three test cases: 1D Burgers' equation, Allen–Cahn equation, and lid driven cavity to showcase the capability of various tools that we have discussed so far. In all the test cases, we used Adam optimiser, with Xavier normal weight initialisation and hyperbolic tangent activation function. We sampled the initial set of collocation points with Sobol sequence in all the test cases.

1.7.1 1D Burgers Equation

The 1D Burgers' equation is a time-dependent problem with details given in Eqn. 1.8.

$$\begin{aligned} &u_t + uu_x - (0.01/\pi)u_{xx} = 0, \quad x \in [-1,1], \quad t \in [0,1], \\ &u(0,x) = -\sin(\pi x), \\ &u(t,-1) = u(t,1) = 0 \end{aligned} \tag{1.8}$$

The Burgers' equation is a second-order nonlinear convection-diffusion problem with an analytical solution available in reference [47]. The presence of a nonlinear convection term uu_x exhibits a discontinuity over time.

We employed an FNN architecture, i.e., a baseline PINN, and trained the model for 10k iterations. Figure 1.8 shows the PINN-predicted solution and the absolute pointwise error between the analytical solution and PINN-predicted solution.

The observed discontinuity in the solution may be attributed to the spectral bias inherent to FNNs, which we have previously discussed. Thus, the network is not able to capture the discontinuity as shown in Figure 1.8. This limitation is reflected in the computed relative L2 error of 5.17%.

1.7.2 1D Allen–Cahn Equation

The Allen–Cahn equation is used to model the process of phase separation and is characterised by a diffusion term and a nonlinear reaction term that drives the system towards minimising its free energy, often resulting in the creation of interfaces between phases over time. Consider the Allen–Cahn equation along with periodic BC (Eqn. 1.9).

$$\begin{gathered} u_t - 0.0001u_{xx} + 5u^3 - 5u = 0, x \in [-1,1], t \in [0,1], \\ u(0,x) = x^2 \cos(\pi x), \\ u(t,-1) = u(t,1), \\ u_x(t,-1) = u_x(t,1) \end{gathered} \tag{1.9}$$

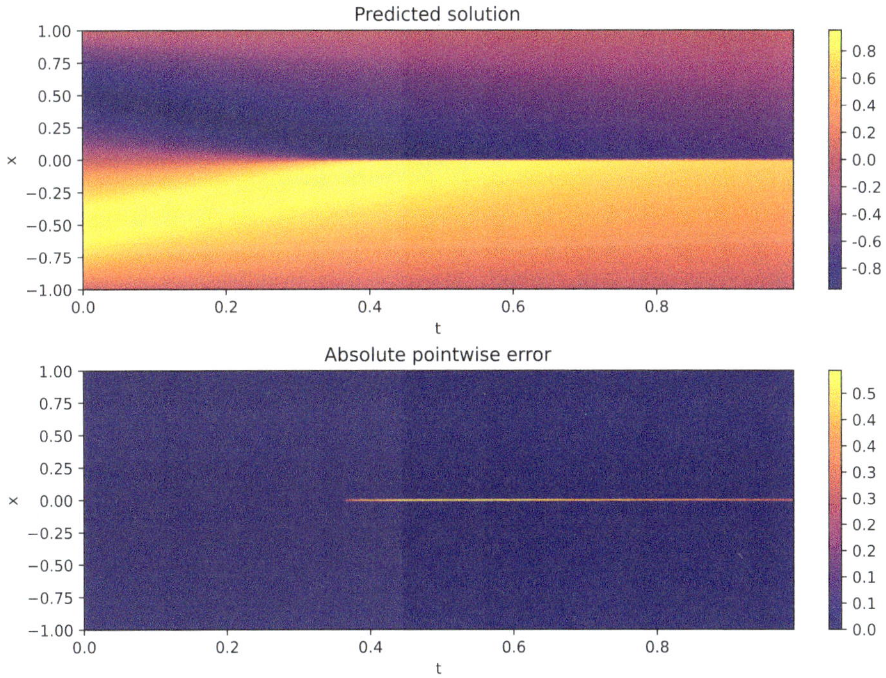

FIGURE 1.8 Baseline PINN-predicted solution of 1D Burgers' equation is shown on the top and absolute pointwise error between the analytical solution and PINN-predicted solution is shown at the bottom.

The ground truth was generated using spectral Fourier discretisation and fourth-order explicit Runge–Kutta time integrator. The solution $u(x,t)$ evolves over time due to the combined effects of diffusion and reaction. The diffusion term $-0.0001u_{xx}$ tends to smooth out variations in u, while the cubic nonlinear reaction term $u^3 - 5u$ can create multiple stable states over time.

The initial condition $u(0,x) = x^2 \cos(\pi x)$ is smooth and contains no discontinuities. However, as time progresses, the solution develops sharp transitions between the phases due to the nonlinear dynamics, which resembles a stiffness in the numerical solution.

Similar to the 1D Burgers' equation, the baseline PINN couldn't capture the regions with stiff solution, as shown in Figure 1.9, leading to a relative L2 error of 1.32%. However, with the application of Fourier NN the PINN was able to accurately capture the stiff regions in the solution, as shown in Figure 1.10, thus reducing the relative L2 error to 0.06%. This example underscores how employing a high-dimensional training dataset can effectively mitigate the spectral bias.

1.7.3 Lid-Driven Cavity

The lid-driven cavity is a well-known benchmark problem in computational fluid dynamics. It consists of a square cavity with of three rigid walls having no-slip

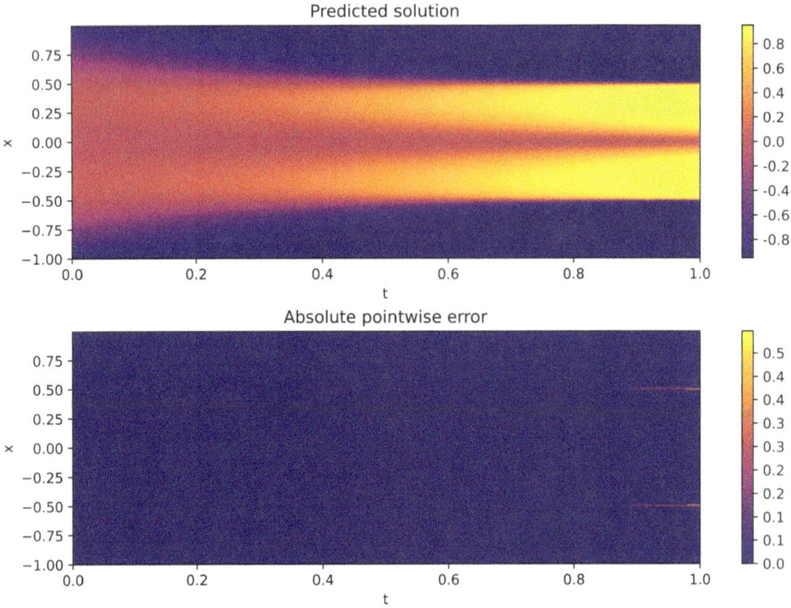

FIGURE 1.9 Baseline PINN-predicted solution of 1D Allen–Cahn equation is shown on the top and absolute pointwise error between the analytical solution and PINN predicted solution is shown at the bottom.

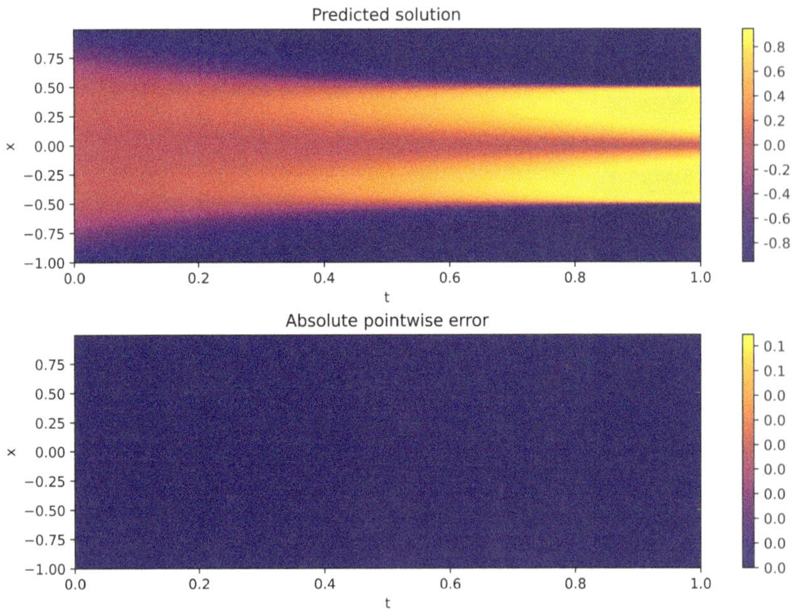

FIGURE 1.10 Fourier NN-predicted solution of 1D Allen–Cahn equation is shown on the top and absolute pointwise error between the analytical solution and PINN predicted solution is shown at the bottom side.

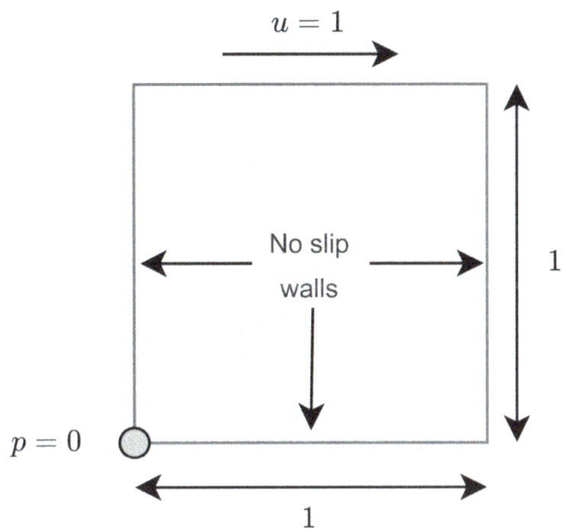

FIGURE 1.11 Geometry of the lid-driven cavity problem.

conditions and the top lid moving with a tangential unit velocity $u = 1$. The lower left corner of the domain has a reference pressure p of 0 as shown in Figure 1.11.

The lid-driven cavity uses the 2D steady-state incompressible Navier–Stokes equations to model fluid flow, as detailed in Eqn. 1.10.

$$\frac{\partial u}{\partial x} + \frac{\partial v}{\partial y} = 0$$
$$u\frac{\partial u}{\partial x} + v\frac{\partial u}{\partial y} = -\frac{\partial p}{\partial x} + \nu\left(\frac{\partial^2 u}{\partial x^2} + \frac{\partial^2 u}{\partial y^2}\right) \tag{1.10}$$
$$u\frac{\partial v}{\partial x} + v\frac{\partial v}{\partial y} = -\frac{\partial p}{\partial y} + \nu\left(\frac{\partial^2 v}{\partial x^2} + \frac{\partial^2 v}{\partial y^2}\right)$$

where u and v are velocities in x and y direction, p is the pressure, ν is the kinematic viscosity, and ρ is the density of the fluid. Focusing on a Reynold's number of 100 to simulate laminar flow conditions, we set $\rho = 1$ and $\nu = 0.01$. The numerical solution was acquired through the semi-implicit method for pressure-linked equation (SIMPLE) algorithm [48], and Figure 1.12 shows this solution.

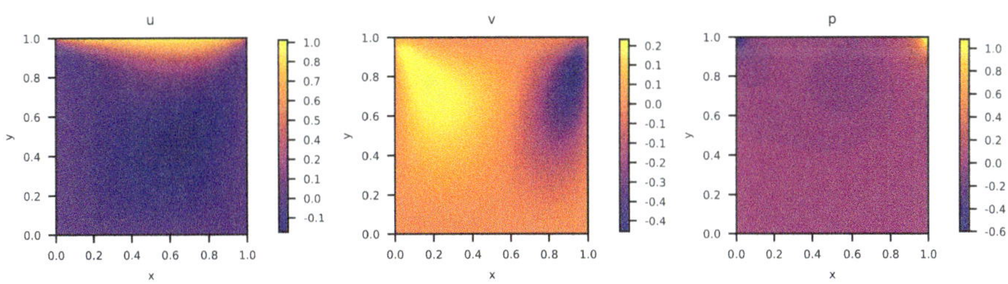

FIGURE 1.12 Numerical solution for the lid-driven cavity problem obtained using the SIMPLE algorithm, showcasing the velocity field and pressure distribution.

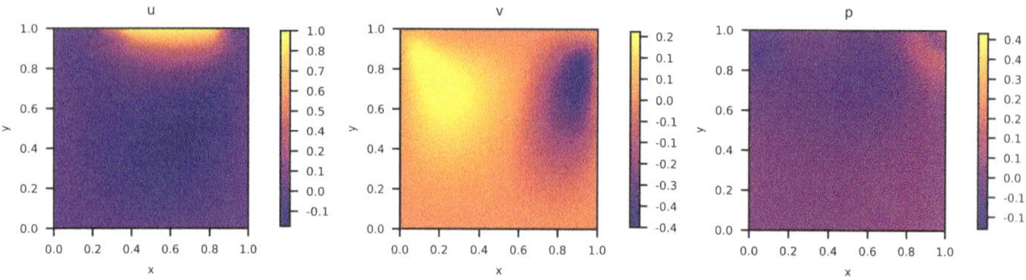

FIGURE 1.13 Baseline PINN-predicted solution of the lid-driven cavity.

The conflicting BCs on top left and top right corners result in sharp discontinuities. Similar to the Burgers' equation, the baseline PINN struggles to accurately capture these discontinuities due to spectral bias. This limitation is shown in Figure 1.13.

To mitigate the issue of spectral bias, we employed the DGM architecture [35], coupled with self-adaptive PINN's weight balancing algorithm, L-LAAF, and importance sampling. This approach resulted in a reduction of the relative L2 error by nearly 50% compared to the baseline PINN, as shown in Table 1.2. Figure 1.14 showcases the solution predicted by DGM, highlighting the improved accuracy.

Following the DGM-based model, we integrated it with the XPINN framework, dividing the domain into three equally spaced sub-domains along the x-direction. This division resulted in three fold reduction in the relative L2 error compared to the DGM-based model, as detailed in Table 1.2 and Figure 1.15. This shows the effectiveness of domain decomposition technique while solving problem involving discontinuities.

TABLE 1.2
Relative L2 Error (in %) for Lid-Driven Cavity

	u	v	p
Baseline PINN	47.7	26.3	80.4
DGM architecture	26.8	18.3	41.3
DGM with three sub-domains	8.6	9.7	12.6

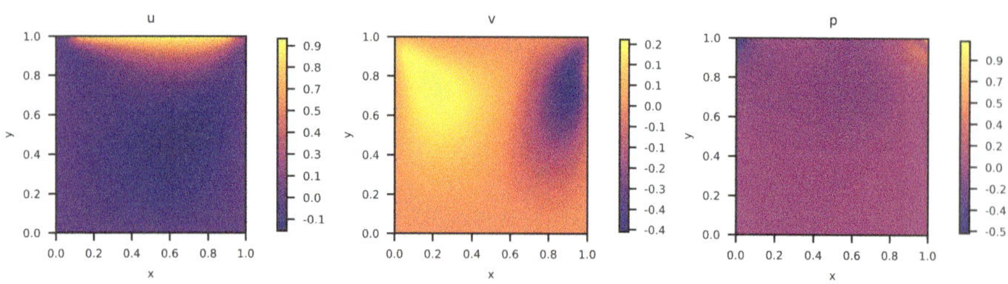

FIGURE 1.14 DGM-predicted solution of the lid-driven cavity.

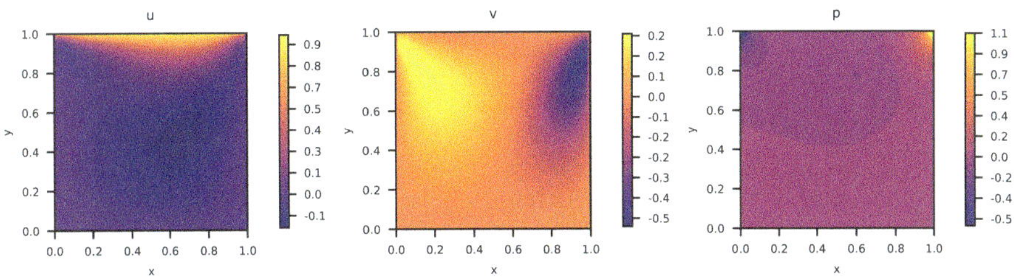

FIGURE 1.15 DGM-predicted solution of the lid-driven cavity with three sub-domains equally spaced along the x-direction.

1.8 CONCLUSIONS

PINNs have emerged as a powerful framework for solving problems involving PDEs. By seamlessly integrating physical laws into deep learning model, they allow exploring a wide range of scientific and engineering challenges. This chapter has provided a comprehensive overview of PINNs, covering the core components, advancements in sampling strategies, multitask loss function challenges, adaptive activation functions, and the issues with spectral bias, enhanced accuracy and efficiency of PINNs. Notably, PINNs offer significant advantages over traditional numerical techniques enabling:

- The application of transfer learning to leverage insights from solved problems on new and similar challenges.
- Efficiently addressing high-dimensional parametric problems.
- Simplified domain decomposition for stiff problems, avoiding complex algorithms required for mesh generation.

We solved the 1D Burgers' equation, 1D Allen–Cahn equation, and the lid-driven cavity problem to showcase the various improvements over the baseline PINN that allowed for effective handling of discontinuities. These improvements mark PINNs as a powerful tool for efficiently solving complex problems. As we enhance PINNs further, we believe their full potential is yet to be realised, promising more innovative solutions in the future.

ACKNOWLEDGEMENTS

This work is part-funded by the United Kingdom Atomic Energy Authority (UKAEA) and the Engineering and Physical Sciences Research Council (EPSRC) under the Grant Agreement Numbers EP/W006839/1, EP/T517987/1, and EP/R012091/1. We acknowledge the support of Supercomputing Wales and AccelerateAI projects, which is part-funded by the European Regional Development Fund (ERDF) via the Welsh Government for giving us access to NVIDIA A100 40GB GPUs for batch training. We also acknowledge the support of NVIDIA academic hardware grant for donating us NVIDIA RTX A5000 24GB for local testing.

REFERENCES

1. Hyuk Lee and In Seok Kang. Neural algorithm for solving differential equations. *J. Comput. Phys.*, 91(1):110–131, November 1990.
2. I.E. Lagaris, A. Likas, and D.I. Fotiadis. Artificial neural networks for solving ordinary and partial differential equations. *IEEE Trans. Neural Netw.*, 9(5):987–1000, September 1998. Conference Name: IEEE Transactions on Neural Networks.
3. I.E. Lagaris, A.C. Likas, and D.G. Papageorgiou. Neural-network methods for boundary value problems with irregular boundaries. *IEEE Trans. Neural Netw.*, 11(5):1041–1049, September 2000. Conference Name: IEEE Transactions on Neural Networks.
4. A. Malek and R. Shekari Beidokhti. Numerical solution for high order differential equations using a hybrid neural network—Optimization method. *Appl. Math. Comput.*, 183(1):260–271, December 2006.
5. Keith Rudd and Silvia Ferrari. A constrained integration (CINT) approach to solving partial differential equations using artificial neural networks. *Neurocomputing*, 155:277–285, May 2015.
6. Maziar Raissi, Paris Perdikaris, and George Em Karniadakis. Numerical Gaussian processes for time-dependent and nonlinear partial differential equations. *Siam J. Sci. Comput.*, January 2018. Publisher: Society for Industrial and Applied Mathematics.
7. Maziar Raissi, Zhicheng Wang, Michael S. Triantafyllou, and George Em Karniadakis. Deep learning of vortex-induced vibrations. *J. Fluid Mech.*, 861:119–137, February 2019. Publisher: Cambridge University Press.
8. M. Raissi, P. Perdikaris, and G. E. Karniadakis. Physics-informed neural networks: A deep learning framework for solving forward and inverse problems involving nonlinear partial differential equations. *J. Comput. Phys.*, 378:686–707, February 2019.
9. Haakon Robinson, Suraj Pawar, Adil Rasheed, and Omer San. Physics guided neural networks for modelling of non-linear dynamics. *Neural Networks*, 154:333–345, 2022.
10. Kristinn Andersen, George E. Cook, Gabor Karsai, and Kumar Ramaswamy. Artificial neural networks applied to arc welding process modeling and control. *IEEE Trans. Industry Appl.*, 26(5):824–830, 1990.
11. Chengping Rao, Hao Sun, and Yang Liu. Hard encoding of physics for learning spatiotemporal dynamics. *arXiv preprint arXiv:2105.00557*, 2021.
12. Ricky TQ Chen, Yulia Rubanova, Jesse Bettencourt, and David K Duvenaud. Neural ordinary differential equations. *Adv. Neural Inf. Process. Syst.*, 31, 2018.
13. O.C. Zienkiewicz, R.L. Taylor, and P. Nithiarasu. *The Finite Element Method for Fluid Dynamics*. Elsevier, Oxford, 7th edition, 2013.
14. P. Nithiarasu, R.W. Lewis, and K.N. Seetharamu. *Fundamentals of the Finite Element Method for Heat and Mass Transfer*. Wiley, 2015.
15. C. Hirsch. *Numerical Computation of Internal and External Flows*, volume 1. John Wiley & Sons, New York, 1992.
16. Yuntian Chen, Dou Huang, Dongxiao Zhang, Junsheng Zeng, Nanzhe Wang, Haoran Zhang, and Jinyue Yan. Theory-guided hard constraint projection (HCP): A knowledge-based data-driven scientific machine learning method. *J. Comput. Phys.*, 445:110624, November 2021.
17. Giorgio Gnecco, Marco Gori, Stefano Melacci, and Marcello Sanguineti. Learning with hard constraints. In Valeri Mladenov, Petia Koprinkova-Hristova, Günther Palm, Alessandro E. P. Villa, Bruno Appollini, and Nikola Kasabov, editors, *Artificial Neural Networks and Machine Learning – ICANN 2013*, Lecture Notes in Computer Science, pages 146–153, Berlin, Heidelberg, 2013. springer.
18. Ange Tato and Roger Nkambou. IMPROVING ADAM OPTIMIZER. page 4, 2018.
19. Persi Diaconis and Mehrdad Shahshahani. On nonlinear functions of linear combinations. *Siam J. Sci. Stat. Comput.*, 5(1):175–191, March 1984. Publisher: Society for Industrial and Applied Mathematics.

20. William J. Morokoff and Russel E. Caflisch. Quasi-Monte Carlo integration. *J. Comput. Phys.*, 122(2):218–230, December 1995.
21. C. Bard and J.C. Dorelli. Neural network reconstruction of plasma space-time. *Front. Astron. Space Sci.*, 8, 2021.
22. Christian P. Robert and George Casella. Monte Carlo integration. In Christian P. Robert and George Casella, editors, *Monte Carlo Statistical Methods, Springer Texts in Statistics*, pages 71–138. Springer, New York, 1999.
23. Luca Martino, Vctor Elvira, and Francisco Louzada. Effective sample size for importance sampling based on discrepancy measures. *Signal Process.*, 131:386–401, February 2017.
24. Mohammad Amin Nabian, Rini Jasmine Gladstone, and Hadi Meidani. Efficient training of physics-informed neural networks via importance sampling. *Comput.-Aided Civ. Infrastruct. Eng.*, 36(8):962–977, 2021. _eprint: https://onlinelibrary.wiley.com/doi/pdf/10.1111/mice.12685.
25. Colby L. Wight and Jia Zhao. Solving Allen-Cahn and Cahn-Hilliard equations using the adaptive physics informed neural networks. *Comm. Comput. Phys.*, 29(3), July 2020.
26. Levi McClenny and Ulisses Braga-Neto. Self-Adaptive Physics-Informed Neural Networks using a Soft Attention Mechanism. Technical Report 68, AAAI-MLPS, February 2019.
27. Shuyan Shi, Ding Liu, and Zhongdan Zhao. Non-Fourier Heat Conduction based on Self-Adaptive Weight Physics-Informed Neural Networks. In *2021 40th Chin. Control Conf. (CCC)*, pages 8451–8456, July 2021. ISSN: 1934-1768.
28. Sifan Wang, Yujun Teng, and Paris Perdikaris. Understanding and mitigating gradient flow pathologies in physics-informed neural networks. *SIAM J. Sci. Comput.*, 43(5):A3055–A3081, January 2021. Publisher: Society for Industrial and Applied Mathematics.
29. Sifan Wang, Xinling Yu, and Paris Perdikaris. When and why PINNs fail to train: A neural tangent kernel perspective. *J. Comput. Phys.*, 449:110768, January 2022.
30. Prakhar Sharma, Llion Evans, Michelle Tindall, and Perumal Nithiarasu. Hyperparameter selection for physics-informed neural networks (PINNS): application to discontinuous heat conduction problems. *Num. Heat Transfer, Part B: Fundamentals*, 1–15, 2023.
31. Ameya D. Jagtap, Kenji Kawaguchi, and George Em Karniadakis. Adaptive activation functions accelerate convergence in deep and physics-informed neural networks. *J. Comput. Phys.*, 404:109136, March 2020.
32. Ameya D Jagtap, Kenji Kawaguchi, and George Em Karniadakis. Locally adaptive activation functions with slope recovery for deep and physics-informed neural networks. *Proc. Royal Soc. A*, 476(2239):20200334, 2020.
33. Matthew Tancik, Pratul Srinivasan, Ben Mildenhall, Sara Fridovich-Keil, Nithin Raghavan, Utkarsh Singhal, Ravi Ramamoorthi, Jonathan Barron, and Ren Ng. Fourier features let networks learn high frequency functions in low dimensional domains. *Adv. Neural Inf. Process. Syst.*, 33: 7537–7547. Curran Assoc., Inc., 2020.
34. Vincent Sitzmann, Julien Martel, Alexander Bergman, David Lindell, and Gordon Wetzstein. Implicit neural representations with periodic activation functions. *Adv. Neural Inf. Process. Syst.*, 33:7462–7473. Curran Assoc., Inc., 2020.
35. Justin Sirignano and Konstantinos Spiliopoulos. DGM: A deep learning algorithm for solving partial differential equations. *J. Comput. Phys.*, 375:1339–1364, December 2018.
36. Prakhar Sharma, Llion Evans, Michelle Tindall, and Perumal Nithiarasu. Stiff-PDEs and physics-informed neural networks. *Arch. Comput. Methods Eng.*, February 2023.
37. Enrico Schiassi, Carl Leake, Mario De Florio, Hunter Johnston, Roberto Furfaro, and Daniele Mortari. Extreme Theory of Functional Connections: A Physics-Informed Neural Network Method for Solving Parametric Differential Equations. Technical Report arXiv:2005.10632, arXiv, May 2020. arXiv:2005.10632 [physics, stat] type: article.

38. Sifan Wang, Hanwen Wang, and Paris Perdikaris. Learning the solution operator of parametric partial differential equations with physics-informed DeepONets. *Sci. Adv.*, 7(40):eabi8605, September 2021. Publisher: American Association for the Advancement of Science.
39. Seid Koric and Diab W Abueidda. Data-driven and physics-informed deep learning operators for solution of heat conduction equation with parametric heat source. *International Journal of Heat and Mass Transfer*, 203:123809, 2023.
40. Ameya D. Jagtap, Ehsan Kharazmi, and George Em Karniadakis. Conservative physics-informed neural networks on discrete domains for conservation laws: Applications to forward and inverse problems. *Comput. Methods Appl. Mech. Eng.*, 365:113028, June 2020.
41. Ameya D. Jagtap and George Em Karniadakis. Extended physics-informed neural networks (xpinns): A generalized space-time domain decomposition based deep learning framework for nonlinear partial differential equations. *Commun. Comput. Phys.*, 28(5):2002–2041, 2020.
42. Khemraj Shukla, Ameya D Jagtap, and George Em Karniadakis. Parallel physics-informed neural networks via domain decomposition. *J. Comput. Phys.*, 447:110683, 2021.
43. Zheyuan Hu, Ameya D. Jagtap, George Em Karniadakis, and Kenji Kawaguchi. When Do Extended Physics-Informed Neural Networks (XPINNs) Improve Generalization? *arXiv:2109.09444 [cs, math, stat]*, December 2021. arXiv: 2109.09444.
44. Ehsan Haghighat, Maziar Raissi, Adrian Moure, Hector Gomez, and Ruben Juanes. A physics-informed deep learning framework for inversion and surrogate modeling in solid mechanics. *Comput. Methods Appl. Mech. Eng.*, 379:113741, June 2021.
45. Chen Xu, Ba Trung Cao, Yong Yuan, and Günther Meschke. Transfer learning based physics-informed neural networks for solving inverse problems in tunneling. Technical Report arXiv:2205.07731, arXiv, May 2022. arXiv:2205.07731 [cs] type: article.
46. Konstantinos Prantikos, Stylianos Chatzidakis, Lefteri H Tsoukalas, and Alexander Heifetz. Physics-informed neural network with transfer learning (TL-PINN) based on domain similarity measure for prediction of nuclear reactor transients. *Scientific Reports*, 13(1):16840, 2023.
47. Mayur P. Bonkile, Ashish Awasthi, C. Lakshmi, Vijitha Mukundan, and V. S. Aswin. A systematic literature review of Burgers' equation with recent advances. *Pramana – J Phys*, 90(6):69, April 2018.
48. Timothy Francis Miller. *Application of Multilevel Methods to a Semi-implicit Pressure-Linked Algorithm*. The Pennsylvania State University, 1987.

2 Multi-objective Optimization of Heat Transfer Problems

Andrea Fragnito, Marcello Iasiello, Gerardo Maria Mauro, Wilson K. S. Chiu, and Nicola Bianco

2.1 INTRODUCTION

The success of the transition from a world of uncontrolled urbanization, energy poverty, and global warming to a world of sustainability nowadays depends in part on whether the optimization of energy systems is performed or not. Luckily, in the past decades the global awareness for climate change provided a quick development of sustainable technologies to manage such problems. Nevertheless, there is still much progress to be made, representing a breeding ground for scientific progress. In this context, the optimization – seen as a process to improve a system from its initial state – is a significant tool in engineering for determining the best, or optimal, value for the decision variables of a system. For various reasons, it is important to optimize processes/systems so that a desired quantity, known as the objective function, is maximized, or minimized, e.g., the output, profit, productivity, product quality, and so on, to be maximized, or the cost per item, investment, energy input, and so on, to be minimized. Therefore, for improved performance, such engineering devices must be optimized for designs that are, e.g., more compact, lighter, with fewer frictional losses, and have increased thermal efficiency. The most crucial and difficult stage in this process is typically formulating a suitable optimization problem, for which there are numerous elements that need to be defined, i.e., system boundaries, optimization criteria, decision variables, and objective functions. Depending on the application, all these elements may notably differ.

The choice of optimization techniques is a crucial decision, with profound implications for the success of a project. Two prominent methodologies, gradient-free and gradient-based techniques [1], have emerged as key players in this domain. Each has its strengths and weaknesses, and the selection between them often hinges on the size and complexity of the system under examination and the desired level of design freedom. This chapter offers considerations surrounding the use of these optimization techniques, emphasizing the preference for genetic algorithms (GA) in large-scale systems while highlighting the advantages of gradient-based methods in small-scale systems that demand high degree of thermo-fluid-dynamic precision. Figure 2.1 gives a visual idea of how – moving from large to small scales – the component complexity increases while the system one decreases. When dealing with micro-channels (tiny fluid passages), attention to small features and geometric intricacies becomes crucial. Micro-channels are commonly used for cooling purposes in

DOI: 10.1201/9781032688121-2

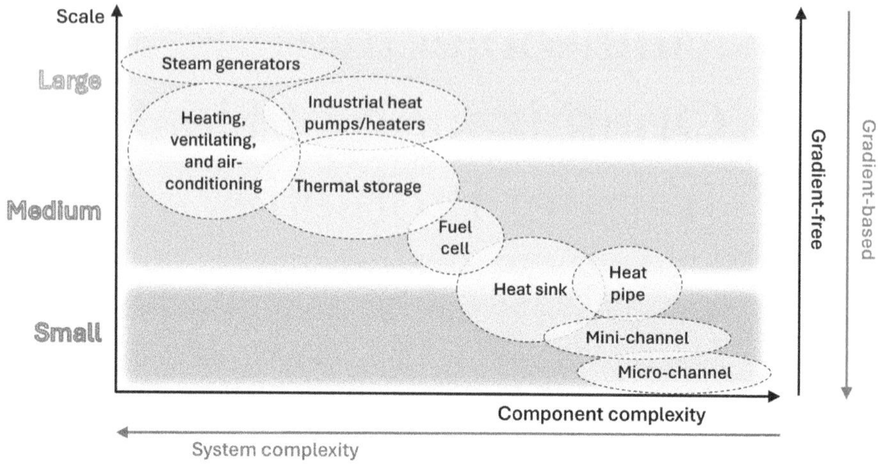

FIGURE 2.1 From small to large scale systems: a trade-off between system and component complexities.

electronics or microfluidic devices. In contrast, thermal storage systems (e.g., large-scale water tanks) do not require the same level of detail. The reason lies in the scale: micro-channels operate at a miniature level, where even minor variations significantly impact the performance. On the other hand, large-scale systems can tolerate some simplification in numerical simulations without compromising accuracy.

Large-scale heat transfer systems, e.g., industrial heat exchangers and power plant components, present typical challenges. These systems often feature complex geometries and detailed thermal interactions, making it impractical to go in depth into the details of the fluid dynamics. In such cases, treating the system as an energy system is a pragmatic approach and GAs are preferred choice. These algorithms typically find a comprehensive solution to optimization challenges, albeit with a slow convergence rate – due to the high number of decision variables – just needing the assessment of cost functions. In addition, they do not require smoothness in the objective function. GAs focus on optimizing parameters that directly impact thermal/fluid performance. They excel in large-scale systems where the computational burden associated with detailed computational fluid dynamics (CFD) simulations or complex numerical formulations is prohibitive. This approach is more scalable, and it maintains a sharp focus on energy efficiency, which is a crucial factor in large-scale heat transfer applications. GAs offers a pragmatic solution, optimizing performance without an exhaustive exploration of all possible configurations.

In contrast, small-scale heat transfer systems, such as microfluidic devices or compact electronic components, require precision and detail. Here, entering the field of fluid dynamics is imperative to take advantage of the system full of thermo-fluid-dynamic potential. This is where gradient-based techniques – particularly for shape and topology optimization – become useful.

Gradient-based optimization methods move towards best solution within the influence of how sensitive the objective and constraints are to the design variables. These sensitivities come at an extra computational expense. Specifically, the adjoint method helps calculate these sensitivities with a cost that is almost unaffected by the quantity of design parameters, achieved by solving a series of adjoint equations matching the number of objective and constraint functions. In this way, they enable a detailed control over design parameters, allowing to optimize fluid pathways and heat transfer surfaces with precision. The approach ensures that every aspect of the system is fine-tuned to achieve maximum heat transfer efficiency.

One of the key advantages of gradient-based techniques, especially if used for shape and topology optimization, is the affordability and robustness. One can explore innovative configurations and geometric patterns, optimizing not only for energy efficiency but also for innovative form factors and structural considerations. This level of design flexibility is invaluable, particularly in small-scale systems where every element must be meticulously tuned for optimal performance.

The choice between GAs and gradient-based techniques in heat transfer analysis is far from arbitrary, as it centers on the size and complexity of the system under investigation. GAs, with their energy-centric approach, are the preferred choice for large-scale systems, where the emphasis is on enhancing heat transfer efficiency without deeply getting into fluid dynamics. Ultimately, the selection of the most suitable optimization technique should be a well-considered decision that aligns with the specific goals and constraints of the heat transfer problem at hand. By carefully weighing the scale, complexity, and design requirements, researchers can control these powerful optimization tools to achieve superior heat transfer performance across a wide range of applications.

2.1.1 The Optimization Problem

The optimization process is a dynamic and iterative path that grows from the initial problem formulation to the achievement of optimality or the fulfillment of termination criteria. It comprises various essential stages, each of which contributes to refining the solutions in pursuit of *an optimal* outcome. At the outset, the optimization begins with a comprehensive problem formulation [2]. This phase entails the explicit delineation of design/decision variables and objective functions that necessitate minimization or maximization, coupled with the assessment of constraints that must be satisfied. If the optimization does not include constraints, it is called *unconstrained optimization*. Design variables are also explicitly defined with their permissible ranges during this critical stage. A general optimization statement (minimization) can be formulated as follows:

$$\begin{aligned} &\underset{x}{\text{minimize}} && f(x) \\ &\text{subject to} && c_i(x) \le 0 \quad i = 1,....,n \\ & && c_j(x) = 0 \quad j = 1,....,m \end{aligned} \tag{2.1}$$

where $f(x)$ is the objective function, and $c_i(x)$ and $c_j(x)$ are the constraints. For instance, a heat exchanger (HX) is a system through which heat goes from one fluid stream to another with or without merging. HX are used to facilitate heat transfer in a myriad industrial processes, from refrigeration to power generation. Within this system lies the heart of the optimization problem challenge: to maximize heat transfer while minimizing energy consumption and cost. Every journey begins with a map, and in optimization that map takes the form of a well-defined problem statement. The task is to schedule the complexities of heat exchanger design into a mathematical framework ready for optimization. Consider the following variables:

- *Geometry*: dimensions of the heat exchanger, including length, diameter, and surface area.
- *Fluid properties*: thermal conductivity, viscosity, and specific heat capacity of the fluids.
- *Operating conditions*: inlet temperatures, flow rates, and pressure differentials.
- *Material selection*: choice of materials for heat exchanger construction, impacting both performance and cost.
- *Constraints*: physical limitations, e.g., space availability, pressure ratings, and material compatibility.

With these variables in hand, the objective function can now be crafted. In the case of a heat exchanger, a suitable but not unique goal should be to maximize heat transfer efficiency while minimizing energy consumption and cost. Here, the heat transfer rate represents the amount of heat exchanged per unit time, while energy consumption and cost encompass the operational and capital expenses associated with the heat exchanger, respectively. These contrasting objectives clearly define the need for a multi-objective approach that aims at balancing this trade-off, offering a set of candidate solutions to be selected. The way performance indicators are linked to design variables is matter of the simulation model: it can be a physically based model, software, or a surrogate model. Figure 2.2 shows this classification with reference to a common case, i.e., the heat rate as a function of volumetric heat rate.

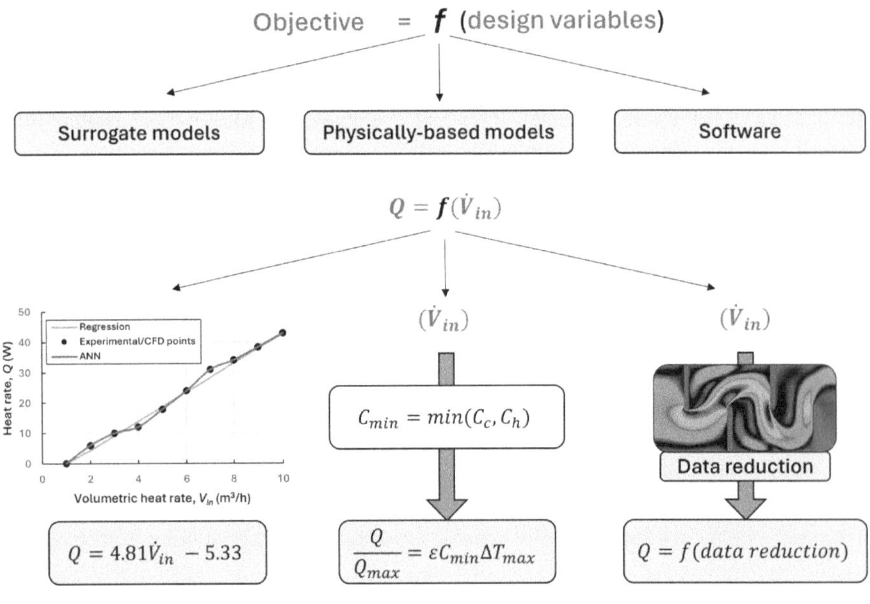

FIGURE 2.2 Simulations model: the link between variables and objectives. Example of heat rate as a function of volumetric heat rate.

Physically based modeling represents a system that uses fundamental physical principles and equations. This method offers a detailed understanding of the system's behavior and relationships between variables. Physically based models are advantageous for accurate predictions within their validity range and for gaining insights into physical phenomena. Nonetheless, they may be impractical for highly nonlinear or complex systems due to their computational demands and need for extensive experimental validation. Software model, on the other hand, often requires complex mathematical equations or simulations – as it performs the resolution of differential equations by means of numerical schemes – making it computationally intensive and time-consuming. Surrogate modeling offers a simplified approximation of the system, typically made with statistical techniques or machine learning algorithms. Surrogate models are built on the basis of input-output data from the original system and can capture complex relationships without explicitly modeling underlying physical principles. They provide a computationally efficient alternative to physically based models, allowing for rapid evaluation and faster optimization. However, surrogate models rely heavily on the quality and representativeness of training data and may lack accuracy outside the range of the training data or in regions of high uncertainty.

2.1.2 Optimization Framework

The definition of the model is crucial for the subsequent selection of optimization technique: going back to the heat exchanger, using a physically based model with a CFD approach with the aim of changing its design to increase thermal performance, a challenge emerges. While computational simulation offers exceptional capabilities for design exploration, achieving the best design necessitates the integration of automated optimization procedures. These tools can drive innovation and create advanced designs, but at what cost? From classical methods such as gradient descent [3] to modern metaheuristics such as GAs [4], a wide spectrum of choices appears as possible tools [1]. Opting for gradient-free techniques could lead to an optimal solution, but with prohibitive computation time, just as a gradient-based approach could quickly lead to a local optimum. So, which one to choose?

As previously specified, classical optimization methods, rooted in the calculus, offer a solid foundation for tackling complex problems. Gradient-based techniques such as gradient descent and Newton's method allow to solve the problem, while trying – without always succeeding – to avoid local optima. However, their reliance on smooth, convex functions may prove inadequate on the difficult ground of real-world optimization problems.

In contrast, metaheuristic algorithms offer multiple solutions, from principles of evolution, swarm behavior, and simulated annealing to explore the optimization possibilities. GAs, for instance, simulate the process of natural selection, breeding successive generations of solutions until an optimal specimen emerges. Similarly, particle swarm optimization (PSO) exploits the collective intelligence of a swarm of particles to converge on promising regions of the search space. Following the problem formulation – which is described in Figure 2.3 – the optimization loop proceeds with an initialization step. Here, a set of candidate solutions is introduced to the process. These initial solutions can be generated

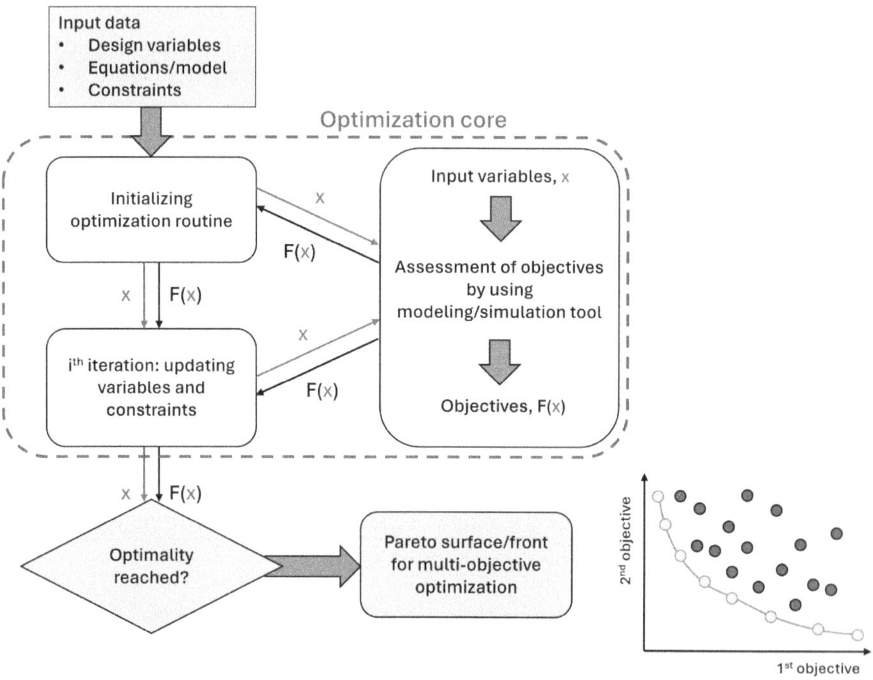

FIGURE 2.3 General scheme of a multi-objective optimization.

randomly or based on prior domain knowledge, setting the stage for subsequent iterations. In the heart of the optimization loop lies the evaluation of the objective function and the associated constraints. In each iteration, the objective function is scrutinized for each candidate solution, along with any pertinent constraint functions.

The objective function serves as the benchmark for assessing the performance of each solution in its quest to achieve the desired goal, while constraint functions determine whether the solution meets the problem's limitations [5]. Once the objective and constraint evaluations are completed, the optimality of each candidate solution is considered. At this point, the objective function values and the degree of constraint violations, if any, are considered. Two possible scenarios emerge: the feasible solution, wherein all constraints are met and the desired objective is reached; or the infeasible or suboptimal solution, which fails to fulfill either or both criteria. In the latter case, the optimization process persists in pursuit of better solutions. The optimization algorithm employs various strategies and methodologies to recalibrate candidate solutions that fall short in terms of feasibility or optimality. These optimization techniques, ranging from gradient-based methods to stochastic methods, such as GAs or simulated annealing, guide the search towards improving the objective function while adhering to the constraints. Following each iteration, a convergence check arises, aimed at determining whether the optimization algorithm is progressing towards an optimal solution. The assessment depends on factors such as changes in objective function values, constraint violation reductions, or the attainment of predefined tolerance levels. The optimization process culminates when convergence is achieved or when predefined termination criteria are met. These termination criteria encompass parameters such as reaching a maximum number of iterations, adhering to a time constraint, or securing a satisfactory solution within an established tolerance range. Once the criteria are met, the optimization loop concludes and the best-found solution is unveiled.

2.1.3 Optimization Challenges: Navigating Local minima/maxima

One of the foremost challenges encountered in the field of optimization is the presence of local minima and maxima [6]. This issue adds a layer of complexity to the optimization process and underscores the importance of a systematic and thoughtful approach to problem-solving.

A local minimum refers to a point within the search space of an optimization problem where the objective function reaches its lowest value within a small neighborhood (see Figure 2.4).

It indicates a solution that outperforms nearby solutions but may not necessarily be the best solution in the entire search space, akin to a dip or valley in the function's landscape. Gradients, which are vectors indicating the direction of steepest ascent or descent of a function at a specific point, play a crucial role in locating local optima in optimization. Gradient descent, a commonly used optimization algorithm, uses gradients to minimize the objective function iteratively. Starting with an initial guess

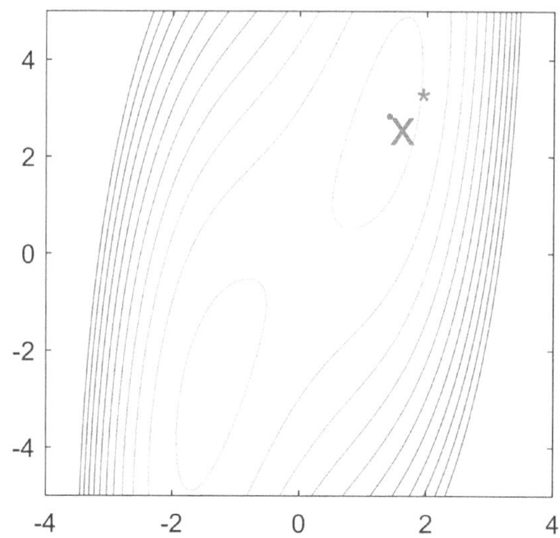

FIGURE 2.4 Example of local minimum for a 2D variables function.

for the parameters, the algorithm calculates the gradient of the objective function with respect to these parameters at the current point (see Figure 2.5). The negative gradient points in the direction of steepest decrease, enabling parameter adjustments to minimize the function. This process continues until convergence to a local minimum. While gradients are fundamental for finding local minima, a potential drawback is that if the optimization starts with an initial guess too close to a local minimum, it may get trapped without exploring the broader parameter space (see Figure 2.5). Consequently, a better global minimum or alternate local minima may remain undiscovered.

To counteract this issue and enhance the likelihood of discovering the global minimum or superior local minima, employing multiple starting points is a recommended strategy [1]. Randomly initializing the optimization algorithm with various initial guesses promotes the exploration of different regions within the parameter space. Each of these starting points may lead to distinct local minima or the global minimum. After optimizing the objective function from each starting point, ensemble methods can be employed to aggregate the results. This may involve selecting the best solution from all runs or creating an ensemble of models trained on the solutions obtained from different starting points. This approach enhances the robustness of the optimization process, diminishing the risk of converging to suboptimal local minima.

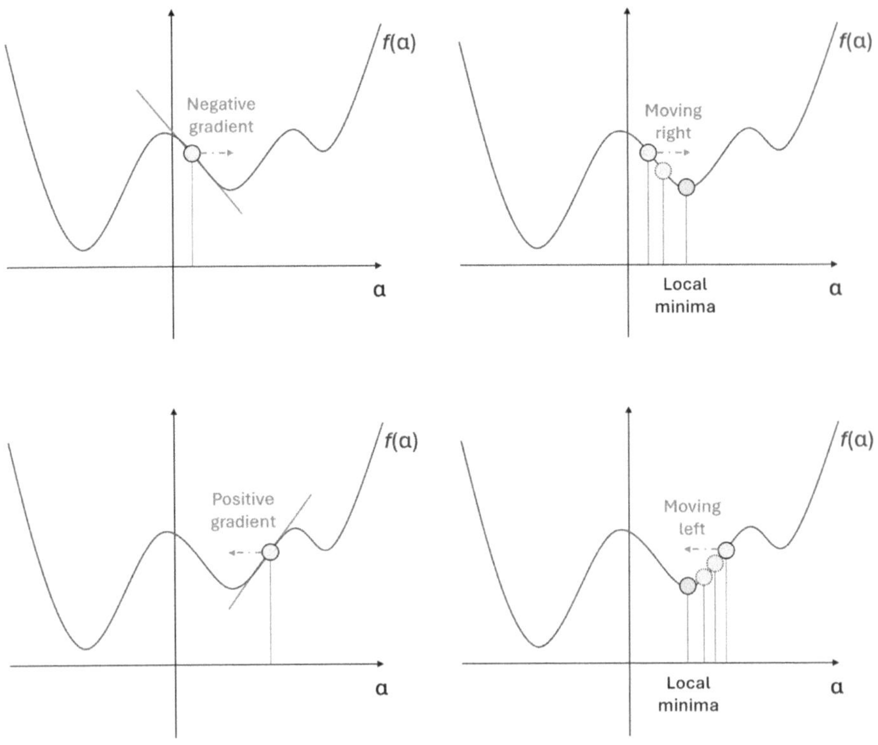

FIGURE 2.5 Gradient descendent search: local minima.

2.1.4 Gradient-free Optimization

Gradient-free optimization, also known as derivative-free optimization (DFO), is a class of optimization techniques used to find the minimum or maximum of a function without relying on its gradient information [7]. Traditional optimization methods often require knowledge of the function's gradient (a vector of partial derivatives with respect to each input variable) to guide the search for the optimal solution. However, gradient-free optimization methods do not make this assumption and are designed to work with functions that may be non-smooth, noisy, or computationally expensive to evaluate. In gradient-free optimization, the primary goal is to find the optimal input values that minimize or maximize the objective function, typically by iteratively exploring the input space and adjusting the input values. These methods rely on various strategies, such as sampling points, approximating the function locally, or using heuristics to make informed decisions about where to search for the optimum. Among the several types of DFO algorithms [7] the following are discussed here:

- *Random search*: it is a straightforward method that randomly samples input values from the search space and evaluates the objective function. Over time, it accumulates information about the function's behavior and refines the search [8].
- *Nelder–Mead algorithm:* it is also known as the downhill simplex method, and is an iterative optimization method that builds a simplex (a geometrical shape) in the input space. It then contracts, expands, or reflects the simplex to explore the space and converge towards an optimal solution [9].
- *Simulated annealing:* it is a probabilistic optimization technique inspired by the annealing process in metallurgy. It explores the input space by accepting moves that reduce the objective function value, even if they occasionally increase it. Over time, it gradually decreases the acceptance probability, allowing it to escape local optima [10].
- *Genetic algorithms:* these are optimization methods inspired by the process of natural selection. They keep a population of candidate solutions and iteratively evolve them through selection, crossover (recombination), and mutation operations. GAs can handle both continuous and discrete search spaces [11].
- *Particle swarm optimization:* it is a population-based optimization algorithm inspired by the social behavior of birds or fish. Particles in the population explore the search space, adjusting their positions based on their own experience and the experiences of their neighbors [12].
- *Bayesian optimization:* it is a probabilistic optimization approach that builds a surrogate model (typically a Gaussian process) to approximate the objective function. It iteratively selects points in the input space based on the surrogate model's uncertainty to balance exploration and exploitation [13].

These gradient-free optimization methods are valuable in scenarios where the gradient of the objective function is unknown, expensive to compute, or noisy. They

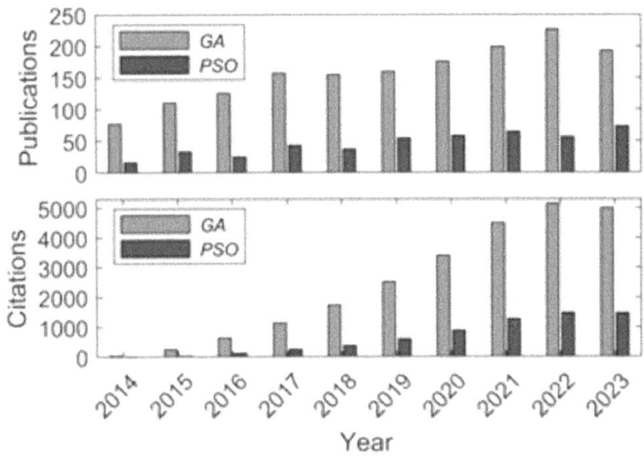

FIGURE 2.6 Comparison between genetic algorithm and particle swarm optimization. Citations (bottom panel) and publications (top panel). (Source: Web of Science™).

have applications in various fields, including machine learning hyperparameter tuning, engineering design, finance, and simulations. A resume of yearly scientific contributions on heat transfer-related optimization problem done with different approach and found via Web of Science™ is reported in Figure 2.6.

The choice of method depends on the problem's characteristics and computational resources available. However, the overall cost of gradient-free optimization is sensitive to the cost of the function evaluations because they require many iterations for convergence, and the number of iterations scales poorly with the number of design variables. In the following section, the primary focus or emphasis is placed on the GA among the various discussed.

2.1.5 Multi-Objective Optimization

Multi-objective optimization (MOO) [14] is a powerful mathematical and computational technique that addresses problems involving multiple, often conflicting, objectives. In the context of engineering and heat transfer, MOO is indispensable for finding optimal solutions that balance various competing goals. This chapter explores the concept of MOO, with a focus on its relevance to heat transfer problems. In addition, it introduces the concept of the Pareto front, which is a fundamental concept in MOO, and highlights the significance of MOO in addressing multi-physics problems in heat transfer.

In heat transfer applications, we often encounter problems where multiple objectives need to be considered simultaneously. Consider a heat sink design as an example. A generic task may want to simultaneously optimize the base temperature of the heat sink (to ensure efficient cooling) and minimize its weight (to reduce material costs and improve portability). These objectives are typically conflicting; decreasing the base temperature may require adding more material, which increases weight.

Therefore, traditional single-objective optimization approaches are inadequate, as they prioritize one goal at the expense of others. Another common scenario is the trade-off between pressure drop and thermal resistance in a heat exchanger/micro-channels design. Reducing pressure drop may require using larger flow channels, which in turn can increase thermal resistance. These competing objectives must be considered together to find an optimal design that suits the specific requirements of the application. In a general view, MOO is particularly crucial in scenarios involving multi-physics problems, which regards the interplay of multiple physical phenomena, such as fluid flow, heat conduction, and radiation, among others. By employing MOO techniques, it becomes possible to explore a vast solution space.

2.1.6 Pareto Front/Surface

Central to MOO is the concept of the Pareto front/surface. The Pareto front represents the set of optimal solutions, where no single solution is superior to another in all objectives [15]. Instead, each solution on the Pareto front represents a trade-off between the conflicting objectives. The term front is used when referring to a bi-objective analysis. Conversely, the term surface is used when three objective functions are involved (Figure 2.7). Advanced visualization techniques ensure the identification of the dominance relationship between solutions when optimizing with more than three objective functions [15]. Once a Pareto front has been defined, the solution space can be divided in two groups:

- *Dominated solutions*: In an MOO problem, a solution is said to be dominant if there exists another solution that is at least as good in all objectives and strictly better in at least one objective. Dominated solutions are not on the Pareto front and can be "virtually" eliminated. Dominated solutions are evidenced in red in Figure 2.7.
- *Non-dominated Solutions:* Solutions that are not dominated by any other solutions are considered non-dominated and lie on the Pareto front. These are the optimal solutions that provide the best trade-offs between the competing objectives that are highlighted in blue in Figure 2.7.

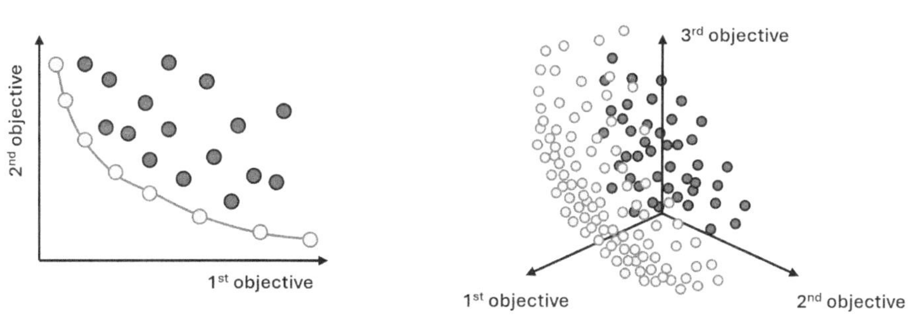

FIGURE 2.7 Pareto front for a two-objective optimization (left). Pareto surface for a three-objective optimization.

2.2 GENETIC ALGORITHMS

The classic GA operates with a collection of candidate solutions that serve as representations for potential solutions to the optimization problem at hand. Each solution represents a possible candidate for achieving an optimum in the optimization problem. The chosen representation of these solutions plays a pivotal role because it dictates the selection of genetic operators. These representations typically take the form of lists of values and are generally founded on sets of symbols. When these representations involve continuous values, they are referred to as vectors, and when composed of binary digits, they are termed bit strings. In the case of combinatorial problems, solutions often consist of symbols that are present within a list. For instance, consider the representation of a path for solving the traveling salesman problem. Genetic operators come into play by generating fresh solutions within the selected representation and facilitating exploration within the solution space. The encoding of a solution as a representation, which undergoes the evolutionary process, is commonly known as the genotype or chromosome (Figure 2.8).

2.2.1 GA Working Principles

The idea is to create the next generation (and the successive ones) in this way:

- Best individuals from the previous generation.
- Individuals resulting from the crossover of individuals with average performance.
- Individuals resulting from the mutation of individuals with mediocre performance.

The idea of the GA was initially conceived by Turing in 1950 [16], who recognized the potential of possible optimization based on bio-evolutionary algorithms. The first successful implementation of a GA was carried out by Holland in 1969 [17], and in the following years, efforts began to be made towards the integration of enhanced mechanics into the typical GA. Each individual is characterized by a specific set of parameters (finite values of variables).

For instance, if one were to optimize a heat exchanger, the relevant design variables could be the material of the tubes, the diameter of the tubes, the number of tubes,

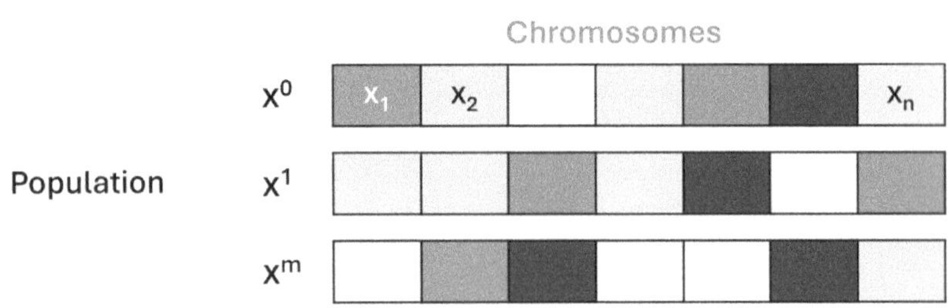

FIGURE 2.8 Individual encoding.

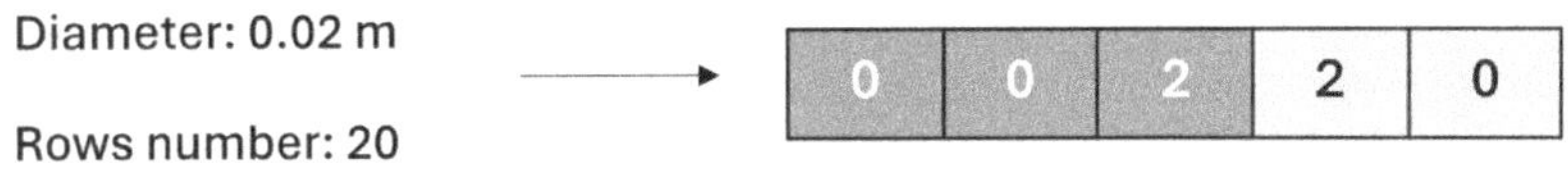

FIGURE 2.9 Example of direct encoding.

etc. Each individual needs encoding in order to be evaluated and to enable the crossover and mutation functions to occur. One possible encoding is *direct vector* encoding where the individual is identified through a vector containing the values of the variables as they are. An example of direct vector encoding is shown in Figure 2.9.

This method has the advantage of being direct and intuitive, but canonical operations such as crossover and mutation may be challenging to implement in the algorithm using this type of encoding. Therefore, *binary vector* encoding [18] is preferred, where each variable is converted into binary format based on the number of values the variable can take. For instance, if the tube diameter can assume four values (0.05, 0.08, 0.01, 0.02), two bits will be needed to represent the individual. If there are eight values, three bits will be required. In general, the number of bits required is equal to $log_2\, n$, where n is the number of values a single variable can assume.

The individual binary value associated with the individual real value will be assigned based on cardinality. In other words, considering this set of values (0.05, 0.08, 0.01, 0.02), the 0.05 value will be associated with the binary value 00, the value 0.08 with the value 01, and so on. This method, even though less intuitive and direct in implementation, offers significant advantages in facilitating crossover and mutation operations. Furthermore, it allows for the encoding of "non-numerical" values. For instance, using direct vector encoding does not allow to directly insert the material of the heat exchanger tubes (what numerical value would it take?), but one would need to insert all its properties instead. However, through binary encoding, one can assign a binary number to each tube material, providing greater flexibility in optimization.

Figure 2.10 shows the classic GA framework, divided in multiple steps, that are now discussed in the following sections.

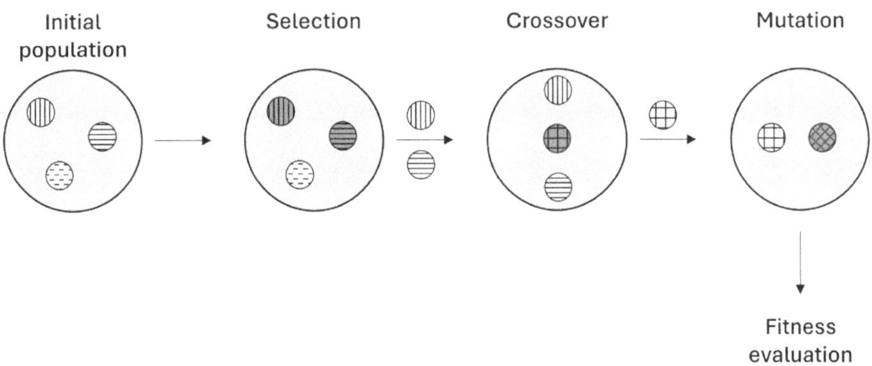

FIGURE 2.10 GA framework.

2.2.2 Initial Population

In a GA the initial population is a crucial starting point for the evolutionary process. It comprises of individual solutions, represented as chromosomes, which hold potential answers to the optimization problem being addressed. Here is an overview of how the initial population is typically established and selected in a GA. First, one needs to decide on the size of the initial population, which is a user-defined parameter. This size can vary depending on the problem's complexity. A larger population size can enhance diversity but might also increase computational requirements. The order of magnitude of the population size has to be one order of magnitude higher than the number of design variables.

The next choice is on how to represent potential solutions as chromosomes. As aforementioned, common representations include binary strings, integer arrays, real-valued vectors, or custom structures tailored to the specific problem domain. To create the initial population, individuals (chromosomes) can be generated *randomly*. Within each chromosome, values are chosen randomly from the problem's search space. This randomization helps ensure a diverse starting population.

Depending on the nature of the problem and optimization goals, various strategies can be employed to create the initial population. These strategies may include completely random selections (uniform random), using domain-specific knowledge or heuristics (heuristic initialization), or starting with promising individuals from previous runs or expert guidance (seed population). In this frame, a suitable option to create an initial population for GA is Latin hypercube sampling (LHS). It was initially developed to create a diverse distribution of parameter values from a multidimensional distribution. It achieves this goal by ensuring that the sample points are arranged in a Latin square, meaning that each row and column contain only one sample. The versatility of LHS becomes evident when applied to problems involving a multitude of dimensions. It generalizes the Latin square concept to handle an arbitrary number of dimensions effectively. When LHS is used to sample a function with k variables, it divides the range of each variable into n equally probable intervals. Subsequently, it selects n sample points in such a way that a Latin hypercube is constructed. One of the key advantages of Latin hypercube sampling is its ability to provide more efficient estimates of desired parameters compared to simple Monte Carlo sampling methods. This efficiency derives from the careful arrangement of sample points, which ensures that the exploration of the parameter space is systematic, diverse, and representative (as in Figure 2.11).

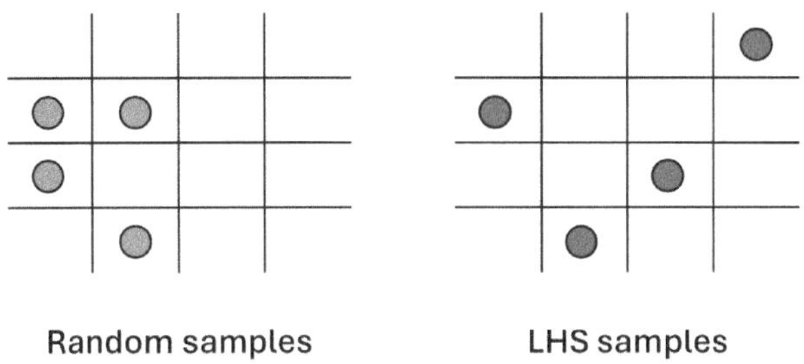

FIGURE 2.11 Latin hypercube sampling principle.

2.2.3 Selection

Among the various techniques used to establish the selection process, one of the most widespread is the *tournament selection*. It is designed to emulate the competitive dynamics of a tournament setting. The following is an outline of its operation. Initially, the tournament size is set as a user-defined parameter. This value dictates how many individuals participate in each tournament round, offering flexibility to adapt to the problem's nature and desired selection intensity. In each round of selection, a random subset of the population, comprising of the specified tournament size, is assembled. This grouping, termed a "tournament," ensures that each individual is involved once and only once. Within each tournament, individuals compete based on their fitness scores. The individual with the highest fitness score (in the context of maximizing fitness) or the lowest fitness score (in minimizing fitness) is declared the tournament winner. The selected tournament champion assumes the role of a parent for the subsequent generation. This preference leans toward individuals with higher fitness scores, aligning with the fundamental principle of "survival of the fittest."

This iterative process repeats to choose multiple parents for reproduction, with the key point that each tournament operates independently. Consequently, the same individual may participate in multiple tournaments, offering numerous opportunities for selection. Tournament selection's appeal lies in its capacity to strike a balance between diversity and selection pressure while introducing an element of randomness into the selection process. This stochastic nature helps prevent premature convergence and promotes exploration of a broader solution space within GAs.

2.2.4 Crossover

The crossover operation involves the interaction between two individuals (parents) to generate a new individual (kid) with characteristics belonging to both parents. An example of crossover using binary vector encoding is shown in Figure 2.12.

The two parents can combine to generate a child characterized by the first variable being the same as that of parent 1 and the second variable being the same as that of parent 2, and vice versa. In addition, there are also combinations where crossover doesn't actually occur, so the kid will be identical to one of the two parents.

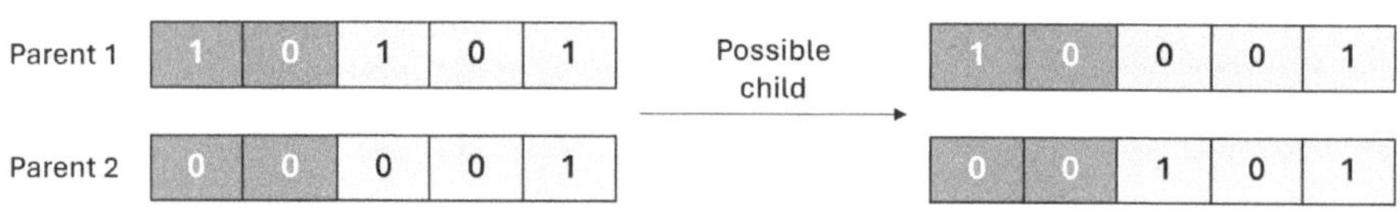

FIGURE 2.12 Example of crossover.

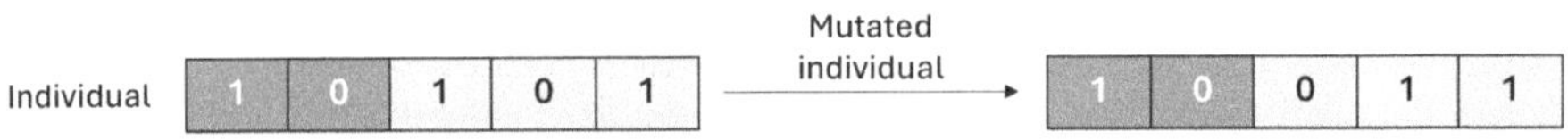

FIGURE 2.13 Example of mutation.

2.2.5 Exploring Design Space: Mutation

In GAs the mutation process is a fundamental genetic operator that introduces small, random changes into the genetic information of individuals (chromosomes) within the population. These changes help diversify the population and explore new regions of the solution space.

During mutation, a random subset of genes within an individual's chromosome is selected. Then, these selected genes undergo random alterations (Figure 2.13).

The nature and extent of these alterations, such as flipping a bit in a binary representation or adding a small random value in a real-valued representation, are determined by the mutation rate, which is a user-defined parameter. Mutation serves as a vital source of genetic diversity in GAs, complementing other genetic operators like crossover. It helps prevent premature convergence by occasionally introducing novel genetic material into the population, enabling the algorithm to discover potentially better solutions. The mutation rate can be adjusted to control the intensity of this diversification, striking a balance between exploration and exploitation.

2.2.6 Elitism

Elitism is a strategy that ensures that the best-performing individuals from one generation are preserved and carried over to the next generation without undergoing any genetic operations like crossover or mutation. These top-performing individuals are often referred to as "elites." By preserving elites, GA aims to maintain the best solutions found so far and prevent them from being lost due to genetic operations that may introduce randomness or errors. Elitism helps improve the convergence and overall performance of a GA by consistently keeping the best individuals in the population.

2.2.7 Convergence

The idea of the algorithm is to let the best individuals, those with a higher or lower objective function value depending on whether the problem is about maximization or minimization, "survive." These individuals will either be taken as they are, or recombined with other individuals through crossover, to form the next generation. However, this logic could present a drawback, which is the convergence towards local optima. Within the context of heat exchangers, if the best individuals all shared a common characteristic, i.e., tube material, for instance, aluminum, only that characteristic would continue. If there were combinations of variables that could make steel or copper heat exchangers efficient as well, they wouldn't be explored. Consequently, the inclusion of mutation within the algorithm is of fundamental importance to increase its variability and make the optimization as global as possible.

As an example, the steps leading to the creation of a new generation from the previous one will be shown. In this case, the chosen population consists of five

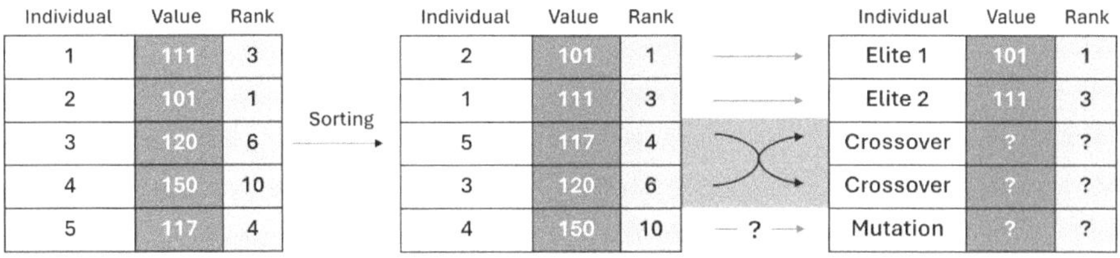

FIGURE 2.14 Sketch of GA evolution.

individuals, the problem requires the minimization of the objective function value, and the new generation will comprise 40% elite individuals, 40% individuals derived from crossover, and 20% mutated individuals. The evaluation of the objective function is carried out using a fitness function, which is responsible for taking the parameters that make up the individual and evaluating it. After calculating the value of the objective function, a ranking operation must be performed to order the population. The next generation will then consist of the best individuals from the previous generation as they are, the intermediate individuals will undergo crossover, and the last individuals will undergo mutation. This is just one of many approaches. It is also possible to have only one individual undergoing mutation and three elite individuals, for example. Finally, a sketch of how a GA evolves is represented in Figure 2.14.

2.3 CASE STUDIES

This section is dedicated to exploring the multifaceted branch of MOO in heat transfer problems, with a specific focus on relevant case studies about heat sinks, heat exchangers, porous channels, and more. In the pursuit of enhancing efficiency, reducing costs, and mitigating environmental impacts, researchers face the challenge of optimizing heat transfer processes while concurrently addressing multiple objectives. This chapter aims to guide readers in the landscape of case studies, to explain useful insights and major complexities that derive from optimization approaches. From the explanation of theoretical frameworks to the application of cutting-edge optimization algorithms, each case study offers a unique vantage point from which insights collected provide the challenges and opportunities inherent in optimizing heat transfer processes across diverse objectives. Table 2.1 outlines – depending on the case study – the involved design variables, relevant constraints, objective functions, and predictive model used.

2.3.1 Heat Sink

The challenges in thermal management of electronic devices, particularly high-performance chips, are nowadays no longer negligible due to the significant increase in power density and non-uniform power dissipation. Old projections for maximum power dissipation and heat flux from high-performance microprocessor chips were expected to reach about 360 W and 190 W/cm^2, respectively by 2020 [19]. Nowadays heat flux generation of many high-performance electronic devices already exceed these projections. In addition, the trend of miniaturization and increasing

TABLE 2.1
Case Studies of Multi-Objective Optimization Problems Solved with Gradient-Free Tools. Only the Work from Shridan et al. [13] Uses PSOA Instead of GA for the Optimization. Predictive Models Labelled with Direct Means that a CFD Solver Is Directly Coupled with the Optimized. In All the Other Cases, Regression Models Have Been Derived from CFD Computations. Variables to Be Minimized Have Been Labelled with the Minus Sign.

Case Studies	Reference	Design Variables	Relevant Constraint	Objective Function	Predictive Model
Heat sink (HS)					
Air	Bianco et al. [20]	$e, PPI, u_{in}, t_f, N_f, d, H$	Plate size	$[Q, -L]$	Direct
Liquid	Radmard et al. [21]	$t_f, s_f, H(y)$	T_w, P_{max}	$[R_{th}, -DP]$	ANN/RA
PCM	Bianco et al. [22]	e, PPI, t_f, N, H	Plate size	$[t_{opt}, -€]$	Direct
	Sridharan et al. [23]	t_f, H, D, d, N	Device and fins volumes		
Fins volumes	$[t_{ch}, -t_{dis}]$	ANN			
	Maleki et al. [24]	N, H, V_f*	Plate size	$[E_{ch}, -t_{ch}]$	GMDH
	Srikanth et al. [25]	L, H	Device and fins volume		
Fins volume	$[t*_{ch}, -t*_{dis}]$	ANN			
Thermal energy storage (TES)					
Sensible	Li et al. [26]	$d_{t,i}*, d_{t,j}*, d_{t,i} - d_{t,j}$	TES height	$[-d, -A]$	SVM
	Liu et al. [27]	$N_f, N_p, T_{in}, u_{in}, T_{ini}$	TES sizes	$[t_{ch}, E_{ch}, h]$	RSM
Latent	Bianco et al. [28]	u_{in}, T_{in}	TES sizes	$[t_{ch}, -t_{dis}]$	Direct
	Bianco et al. [28]	d_p, L_p, D	PCM amount	$[A, Dt]$	Direct
	Bianco et al. [29]	$d_{ext}, u_{in}, p_f, t, T_{melt}$	Max. velocity,		
Time interval	$[Q, -€, -DP]$	Direct			
Heat exchanger (HX)					
Plate fin	Liu et al. [30]	H_f, p_f, s_f, t_f	HX sizes	$[j, -f]$	KRI
Plate fin	Han et al. [31]	$t_{f,i}, t_{f,j}, p_{f,i}, p_{f,j}, q_f$	HX sizes	$[Q, -DP]$	KRI
Shell and tube	Daneshparvar et al. [32]	u_{in}, q_f, p_f	HX size	$[h, -DP]$	RA
Shell and tube	Wen et al. [33]	H_f*, q_f, u_{in}	HX size	$[h, -DP]$	RSM
Tubular	Zhang et al. [34]	Re, d_t*	HX size	$[h, -DP]$	ANN
Porous channel (PC)					
Channel	Shi et al. [35]	$PPI(y), e(y)$	Channel size, e, PPI	$[Nu, -f]$	KRI
Channel	Mauro et al. [36]	$PPI(y), e(y), H*, Re, k*$	Channel size	$[Nu, -f]$	Direct
Tube	Ge et al. [37]	$e(y), Da, H_1, H_2$		$[Nu, -f]$	Direct
Tube	Keykhah et al. [38]	$e(y), Da, H_1, H_2$		$[Nu, -f]$	Direct

Abbreviations: ANN, artificial neural network; GMDH, grouped method of data handling; KRI, kriging model; RA, regression analysis; RSM, response surface model; SVM, support vector machine.

transistor density leads to high power densities and operating temperatures in electronic devices. The impact of the classical Moore's law progression toward shrinking feature size, increasing transistor density, faster circuit speeds, and higher chip performance, emphasizes that while the trend of smaller and faster electronic devices is desirable, it also leads to high power densities, high operating temperatures, and reduced performance and longevity of the devices. This aspect is strictly associated with the failure rate of electronic devices, which increases exponentially with rising operating temperatures, highlighting the critical importance of efficient heat removal to maintain device reliability and performance. When performing optimization on heat sink designs, there is usually a trade-off to balance the need for high heat transfer performance and reducing power consumption to move heat transfer fluid. An example of balancing fluid flow and heat transfer in a heat sink by performing an MOO on is shown in Figure 2.15. As design variables are inlet velocity, solid material displacement and quantity, objective functions change accordingly.

Furthermore, limitations of traditional coolants such as air, oil, and water in meeting the escalating cooling demands of modern electronic devices due to poor heat transfer properties are now overcome by consolidated cooling techniques, such as microchannel-based liquid cooling, two-phase liquid cooling, heat pipe cooling systems, and thermoelectric cooling, as promising solutions to address the thermal challenges of high-performance computing architectures.

2.3.1.1 Air

When dealing with the task of enhancing thermal management in electronic component miniaturization, a viable solution is to propose metal foam heat sinks. The main challenge in this case is to improve heat transfer while minimizing pressure drop, which is a typical MOO problem [20]. The investigation can be on both non-finned and finned metal foam heat sinks, through numerical thermo-fluid dynamics analysis, to maximize heat dissipation and minimize pumping power. In the specific case [20], governing

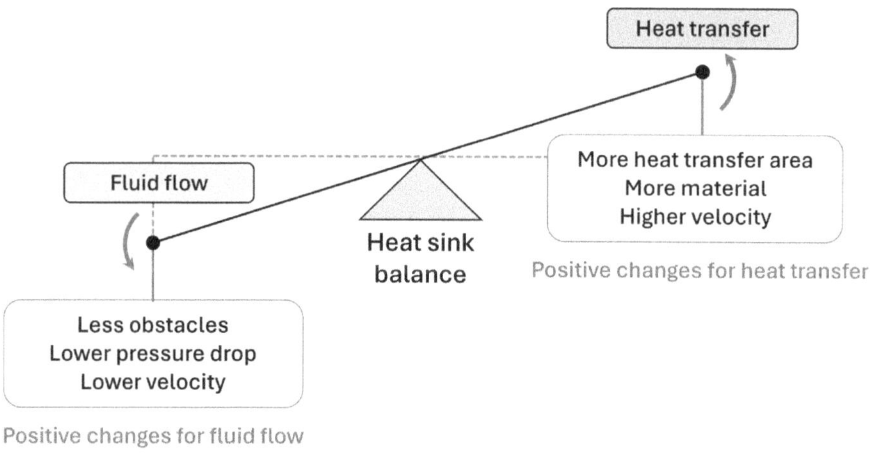

FIGURE 2.15 Example of a heat sink balance when performing multi-objective optimization on fluid flow and heat transfer.

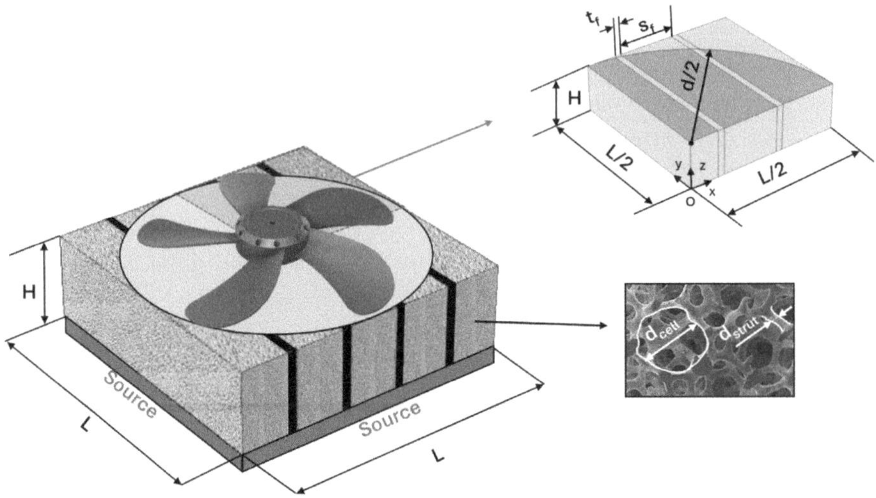

FIGURE 2.16 Sketch of the heatsink concept and computational domain: heat that comes from the plate via foam struts and/or fins (tf = fin thickness, sf = fin spacing) is dissipated with an air flow that comes via a circular nozzle inlet section of diameter d [19].

equations, based on porous media theory with local thermal non-equilibrium assumptions, are solved using a finite element tool. The numerical model is coupled with a multi-objective genetic algorithm (MOGA) to find optimal solutions considering various design parameters. Figure 2.16 shows a scheme of the investigated domain.

Morphological and geometrical variables, along with inlet flow velocity, are treated as design parameters. Specifically, for the finned configuration foam porosity, pores per inch (PPI), dimensionless inlet section diameter inlet velocity, heat sink height, fin numbers, and fin thickness are considered decision variables. Values that can take are shown in Figure 2.17. The objective functions in the design of porous

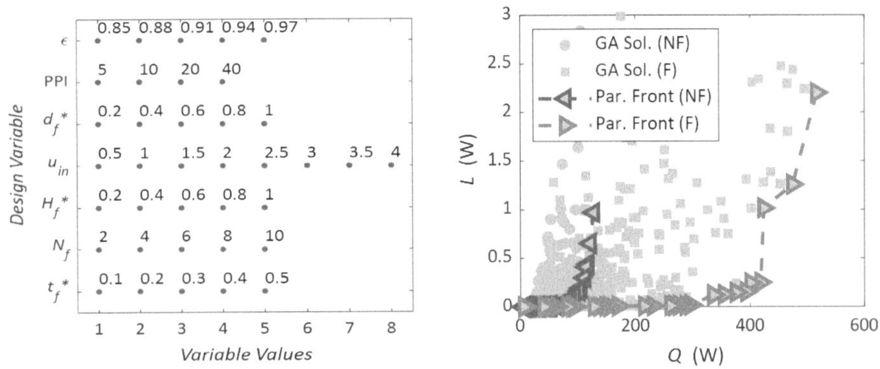

FIGURE 2.17 Comparison between investigated solutions for the finned and non-finned systems [19].

media heat sinks pose a critical challenge. On the one hand, we aim to maximize the heat rate (Q), for efficient heat transfer augmentation. On the other hand, to minimize the pumping power (L_p) to mitigate pressure drop limitations. Balancing these two objectives is essential for optimal performance.

$$Q = 4\left[\iint_{A_{foam}} \left(-k_{eff,s}\frac{\partial \langle T_s\rangle^s}{\partial z} - k_{eff,f}\frac{\partial \langle T_s\rangle^s}{\partial z}\right) dxdy + \iint_{A_{fin}} -k_t\frac{\partial T_t}{\partial z} dx\,dy\right] \tag{2.2}$$

$$L = 4\left[\frac{1}{A_{in}}\iint_{A_{in}} u_{in} P\,dx\,dy\right] \tag{2.3}$$

Results indicate that the heat rate of the finned foam heat sink is approximately 3.3–3.5 times higher than that of the non-finned counterpart under equal pumping power (see Figure 2.17). Comparison with experimental data suggests that optimization can increase heat rates by about 2.5-3 times for non-finned heat sinks and up to 5-6 times for finned ones under equal pumping power conditions.

According to existing knowledge, there are limited papers [39] that offer correlations of Pareto fronts for similar applications. To address this, researchers have proposed two types of correlations for comparing dimensionless heat transfer and pressure drop. These correlations are expressed in a dimensionless form to enhance the overall validity of the investigated problem, and their trends are showed in Figure 2.18. Specifically, the Nusselt number is presented as a function of the Poiseuille number, allowing for a correlation in terms of dimensionless heat rate based on the dimensionless pressure drop (Eqs. 2.4 and 2.5a.).

$$Nu = \frac{Q}{(A_{fin} + A_{foam})(T_W - T_0)}\frac{D}{k_f} \tag{2.4}$$

$$f\,\mathrm{Re} = 2\frac{\Delta P}{H}\frac{D^2}{\mu w_0} \tag{2.5}$$

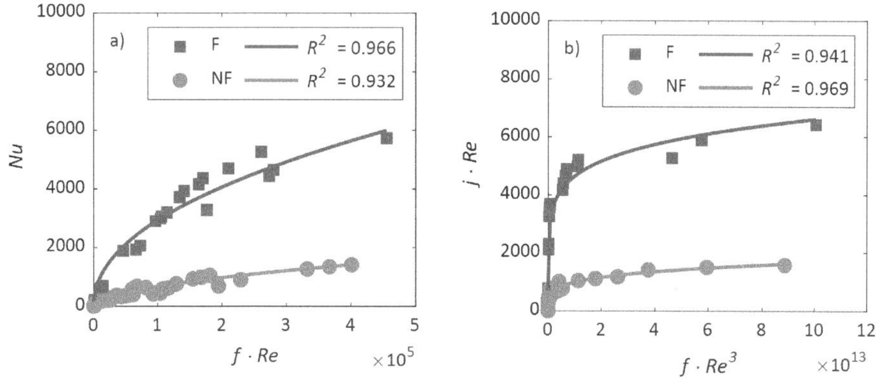

FIGURE 2.18 Correlations of Pareto fronts. On the left Nusselt vs. $f\,\mathrm{Re}$; on the right dimensionless heat rate $j\,\mathrm{Re}$ vs. $f\,\mathrm{Re}^3$ [19].

In addition, other correlations are related to the Colburn j-factor, which represents a dimensionless heat rate concept similar to the Nusselt number. These correlations are linked to the dimensionless pumping power $f\,\mathrm{Re}^3$, where the nozzle diameter serves as the characteristic length for the Reynolds number. The dimensionless heat rate (j Re) and the dimensionless pumping power factor ($f\,\mathrm{Re}^3$) are shown in Eqs. 2.6 and 2.7:

$$j\,\mathrm{Re} = \frac{Q}{(A_{fin} + A_{foam})(T_W - T_0)} \frac{\mathrm{Pr}^{2/3}}{c_p} \frac{D}{\mu} \tag{2.6}$$

$$f\,\mathrm{Re}^3 = \frac{L}{w_0 A_{in}} \frac{D}{H} \frac{2}{\rho w_0^2} \left(\frac{\rho u_0 D}{\mu} \right)^3 \tag{2.7}$$

These dimensionless relationships provide a comprehensive understanding of the heat transfer and pressure drop phenomena in the investigated context.

The findings suggest that optimization techniques significantly enhance the performance of such devices. Furthermore, results indicate that finned metal foam heat sinks significantly enhance heat dissipation compared to non-finned ones. Optimization improves heat rates while minimizing pumping power, yielding substantial performance enhancements. Pareto front correlations aid in designing optimized devices. Overall, the study demonstrates the effectiveness of optimization procedures in enhancing the performance of metal foam heat sinks for various applications.

2.3.1.2 Liquid

The bottleneck of using air as heat transfer fluid is the relatively low heat transfer coefficient compared to liquids. Therefore, many works focused on investigating solutions employing liquids, for instance analyzing heat transfer in a chip-attached micro pin fin cooling device under water jet impingement, aiming to reduce thermal resistance and pressure drop [21]. By directly attaching fins to the chip, the traditional thermal interface resistance is reduced. MOO techniques are employed to minimize thermal resistance and pressure drop, considering various design parameters like fin spacing and height. The heat flux from the chip is assumed of 125 W/cm^2. CFD and experimentation are used to analyze the impact of these parameters. Geometric parameters such as the pin fin cross-section, fin spacing, and fin height profile are analyzed for their impact on the thermal and hydraulic performance of a cooling device through the full factorial design of experiments (FFD). A multi-objective geometric optimization is conducted in two parts: (1) Optimizing pin fin cross-section and spacing to minimize thermal resistance and pressure drop, using both traditional and modern optimization algorithms; and (2) Fin height profile optimization is performed using an artificial neural network (ANN) with an elitist NSGA-II algorithm. Shape optimization is used to reduce pressure drop without compromising thermal resistance. Compared to conventional cold plate designs, the chip-attached configuration significantly decreases thermal resistance.

2.3.1.3 PCM

Liquid cooling systems, despite having superior heat transfer coefficients, entail additional complexity due to the requirement of pumps and associated components. These pumps are necessary to circulate the coolant through the system, adding an extra layer of complexity and potential points of failure. Moreover, the compatibility of liquid cooling setups can be limited by the need for sufficient space to accommodate the components and potential modifications to the hardware. This complexity and dependence on pumps and space requirements further add to the drawbacks of liquid cooling, making alternatives like phase change material (PCM) more appealing for thermal management in certain scenarios. Therefore, a thermal management approach using PCM integrated into heat sinks emerges as a viable solution for controlling temperature peaks without the added complexity and maintenance demands of liquid cooling systems. For instance, the optimization of finned heat sinks integrated with PCMs and metal foams for enhanced thermal management has been exhaustively explored [22].

A mathematical model based on volume-averaged porous media equations is used to predict a transient process shown in Figure 2.19. Multi-objective Pareto

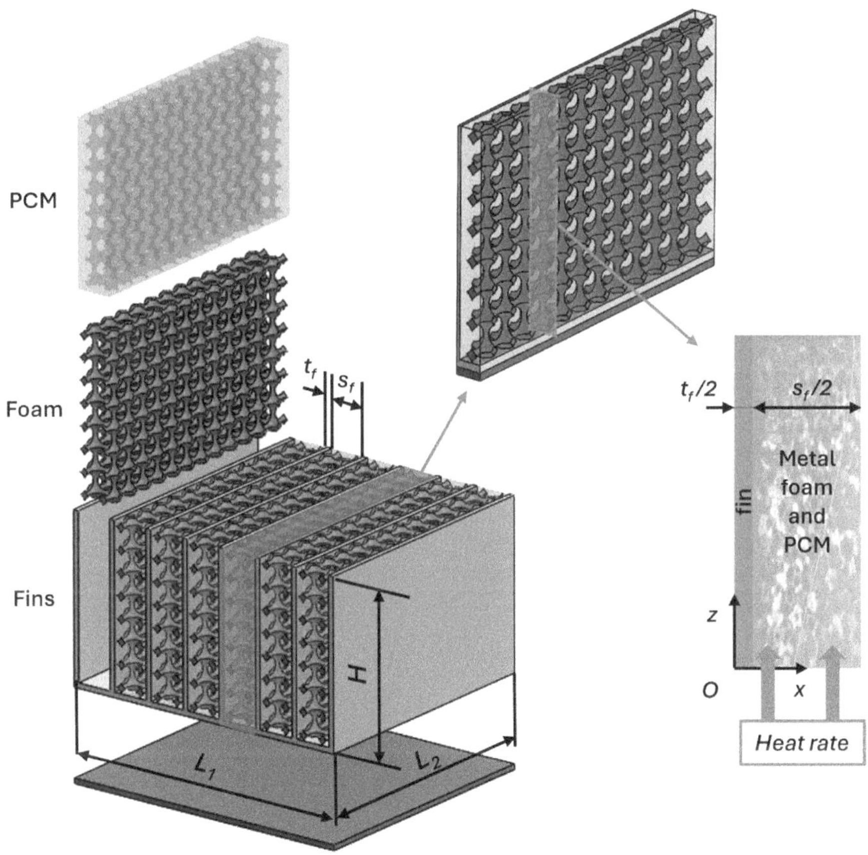

FIGURE 2.19 Investigated domain: heat sink features metallic fins and is equipped with both phase change material (PCM), specifically docosane (a type of paraffin wax), and metal foam. Heat is delivered from the bottom of the device and dissipated through the combination of PCM, metal foam, and the fins [21].

optimization employing a GA balances cost minimization and operation time maximization. The maximum allowable temperature is assumed equal to 90°C. The time required for the device to reach this temperature is represented as t_{opt}. Consequently, the optimization efforts focus on maximizing the operational time to extend the period during which the device can function effectively without surpassing the specified temperature limit and minimizing the total investment cost per device. This cost includes both the expenses related to foam (€$_{foam}$) and the PCM (€$_{PCM}$). Results presented in Figure 2.20 reveal a trade-off between operation time and device cost, with Pareto front solutions ranging from 2000 to 6000 seconds and €200 to €275 per device, respectively. The utopia criterion identifies an optimal solution with an operation time of 3780 seconds and a cost of €235 per device. Lower porosity, thicker fins, and reduced device height contribute to a decreased operation time.

The study highlights the effectiveness of fins in aiding the melting process and emphasizes the importance of natural convection effects. Overall, the findings provide valuable insight for designing cost-effective and efficient heat management devices, with implications for future research focusing on complete charging/discharging cycles and experimental validation of optimized solutions.

As mentioned above, a bottleneck of PCMs is the exploration of solutions to increase their thermal diffusivity. Various shape and materials have been explored to optimize the geometric configuration of a cylindrical heat sink containing a PCM and thermal conductivity enhancers (TCEs) [23]. The aim is to maximize the charging time and minimize the discharging time while maintaining a constant volume for both the TCEs and the cavity. Experimental and numerical simulations were conducted to assess the heat transfer coefficient and charging/discharging times for various geometric configurations of TCEs. The results indicated that MOO is feasible, as charging and discharging times are orthogonal, allowing for simultaneous maximization and minimization, respectively. The multi-objective PSO (MOPSO)

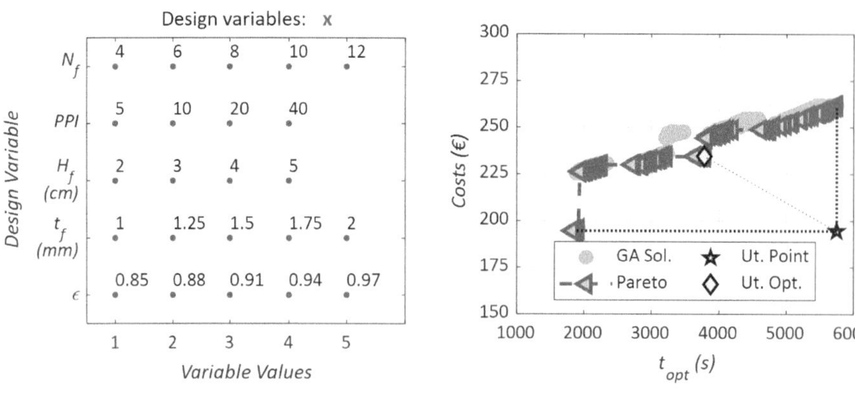

FIGURE 2.20 Resulting Pareto front: trade-off between costs to be minimized and operation times to be maximized [21].

algorithm demonstrated stability in obtaining optimal solutions. Furthermore, the study revealed that the obtained optimal solutions represent a trade-off between charging and discharging times, providing flexibility in selection based on user priorities. Significant improvements were observed in the discharging cycle performance compared to the charging cycle, highlighting the practical importance given the intermittent nature of heat sink operation. Overall, the findings of this study serve as a benchmark for future optimization studies on cylindrical heat sinks and can potentially be extended to address different constraints and objectives in similar optimization problems.

Overall, the high complexity involved in the simulation of PCMs means that it is often convenient to use several tools simultaneously to facilitate the optimization process and obtain reliable results in a reasonable timeframe. An example is the optimization of a finned latent heat thermal energy storage (LHTES) unit to minimize phase change time and maximize stored energy [24]. This is achieved through a hybrid approach involving CFD, grouped method of data handling (GMDH) type of ANN, non-dominated sorting GAs (NSGA-II), and multi-criteria decision-making (MCDM) methods.

The design variables considered are geometrical parameters of fins, including number, length, and volume fraction. CFD simulations are first used to study the effect of these variables on objective functions. Then, GMDH-type ANN models are employed to predict objective function response based on design variables. Pareto optimal points are determined using NSGA-II and GMDH predictive models. MCDM methods like TOPSIS and VIKOR are used to suggest design points per weight of objectives. Results show that the proposed hybrid approach is effective in optimizing TES systems, with most of the optimal cases having specific characteristics such as high dimensionless fin length and low volume fraction of fins. Future research will explore other geometries and boundary conditions for LHTES units.

Finally, the optimization of a PCM based composite pin fin matrix heat sink [25] is presented to extend the operation time during the heating cycle and minimize it during the discharging cycle. The heat sink is made of aluminum with n-eicosane PCM. Through numerical simulations using CFD software, temperature time histories for charging and discharging cycles are obtained for 40 different geometric configurations. These results are then used as inputs for a neural network for MOO to determine the optimal heat sink configuration. Experimental validation is conducted to match numerical simulations under uniform heat input conditions.

Authors reported that NSGA II MOO effectively extends the charging cycle by 5% and reduces the discharging cycle by 12.5% simultaneously. Non-dominated solutions reveal a trade-off between objectives, aiding in an understanding of the impact of input variables on heat sink performance.

2.3.2 Thermal Energy Storage

The rising demand for thermal energy driven by modern lifestyle applications necessitates optimized systems to address global energy supply and demand challenges.

Recent years have seen a growing focus on designing new systems to meet increasing competition and the imperative for improved efficiency. Innovation aims for high-quality, environmentally friendly products, and multiple viable designs must be explored, emphasizing process optimization to maximize or minimize objective functions under constraints. This optimization is particularly crucial for components that enhance thermal system performance, such as air conditioning and solar energy systems. Well-established and emerging approaches aim to boost thermal energy system efficiency through optimized thermal storage, notably sensible thermal energy storage (STES) and LHTES. These options differ depending on whether the latent heat of the storage material is exploited or not. The latter uses PCMs, which undergo phase transition – typically solid/liquid – to store/release thermal energy. The selection of PCM is critical, requiring materials with specific thermophysical properties. Studies employ experimental and numerical analyses to explore various PCM types in different contexts. There exists a challenge between the need for high storage capacity and thermal performance, specifically related to thermal diffusivity. Storage capacity refers to the amount of thermal energy that a system can store, typically measured in terms of energy density or total energy stored. A higher storage capacity can be beneficial for applications requiring large amounts of energy over extended periods. On the other hand, thermal performance, often quantified by thermal diffusivity, measures how quickly a material can spread heat. Higher thermal diffusivity can lead to faster heat transfer rates. These two aspects require – depending on the application – optimization techniques to modify the system design/operation. An example showing a possible balance between energy storage and thermal power for a heat sink is presented in Figure 2.21.

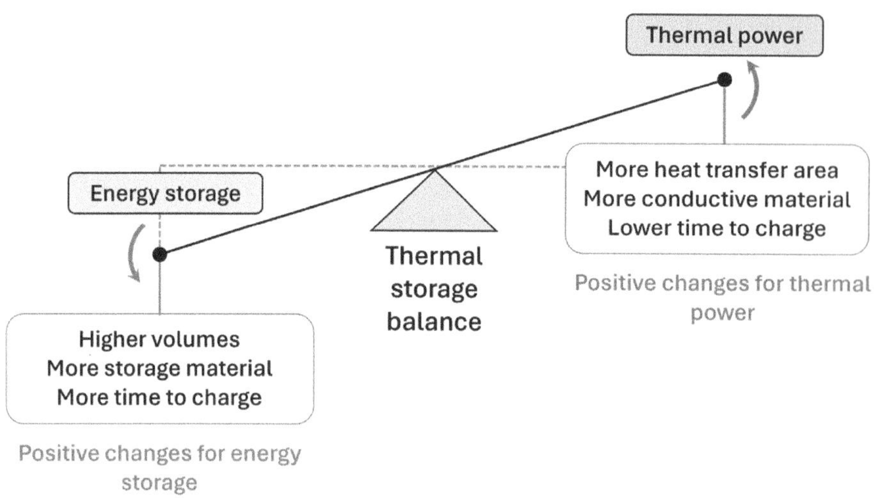

FIGURE 2.21 Example of a thermal storage balance when performing multi-objective optimization on energy storage and thermal power.

2.3.2.1 Sensible Heat

Optimization of a STES system involves exploring and evaluating various designs to select an optimal one. Each design is characterized by operational, geometrical, thermophysical, and performance parameters [40]. Operational parameters are influenced by the specific application. Geometrical parameters are typically arbitrary but must adhere to structural constraints. Thermophysical parameters depend on the materials used and are influenced by temperature and pressure, while performance parameters are used for assessing and comparing different designs. The design space becomes highly dimensional due to the numerous geometric parameters. Given the high cost and time associated with experimental investigations at an industrial scale, TES designs are often evaluated using simulations. However, brute-force evaluations of all possible designs are computationally expensive. To address this, the reduction of design variables is essential.

For instance, optimizing a set of unequal diameter radial diffusers in stratified thermal energy storage (TES) tanks is crucial for enhancing electric power peak load shifting and integrating renewable energy sources [26]. Initially, numerical experiments are conducted based on an orthogonal experimental design to build a performance database. A support vector machine (SVM) is then employed to create a data-driven surrogate model from this database. Subsequently, mono-objective, and MOO are carried out using a GA, evaluating fitness functions via the surrogate model. The surrogate model based on SVM exhibits high accuracy (squared correlation coefficient of 0.96 and a mean squared error near zero), with computational efficiency comparable to numerical simulations. Single-objective optimization reveals an optimized thermocline thickness of 0.829 m, resulting in a 4.2% improvement in thermal stratification performance and an increase in heat storage space by 11.6 m^3. MOO yields Pareto optimal solutions that predict a thermocline thickness of 0.866 m and reduced steel cost by 88.1%, while maintaining comparable thermocline thickness to the benchmark. Overall, these findings aid in designing efficient and economical unequal diameter radial diffusers for TES tanks, enhancing their performance while minimizing costs.

A similar approach has been used for the numerical investigation of a concrete thermal energy storage (CTES) system utilizing air as the heat transfer fluid (HTF) [27]. To efficiently analyze the complex interactions between parameters, response surface models (RSMs) for multiple responses are established based on 27 specific design points determined by a central composite rotation design (CCRD). The effect of design parameters on the CTES system's performance are analyzed using these models. The results highlight the significance of HTF velocity on charging time and efficiency, and the substantial influence of HTF inlet temperature on energy storage. Interactions among parameters also play a crucial role in performance indices. Optimization is conducted to simultaneously minimize charging time and maximize energy storage and charging energy efficiency. The optimal parameter combination is determined using a desirability function. Additionally, the discharging process is considered to optimize the overall cycle performance. The study concludes that a configuration with 22 tubes and 4 fins under selected operating conditions is appropriate for the CTES unit. This method offers a viable approach to CTES unit design.

Temperature profiles of the CTES and charging/discharging completion status are observed, with heat transfer being more uniform radially than axially. The heat transfer rate decreases along the flow direction with decreasing temperature between air and concrete.

RSMs are established for charging time, energy storage, and charging energy efficiency, requiring fewer simulations. These models demonstrate reasonable accuracy, effectiveness, and good prediction capability, as verified through ANOVA. Various factors influence different response indices, with interactions playing a significant role. Optimization considering the charging process yields an optimal parameter combination, including 4 fins and 22 tubes, an HTF inlet temperature of 672.8 K, HTF velocity of 4 m/s, and initial temperature of 463 K. The optimal number of tubes and fins align with considerations from the charging process, optimizing the overall cycle performance.

2.3.2.2 Latent Heat

Optimizing a thermal system that operates depending on the transient response of PCM strongly depends on the application. One noticeable constraint observed in many referenced studies is the limited analysis of comprehensive enhancement techniques for systems that employs PCM in thermal storage. Typically, these studies focus on individual techniques or combine only two methods, without comprehensively explore the synergistic effects of various enhancement strategies. Moreover, a noticeable gap exists in research concerning the comparison of PCM melting times for different geometric configurations of LHTES units. This lack of comprehensive analysis poses a significant challenge in understanding and optimizing the overall storage performance of PCM-based systems. Recognizing these limitations, recent efforts have been directed toward addressing this gap to search for the combined impact of PCM heat transfer enhancement and geometric optimization. The goal is to develop a robust optimization framework aimed at improving the thermal energy storage performance of PCM-based shell-and-tube heat exchangers (STHXs) [28].

The proposed methodology represents a significant departure from traditional approaches, introducing a novel MOO framework. Applied to a chiller system serving a single-family cooling load, this simulation methodology integrates GAs and experimental validation to offer insight into PCM behavior and prompting further optimization stages to increase PCM utilization. At the heart of this effort lies a detailed investigation into the effects of varying chiller operating conditions and key geometric parameters through a detailed parametric analysis. This analysis not only identifies feasible objective functions but also aims to accelerate the PCM melting process, thus driving optimization efforts. Central to the optimization process is the selection of solutions that minimize both charging and discharging times while ensuring compatibility with the utopia point. This multi-step approach involves assessing the final configuration to reach minimum heat transfer area, thereby addressing the limitations in PCM exploitation. The benefits of this proposed framework extend beyond classic optimization. By optimizing PCM utilization and reducing system size, driven by a multi-objective approach, this framework holds the potential to contribute significantly to energy efficiency and sustainability efforts.

The selection of objective functions for systems employing PCM for high thermal storage potential is strongly influenced by the application. For instance, many strategies include reducing heat loss via PCM layers, implementing heat recovery and ventilation (HRV) systems to maintain air quality, and integrating thermal storage to minimize energy consumption. Despite numerous proposals for optimizing heat recovery systems with embedded PCMs, a significant knowledge gap exists regarding the optimization of wall-type HRV systems. These systems, distinguished by their compactness and inclusion of various storage media such as PCMs, require a tailored optimization approach. One viable option is to address this gap by developing an optimization methodology focusing on factors like fan speed, PCM phase change temperature, and tube arrangement [29]. By means of numerical simulation and optimization techniques, the study introduces a framework for optimizing a decentralized PCM-based HRV unit. The investigated unit is shown in Figure 2.22.

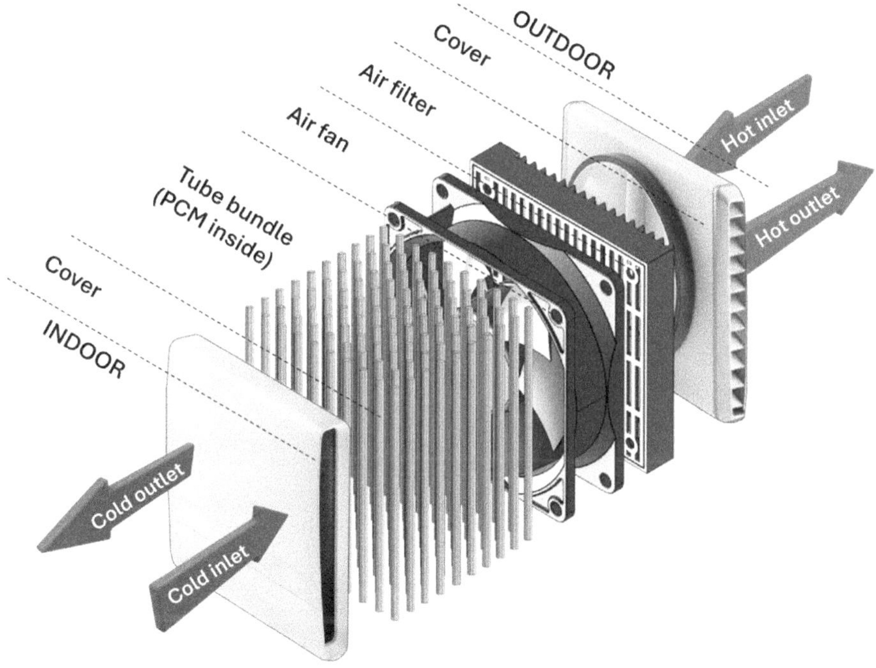

FIGURE 2.22 Main heat recovery and ventilation (HRV) system components: when warm air from the outside (hot inlet) passes over the heat exchanger, it cools down (cold outlet) because it transfers heat to the PCM inside the tubes. When air from the inside (cold inlet) flows around the tubes, it cools the PCM since its temperature is lower than the PCM's melting point, and then exits the system as warmer air (hot outlet) [28].

Through a MOGA, the study achieves enhanced thermal performance while simultaneously reducing investment costs and pressure drop, with results obtained in a 3D space shown in Figure 2.23. The validation of the model against experimental data demonstrates its accuracy. This study demonstrates a constrained MOO with the following procedure:

- *Problem definition*: the process begins by defining design variables, equations, and constraints.
- *Multi-objective genetic algorithm (MOGA)*: a random initial population is generated. Then, each individual in the population is evaluated against an objective function by using the governing equations using a finite element tool. Thus, a MATLAB® code produce individuals in each generation, which are fed into the finite element tool. The finite element model (FEM) analysis provides the fitness values, transferred to MOGA, which performs evolution through conversion, mutation, and selection functions. The GA iteratively improves tentative solutions.

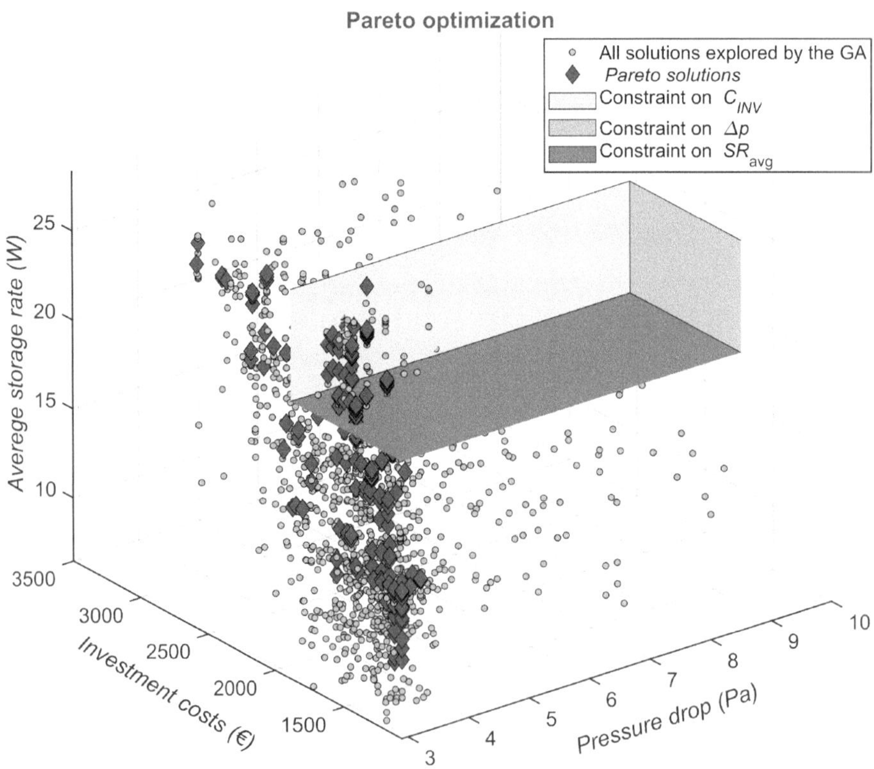

FIGURE 2.23 Constrained 3D Pareto front: solutions with higher pressure drop than 8.5 Pa and higher investment costs than 1880 €, and lower average storage rate than 22.05 W (reference values) are excluded [28].

Key findings include the ability to model the reference design accurately using a 2D conduction-based model and the outperformance of the optimal design in terms of average storage rate.

The study also identifies economically viable solutions with low pressure drop and heat storage rates. Moreover, improvements in HRV efficiency are observed through this optimization approach. Overall, the study concludes that the optimized configuration offers comparable operating costs to household appliances and underscores the growing viability of HRV systems with PCMs. Future research directions include exploring various PCMs and system configurations to address seasonal requirements and improve heat diffusion.

2.3.3 Heat Exchangers

Heat exchangers, e.g., shell and tube heat exchangers (STHEs) and plate-fin heat exchangers (PFHEs), are in widespread use due to their capacity to facilitate effective heat transfer across diverse applications. However, the design of heat exchangers is a complex task, requiring a comprehensive understanding of principles spanning thermodynamics, fluid dynamics, cost estimation, and optimization. The primary objectives of heat exchanger design are dual: maximizing thermodynamic efficiency while minimizing costs. Traditionally, the design process for heat exchangers has been labor-intensive, involving the evaluation of numerous geometric configurations against a backdrop of operational constraints. While thorough, these methods often fall short of achieving optimal solutions and consume significant time and resources. To address these challenges, researchers have turned to nature-inspired optimization algorithms such as GAs and PSO. In this context, the balance between thermal power and pumping power (see Figure 2.24) depends on several factors,

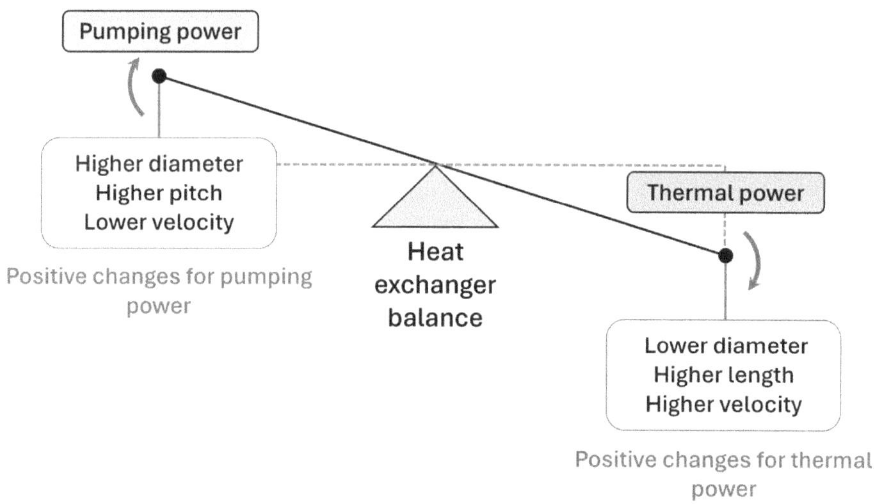

FIGURE 2.24 Example of a heat exchanger balance when performing multi-objective optimization on pumping power and thermal power.

e.g., geometrical ones as pitches, diameters, length and so on, or external ones, e.g., velocity, heat source, etc.

A theoretical optimization method using 3D CFD simulations combined with MOO can significantly enhance design capabilities compared to costly physical experiments. While traditional methods relying on experiments can be time-consuming and may not guarantee the best optimization program, various research have demonstrated a more efficient and reliable optimization process. For instance, a multi-objective framework has been used on optimizing the performance of a PFHE named 95JC1803 for use in a hydraulic retarder [30]. Initially, the heat transfer factor increased by 12.83% and friction factor decreased by 26.91% after optimization, indicating significant performance improvements. Qualitative comparisons of internal flow fields involving temperature, pressure, and velocity further highlight the optimization effects, resulting in reduction of the hot flow temperature, increases in cold flow temperature, and significant improvements in pressure drops and velocities. The study emphasizes the importance of MOO in heat exchanger design due to conflicting objectives and uncorrelated mathematical relationships between parameters and thermal performance. MOO aims to balance various target parameters efficiently, and it is considered a development trend in optimization.

The concept of the field synergy principle, used to explain local heat transfer augmentation, was employed in the study to elucidate optimization results. In this research, an MOO based on CFD simulations was conducted to optimize the structure of an oil-to-water PFHE. The maximum Colburn factor (j) and the minimum friction factor (f) were treated as conflicting objectives. Design parameter variables such as plate-fin height, pitch, spacing, and thickness were optimized, and their influence on PFHE performance was evaluated. The field synergy principle was then analyzed to validate the optimization results based on the studied geometrical parameters and flow field considerations. However, it is noted that the optimization focused solely on improving heat transfer performance without considering economic factors such as cost, technology, and maintenance.

The use of polymeric materials in heat exchanger design offers advantages such as lightweight construction, resistance to corrosion, and low production costs. However, the low thermal conductivity of polymers has traditionally hindered their performance in heat transfer applications. To address this limitation, the authors propose a novel polymer heat exchanger design with five optimal design variables [31]. By optimizing the geometry and arrangement of the heat exchanger components, they aim to improve thermal performance while minimizing airside pressure drop. The design process involves employing analytical and CFD models to analyze the impact of material changes on heat transfer efficiency and other performance metrics. The proposed design incorporates modifications such as altering the shape of the heat exchange tubes to a teardrop-shaped, which increases heat transfer area while minimizing airside pressure drop.

Optimization is performed using a MOGA, considering both thermal and hydraulic performance criteria. Through a combination of design of experiments (DOE), metamodeling, and optimization techniques, the authors identify Pareto optimal design points that balance competing objectives. The results show significant improvements in thermal-hydraulic performance compared to baseline heat exchangers, with the newly designed polymer heat exchanger achieving 92.7% to 162.8% of the baseline's performance. However, challenges remain, particularly regarding the higher airside pressure drop associated with the new design. Further research is needed to address this issue and to explore manufacturing techniques, such as 3D printing, for producing polymer heat exchangers with extremely thin profiles. In conclusion, the study demonstrates the feasibility of polymer heat exchanger application in industrial settings, showing promising thermal performance improvements.

When faced with optimization of a heat exchanger, the main challenge is to realize a change in the design that allows to increase overall performance, while preserving its functionality and satisfying the relevant constraints. For STHXs, a fundamental feature is the baffle shape and location. Its design strictly influences fluid motion and heat transfer, thus has been subject of extensive studies, for which the focus lies on optimizing the performance of STHXs featuring the baffle shape, i.e., helical baffles. These components are crucial for regulating flow distribution and enhancing heat transfer within the exchanger.

One approach is to evaluate the thermal-hydrodynamic performance of such heat exchangers using CFD modeling [32]. Key geometric parameters like baffle pitch and baffle angle are investigated for their effect on heat transfer coefficient and pressure drop. From an experimentally validated CFD model, predictive correlations are developed to estimate the heat transfer coefficient and pressure drop. Optimization using MOGA is then applied to determine optimal baffle geometries, considering the trade-offs between the heat transfer coefficient and pressure drop.

A second approach is to the combine benefits of MOGA optimization coupled with the numerical simulation of a central composite design (CCD) and response surface. The aim is to optimize the performance of STHXs with helical baffles [33]. The CCD is used to generate test points, ensuring precision and efficiency. CFD simulations are then conducted based on these points to acquire objective data. Subsequently, a response surface, derived from CFD results, predicts heat transfer coefficients and pressure drops. This combination of CCD and response surfaces enhance optimization by reducing reliance on empirical correlations. Validation of the surrogate model is crucial and is achieved by comparing predicted values from the response surface with CFD simulation results. Structural parameters such as the helix angle and overlapped degree are optimized, considering conflicting objectives of maximizing the overall heat transfer coefficient while minimizing the shell-side pressure drop. Sensitivity analysis reveals a correlation between these parameters and the performance metrics. Through optimization, improvements are achieved in both overall heat transfer coefficient and pressure drop, showcasing the effectiveness of the optimized heat exchanger configuration.

Finally, the application of machine learning and GAs has been tested in the MOO of tube fin heat exchangers (TFHE) [34]. CFD simulations are conducted across various Reynolds numbers and tube ellipticities to optimize heat transfer and pressure drop performance. Back-Propagation neural networks (NNs) are trained using simulation data to predict heat transfer coefficients and pressure drops. The non-dominated MOGA NSGA-II is then employed to optimize the NN predictions, resulting in optimal Pareto fronts. Findings indicate that elliptical tubes in TFHEs enhance thermal-hydraulic characteristics, with NNs effectively establishing relationships between design variables and performance. The combination of CFD, NNs, and GAs offer efficient optimization, yielding a range of optimal solutions. Through MOO, a significant enhancement in performance is achieved, with a 20% decrease in flow resistance observed at specific Reynolds numbers and tube ellipticities. This approach offers insight into improving TFHE design and underscores the potential of combining artificial intelligence techniques with optimization algorithms in heat exchanger design.

2.3.4 Porous Channels

In porous channels, optimizing heat transfer and minimizing pressure drop pose conflicting objectives due to inherent trade-offs shown in Figure 2.25. Increasing heat transfer often involves maximizing surface area or enhancing fluid-solid interactions, which in turn typically increases flow resistance and thus pressure drop. Conversely, reducing pressure drop usually entails enlarging flow area or minimizing resistance, which may decrease surface area and compromise heat transfer efficiency. This conflict arises from the fundamental relationship between fluid flow and heat transfer within the porous medium. A challenge for porous channels is to find a balance between these objectives to achieve optimal channel performance.

Various optimization techniques, e.g., MOO, can help explore the design space and identify solutions that best satisfy the competing demands of pressure drop and heat transfer. Effective design strategies aim to achieve an optimal compromise, considering the specific requirements and constraints of the application at hand, as for the investigation of the effects of gradient porous media on flow and heat transfer characteristics in a tube [35]. Gradient porous media are materials where properties such as pore size and porosity vary along a spatial gradient. The study employs

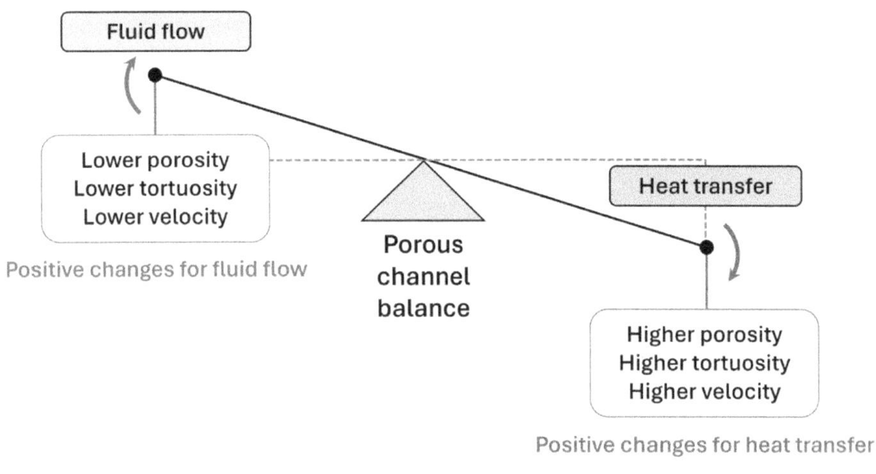

FIGURE 2.25 Example of a porous channel balance when performing multi-objective optimization on fluid flow and heat transfer.

numerical simulations to analyze fully developed flow in a tube partially filled with such gradient porous media. Different configurations of gradient porous media, characterized by variations in pore size and porosity along the radial direction, are compared [35]. The fluid enters the pipe with a velocity profile that follows a parabolic distribution. It is observed that configurations with a simultaneous increase in porosity and pore size along the radius exhibit the best heat transfer performance (Nusselt number), while configurations with simultaneous decrease show minimum flow resistance (friction factor). Configurations with opposing gradients (increasing pore size and decreasing porosity) perform similarly to homogeneous porous media. For a given spatial gradient in pore size, both the friction factor (flow resistance) and the Nusselt number (heat transfer coefficient) decrease as the porosity gradient increases.

However, this effect becomes less significant as the pore size gradient decreases. Increasing the filling ratio (radius of porous inserts/radius of tube) of gradient porous media enhances both flow resistance and heat transfer efficiency. Moreover, the sensitivity of these parameters to gradient variations also increases with the filling ratio. The study performs MOO considering maximum Nusselt number and minimum friction factor. Optimal configurations are obtained, showing improvements in flow resistance reduction and heat transfer efficiency compared to homogeneous porous media, especially at higher filling ratios.

In general, graded foam-filled channels are promising for improving heat sink efficiency due to their customizable structures that facilitate heat transfer enhancement. A viable and effective approach can be an optimization framework that considers both mono- and multi-objective approaches to identify the best combinations of design variables for the foam's fluid dynamics, geometry (see Figure 2.26), and morphology [36].

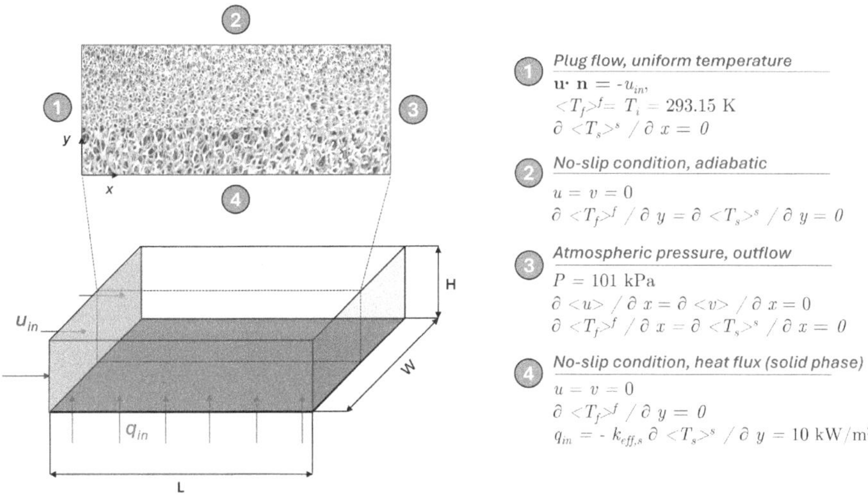

FIGURE 2.26 The domain of a graded foam, with PPI increasing along the height, and the problem boundary conditions [35].

In the mono-objective optimization, the performance evaluation criterion (PEC) is maximized to enhance the thermal efficiency of graded foams. The PEC serves as the objective function, and optimizations are performed by parameterizing the spatial distributions of porosity and/or PPI of the foam. For MOO, Pareto optimization is used to find optimal trade-off solutions between heat transfer enhancement and pressure drop reduction. Different objective functions are considered to explore various aspects of the foam's performance. The optimizations assume a local thermal non-equilibrium in the foam, allowing for more accurate modeling of heat transfer processes. The optimal design variables, such as dimensionless foam height (H^*), Reynolds number (Re), porosity distributions (ε), PPI variations, and thermal conductivity (k^*), are determined through the optimization process. Design variables include fluid velocity, foam morphology, and channel geometry. A resume of design variables is represented in Figure 2.27 and other variables shown there will be introduced later through the text here.

Authors assume porosity and PPI varying according to a power law with indices i_ε and i_{PPI}, varying in ranges typical of commercial metal foams. Graded foams are also compared to uniform foams with porosity and PPI the same as their averaged values, # and PPI, computed via the following correlations:

$$\bar{\varepsilon} = \frac{\varepsilon(1) - \varepsilon(0)}{i_\varepsilon + 1} + \varepsilon(0) \tag{2.8}$$

$$\overline{\text{PPI}} = \frac{\text{PPI}(1) - \text{PPI}(0)}{i_{\text{PPI}} + 1} + \text{PPI}(0) \tag{2.9}$$

Mono-objective optimization aims to maximize the PEC, while MOO explores various objective functions related to heat transfer and pressure drop. A GA (NSGA-II) coupled with MATLAB® and COMSOL Multiphysics® is used for optimization. The study reports significant improvements in heat transfer efficiency

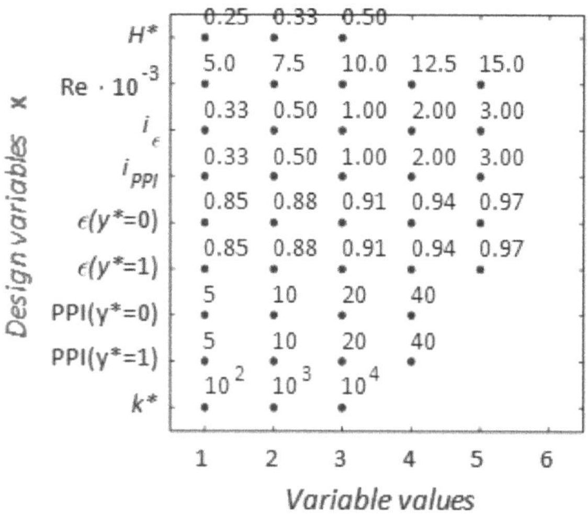

FIGURE 2.27 Design variables and their ranges [35].

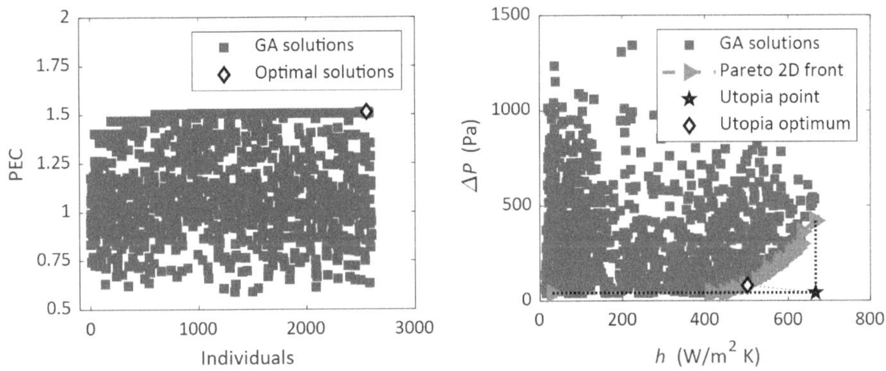

FIGURE 2.28 Mono- and multi-objective optimization results [35].

compared to uniform foam configurations. For instance – see Figure 2.28 – a maximum PEC value of 1.51 is achieved, representing a 51% increase in heat transfer efficiency compared to uniform foam (through the mono-objective optimization). The Pareto front for the multi-objective case (Figure 2.28) ranges from 413 W/m^2 K to 667 W/m^2 K in the averaged heat transfer coefficient and from 42 Pa to 420 Pa in the pressure drop. Utopia optimum values are h = 502 W/m^2 K and ΔP = 80 Pa.

The considered performance indicators of the optimal solutions undergo significant variation as a function of the employed optimization approach. The same occurs for the optimal values of the design variables. Therefore, the reader is strongly recommended to carefully choose the objective function depending on actual needs and wills.

The convergence of the Pareto front and various percentages of the generations number (GEN) reveals how optimal solutions are achieved gradually along the optimization (Figure 2.29). The authors show how solutions reach the final non-dominated

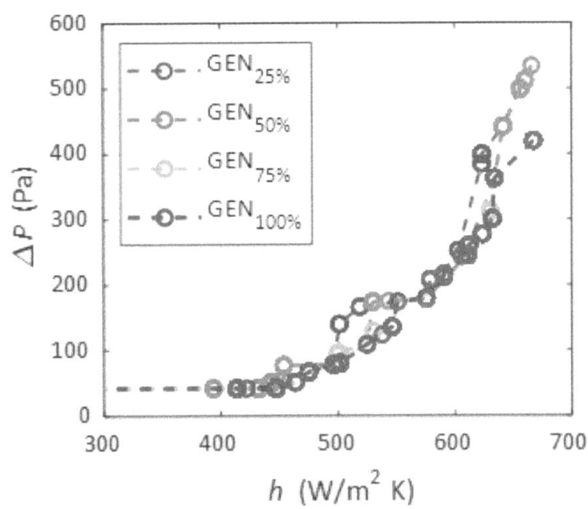

FIGURE 2.29 Multi-objective optimization convergence with generations [35].

ones after $GEN_{75\%}$ (75% of individuals analyzed before reaching convergence). Finally, the utopia point criterion is employed for multi-criteria decision making to determine the main performance indicators for optimal solutions, balancing heat transfer coefficient and pressure drop. When employing the same set of design variables for different objective functions, distinct optimal performance indicators are obtained. This emphasizes that the utopia optimal solution is contingent upon the chosen objective function, necessitating careful consideration for its selection to achieve the desired outcome.

2.4 CONCLUDING REMARKS

This chapter presents an overview of MOO of heat transfer problems, with a special emphasis on GAs. The main concepts and methods of MOO have been introduced, such as Pareto optimality, dominance, preference articulation, and performance indicators. The advantages and challenges of using GAs for MOO have been discussed, such as diversity preservation, convergence speed, and computational cost.

We have also reviewed the application of these algorithms to several case studies involving heat transfer problems, *e.g.*, heat sink, thermal storage, heat sink optimization, and thermal management of electronic devices. The results have shown that GAs can effectively find a set of Pareto optimal solutions that trade off different objectives, such as minimizing the heat transfer area, maximizing the heat transfer rate, or minimizing the entropy generation. These Pareto optimal solutions can be used to guide engineers and researchers in making informed decisions regarding the design and optimization of heat transfer systems. By providing a comprehensive understanding of the trade-offs between various objectives, *e.g.*, thermal efficiency, cost, these Pareto optimal solutions enable stakeholders to explore the design space more effectively. Overall, we have provided a comprehensive and up-to-date introduction to the field of MOO of heat transfer problems using GAs. However, several challenges remain, including the proper fine-tuning of algorithm parameters, the handling constraints, and uncertainties of the prediction require further investigation, especially when integrated with the well-established machine learning techniques and hybrid optimization methods.

This chapter is intended to serve as a catalyst for deeper exploration of MOO of heat transfer problems, inspiring future researcher, and practitioners within interested in this challenging area.

2.5 NOMENCLATURE

2.5.1 SYMBOLS

A	area (m^2)
c	specific heat (J/kg/K)
d	inner diameter (m)
D	outer diameter (m)
E	energy (J)
f	friction factor

F	finned
g	gravitational acceleration (m/s^2)
h	heat transfer coefficient (W/m^2 K)
H	height (m)
i, j	counter
j	Colburn j-factor
k	thermal conductivity (W/m/K)
L	length (m)/pumping power (W)
N	number
Nu	Nusselt number
p	pitch (m)
P	pressure (Pa)
Q	heat rate (W)
R_{th}	thermal resistance (K/m^2 W)
s	spacing (m)
t	thickness (m)
u	velocity (m/s)
T	temperature (K)
V	volume (m^3)
$\dot{V}$	volumetric flow rate (m^3/s)
x, y, z	rectangular coordinates (m)

Greek Symbols

e	porosity
d	thermocline thickness
Δ	Interval
q	theta
h	efficiency
μ	dynamic viscosity (Pa s)
ρ	density (kg/m^3)
t	time (s)

Subscripts

c	cold
ch	charging
dis	discharging
eff	effective
f	fin/baffles/fluid
h	hot
in	inlet
ini	initial
opt	operation
s	solid
t	tubes

Superscripts

*	dimensionless

Acronyms

CFD	computational fluid dynamics
GA	genetic algorithm
HRV	heat recovery and ventilation unit
HS	heat sink
HX	heat exchanger
KRI	kriging model
LHTES	latent heat thermal energy storage
MCDM	multi-criteria decision making
MOO	multi-objective optimization
NN/ANN	neural network/artificial neural network
NSGA-II	non-dominated sorting genetic algorithm
PCM	phase change material
PEC	performance evaluation criteria
PPI	pore per inches
PSO	particle swarm optimization
RA	Regression analysis
RSM	response surface method
STHE	shell-and-tube heat exchanger
SVM	Support vector machines
TES	thermal energy storage
TFHE	tube fin heat exchanger

REFERENCES

1. Martins J R R A and Ning A 2021 *Engineering Design Optimization* (Cambridge University Press)
2. Patel V K, Savsani V J and Tawhid M A 2019 *Thermal System Optimization* (Cham: Springer International Publishing)
3. Dogo E M, Afolabi O J, Nwulu N I, Twala B and Aigbavboa C O 2018 A Comparative Analysis of Gradient Descent-Based Optimization Algorithms on Convolutional Neural Networks *2018 International Conference on Computational Techniques, Electronics and Mechanical Systems (CTEMS)* (IEEE) pp 92–9
4. Sivanandam S N and Deepa S N Genetic Algorithm Optimization Problems *Introduction to Genetic Algorithms* (Berlin, Heidelberg: Springer Berlin Heidelberg) pp 165–209
5. Deb K 2001 *Multi-Objective Optimization Using Evolutionary Algorithms* (Chichester: Wiley)
6. Floudas C A, Pardalos P M, Adjiman C S, Esposito W R, Gümüş Z H, Harding S T, Klepeis J L, Meyer C A and Schweiger C A 1999 *Handbook of Test Problems in Local and Global Optimization* vol 33 (Boston, MA: Springer)
7. Larson J, Menickelly M and Wild S M 2019 Derivative-free optimization methods *Acta Numer.* **28** 287–404
8. Solis F J and Wets R J-B 1981 Minimization by random search techniques *Math. Oper. Res.* **6** 19–30
9. Gao F and Han L 2012 Implementing the Nelder-Mead simplex algorithm with adaptive parameters *Comput. Optim. Appl.* **51** 259–77
10. van Laarhoven P J M and Aarts E H L 1987 Simulated annealing *Simulated Annealing: Theory and Applications* (Dordrecht: Springer Netherlands) pp 7–15

11. Srinivas M and Patnaik L M 1994 Genetic algorithms: A survey *Computer (Long. Beach. Calif).* **27** 17–26
12. Zhang Y, Wang S and Ji G 2015 A comprehensive survey on particle swarm optimization algorithm and its applications *Math. Probl. Eng.* **2015** 1–38
13. Pelikan M 2005 Hierarchical Bayesian Optimization Algorithm pp 105–29
14. Deb K Multi-objective optimization *Search Methodologies* (Boston, MA: Springer US) pp 273–316
15. Tusar T and Filipic B 2015 Visualization of Pareto front approximations in evolutionary multiobjective optimization: A critical review and the prosection method *IEEE Trans. Evol. Comput.* **19** 225–45
16. Turing A 2004 Intelligent Machinery (1948) *The Essential Turing* (Oxford University Press)
17. De J K 2005 Genetic Algorithms: A 30-Year Perspective *Perspectives on Adaptation in Natural and Artificial Systems* (Oxford University Press)
18. Kramer O 2017 Genetic Algorithms pp 11–9
19. Sohel Murshed S M and Nieto de Castro C A 2017 A critical review of traditional and emerging techniques and fluids for electronics cooling *Renew. Sustain. Energy Rev.* **78** 821–33
20. Bianco N, Iasiello M, Mauro G M and Pagano L 2021 Multi-objective optimization of finned metal foam heat sinks: Tradeoff between heat transfer and pressure drop *Appl. Therm. Eng.* **182** 116058
21. Radmard V, Hadad Y, Rangarajan S, Hoang C H, Fallahtafti N, Arvin C L, Sikka K, Schiffres S N and Sammakia B G 2021 Multi-objective optimization of a chip-attached micro pin fin liquid cooling system *Appl. Therm. Eng.* **195** 117187
22. Bianco N, Busiello S, Iasiello M and Mauro G M 2021 Finned heat sinks with phase change materials and metal foams: Pareto optimization to address cost and operation time *Appl. Therm. Eng.* **197** 117436
23. Sridharan S, Srikanth R and Balaji C 2018 Multi objective geometric optimization of phase change material based cylindrical heat sinks with internal stem and radial fins *Therm. Sci. Eng. Prog.* **5** 238–51
24. Maleki H, Ashrafi M, Ilghani N Z, Goodarzi M and Muhammad T 2021 Pareto optimal design of a finned latent heat thermal energy storage unit using a novel hybrid technique *J. Energy Storage* **44** 103310
25. Srikanth R, Nemani P and Balaji C 2015 Multi-objective geometric optimization of a PCM based matrix type composite heat sink *Appl. Energy* **156** 703–14
26. Li Y, Deng Y, Niu M, Sun D and Yu B 2022 Surrogate modeling and optimization for the unequal diameter radial diffuser of stratified thermal energy storage tanks *Energy Sci. Eng.* **10** 2497–508
27. Liu C and Yang H 2022 Multi-objective optimization of a concrete thermal energy storage system based on response surface methodology *Appl. Therm. Eng.* **202** 117847
28. Bianco N, Fragnito A, Iasiello M, Mauro G M and Mongibello L 2022 Multi-objective optimization of a phase change material-based shell-and-tube heat exchanger for cold thermal energy storage: Experiments and numerical modeling *Appl. Therm. Eng.* **215** 119047
29. Bianco N, Fragnito A, Iasiello M and Mauro G M 2023 A CFD multi-objective optimization framework to design a wall-type heat recovery and ventilation unit with phase change material *Appl. Energy* **347** 121368
30. Liu C, Bu W and Xu D 2017 Multi-objective shape optimization of a plate-fin heat exchanger using CFD and multi-objective genetic algorithm *Int. J. Heat Mass Transf.* **111** 65–82
31. Han U, Kang H, Lim H, Han J and Lee H 2019 Development and design optimization of novel polymer heat exchanger using the multi-objective genetic algorithm *Int. J. Heat Mass Transf.* **144** 118589

32. Daneshparvar M R and Beigzadeh R 2022 Multi-objective optimization of helical baffles in the shell-and-tube heat exchanger by computational fluid dynamics and genetic algorithm *Energy Reports* **8** 11064–77
33. Wen J, Gu X, Wang M, Wang S and Tu J 2017 Numerical investigation on the multi-objective optimization of a shell-and-tube heat exchanger with helical baffles *Int. Commun. Heat Mass Transf.* **89** 91–7
34. Zhang T, Chen L and Wang J 2023 Multi-objective optimization of elliptical tube fin heat exchangers based on neural networks and genetic algorithm *Energy* **269** 126729
35. Shi C, Wang M, Yang J, Liu W and Liu Z 2021 Performance analysis and multi-objective optimization for tubes partially filled with gradient porous media *Appl. Therm. Eng.* **188** 116530
36. Mauro G M, Iasiello M, Bianco N, Chiu W K S and Naso V 2022 Mono- and multi-objective CFD optimization of graded foam-filled channels *Materials (Basel).* **15** 968
37. Ge Y, Liu Z and Liu W 2016 Multi-objective genetic optimization of the heat transfer for tube inserted with porous media *Int. J. Heat Mass Transf.* **101** 981–7
38. Keykhah S, Assareh E, Moltames R, Izadi M and Ali H M 2020 Heat transfer and fluid flow for tube included a porous media: Assessment and multi-objective optimization using particle swarm optimization (PSO) algorithm *Phys. A Stat. Mech. Appl.* **545** 123804
39. Sanaye S and Hajabdollahi H 2010 Thermal-economic multi-objective optimization of plate fin heat exchanger using genetic algorithm *Appl. Energy* **87** 1893–902
40. Marti J, Geissbühler L, Becattini V, Haselbacher A and Steinfeld A 2018 Constrained multi-objective optimization of thermocline packed-bed thermal-energy storage *Appl. Energy* **216** 694–708

3 CFD/HT Simulations and DNN Modelling of Conjugate Heat Transfer in Metal Foams

Ubade Kemerli, Muhsin Gokhan Gunay, and Yogendra Joshi

3.1 INTRODUCTION

3.1.1 METAL FOAMS

Stochastic metal foams (MF) have been the focus of significant research due to their superior heat transfer performance compared to conventional finned designs [1, 2] due to higher surface-area-to-volume ratio [3] and continuous disruption of viscous and thermal boundary layers on the structure [4]. Furthermore, the tortuous fluid flow path within the structure contributes to thermal dispersion in the fluid [5]. The research on MFs goes back to the 1990s when manufacturing these structures became possible [6]. Many studies on modeling the flow and heat transfer in MFs are based on porous media transport equations. The physical parameters such as porosity, filament diameter, and pore diameter determine the parameters defined in the porous media equations, including permeability and interfacial heat transfer coefficient. Therefore, many studies have attempted to reveal the relationships between these parameters.

Du Plessis et al. [7] defined permeability and form factor, depending on porosity and pore diameter, with the help of a cubic representative unit cell. Calmidi [8] defined these based on porosity, filament diameter, and pore diameter. Bhattacharya et al. [9] conducted a more comprehensive study and proposed correlations for effective thermal conductivity, permeability, and form coefficient. They also pointed out that the cross-section of the filaments transforms from circular to concave shape with the increase of the porosity.

Some studies simplified and parametrized the foam structure using representative geometries to study their micro-level fluid flow and heat transfer characteristics. Yang et al. [10] investigated pore-scale heat transfer by modeling MFs with the Weaire-Phelan model. Boomsma et al. [11] created an idealized periodic cell structure to represent MF, and numerically investigated the flow and heat transfer characteristics. Kopanidis et al. [12] numerically simulated the MF geometry extracted from Surface Evolver software, and modeled the foam geometry as a combination

DOI: 10.1201/9781032688121-3

of filaments, without modeling the nodes between the filaments. Although this simplified model made the simulations easier and facilitated the parametrization of the foam geometries, some of the critical aspects of the geometry were missing, such as the effects of nodes on the fluid flow.

Few studies have investigated the complex micro-level flow and heat transfer around stochastic structured MFs. Al-Athel [13] modeled the foam geometry, and investigated the internal temperature distribution using a thermal model, defining the efficiency of the MF fins. Ranut et al. [14] studied the permeability and effective thermal conductivity by modeling scanned MFs via high-resolution X-ray micro-computed tomography (μCT) and performing computational fluid dynamics/heat transfer (CFD/HT) simulations. Dixit and Ghosh [15] simulated the airflow through 10 pores per inch (PPI) aluminum foam using images scanned with μCT and identified the intricacies related to the simulations. Iasiello et al. [16] investigated the anisotropy of aluminum MFs by conducting simulations on the μCT scanned geometries. Park et al. [17] experimentally and numerically investigated flow, heat transfer, and the transition from Darcy to the Forchheimer regime in MFs using μCT. Yu et al. [18] investigated the effect of the anisotropy of MF structures on the flow and heat transfer characteristics. In a subsequent study, Yang et al. [19] investigated the flow and heat transfer in MFs having micropores in different directions.

Limited studies in the literature compare MFs having different porosities by direct CFD/HT modeling and provide micro-level physical insights into flow and heat transfer with the change of porosity and filament diameter. In the present study, two MF structures with the same PPI values but different porosities and filament diameters are investigated experimentally and numerically. Numerical simulations were made by using actual geometries obtained by μCT scans. This study investigates the effect of porosity, filament diameter, and nodes on fluid flow and heat transfer. Accordingly, results are analyzed through detailed computational visualization of local heat transfer coefficients and local temperature differences on the solid–fluid interface. New insights are obtained by comparing micro-level transport physics within actual geometries having varied porosity and filament diameter. A DNN machine learning (ML) algorithm is used to predict the heat transfer coefficient, solid and fluid temperatures, and heat flux between solid and fluid, significantly reducing computational time. The results are also helpful for designing heat exchanger channel geometries through an improved understanding of the regions where the heat exchange between the solid and the fluid domains effectively takes place.

3.1.2 Heat Transfer and AI/ML

In recent years, the use of artificial intelligence/machine learning (AI/ML) techniques in heat transfer simulations has gained attention. AI is an umbrella term that refers to software that mimics how humans think and, analyze, reason, and learn. On the other hand, ML is a subset of AI that can employ algorithms that are able to learn from datasets. The main attractiveness of AI/ML comes from the fact that it has been successfully implemented and has the potential to save computational time and energy in different fields, such as healthcare, education, marketing, and

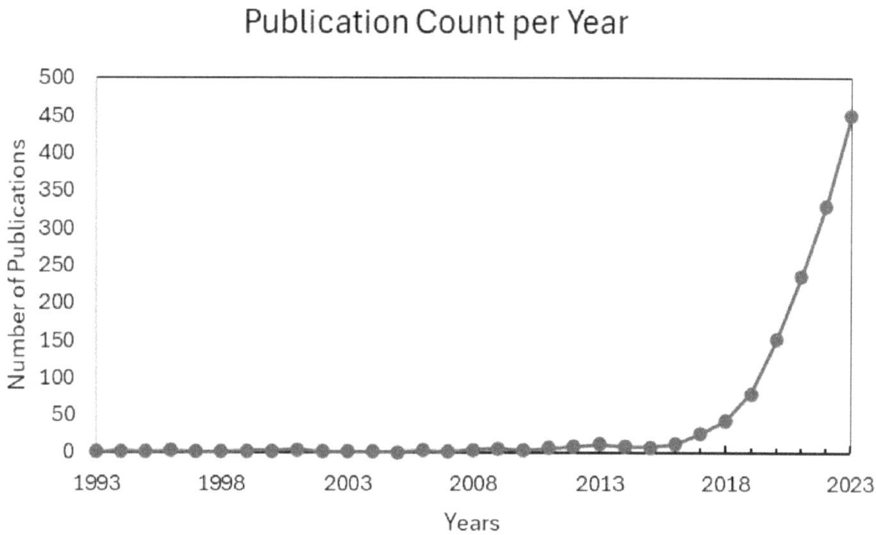

FIGURE 3.1 Publication counts per year between 1993 and 2024 [20].

natural language processing. AI/ML is being implemented in engineering fields, such as mechanical, civil, electrical, biomedical, and materials. These advancements gave rise to the idea that AI/ML techniques can also be implemented in heat transfer applications such as thermal management. Therefore, many studies have been carried out on the use of AI/ML in heat transfer, especially after 2015, as seen in Figure 3.1. As can be seen, there will be continuing exponential growth in the number of studies at the start of 2024. The application areas of the studies on AI/ML can be seen in Figure 3.2. Different topics in fundamental areas of heat transfer

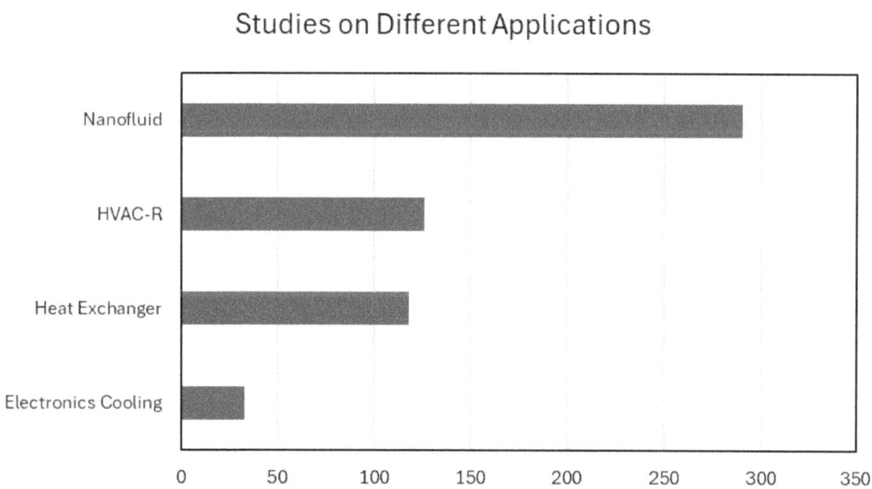

FIGURE 3.2 AI/ML studies on different application areas between 1993 and 2024 [20].

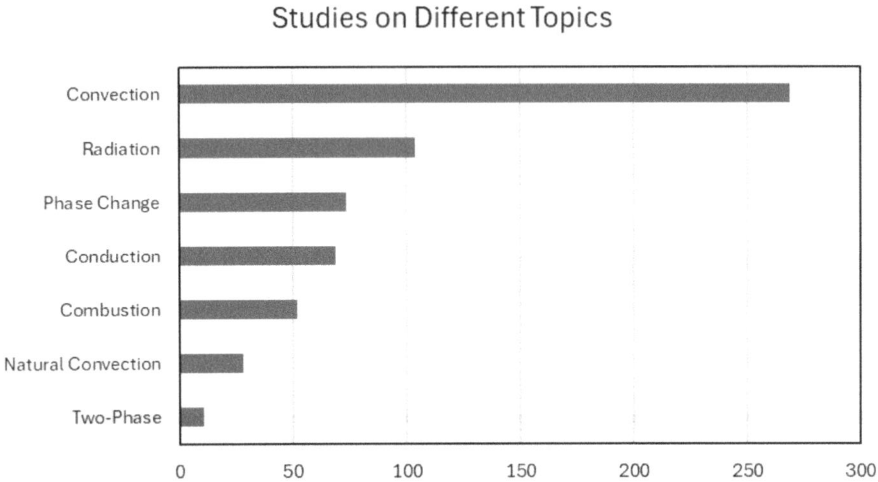

FIGURE 3.3 AI/ML studies on different fundamental topics between 1993 and 2024 [20].

can be seen in Figure 3.3, where most of the studies are on convection heat transfer. This chapter focuses on the use of AI/ML on convection heat transfer, and excludes discussions of nanofluids studies.

3.1.3 Studies on AI/ML Use in Heat Transfer

One of the most common AI/ML methods is neural networks. Neural networks were first proposed in 1943 by Warren McCullough and Walter Pitts [21]. Today, artificial neural networks (ANN) is a broader term for different neural network models. ANNs consist of interconnected nodes called neurons, inspired by the neurons in a human brain. The data is introduced into the algorithm via input neurons. Neurons in different layers are connected to each other. Each connection has its own weight, representing the strength of the connection. Therefore, data is processed through different layers. Finally, it reaches the output layer, which is called the prediction of the algorithm. In the training process, the algorithm trains itself by using the given data by adjusting the weights between neurons to minimize the loss function, which, in simple terms, is the difference between the predicted and actual data. A summary of AI/ML studies on convective heat transfer is provided in Table 3.1.

3.1.3.1 Neural Networks-Based Algorithms

Pacheco-Vega et al. [22] used ANN to predict the heat transfer of humid air–water heat exchangers, and reported more accurate predictions than conventional correlations. They emphasized that the network can capture the complex physics in a heat exchanger very well, but does not provide physical insight into the phenomena. Athavale et al. [23] used ANN to predict the temperature and flow profile in data centers. They reported good agreement between the ANN predictions and CFD/HT simulations with an average error below 0.6°C for rack inlet temperatures and 0.7% for tile flow rate. Souayeh et al. [24] used ANN to predict the *Nu*, *f*, and

TABLE 3.1
AI/ML Studies on Convective Heat Transfer

Algorithm	Task	Objective	Reference	Year
Artificial neural network (ANN)	Prediction of the heat transfer of humid air–water heat exchangers	Regression	Pacheco-Vega et al. [22]	2001
	Prediction of temperature and flow profile in data centers	Regression	Athavale et al. [23]	2018
	Prediction of *Nu*, *f*, and thermohydraulic efficiency of a flow in a circular channel with corrugated spring tape inserts	Regression	Souayeh et al. [24]	2021
	Prediction of friction factor, Nusselt number, irreversibility, exergy efficiency, and thermohydraulic efficiency	Regression	Bhattacharyya et al. [25]	2021
	Prediction of *Nu* in a rectangular channel with a curved winglet vortex generator	Regression	Berber and Gurdal [26]	2023
	Prediction of *Nu* on microscale pin fin heat sinks	Regression	Oh and Guo [27]	2023
	Prediction of pressure drop and heat transfer rate in microchannels and optimization	Regression Optimization	Huang et al. [28]	2023
	Prediction of thermal resistance and pumping power in microchannels	Regression	Sikirica et al. [29]	2023
	Prediction of the skin friction coefficient in a non-Newtonian stagnation point fluid flow towards two different inclined heated surfaces	Regression	Rehman and Shatanawi [30]	2023
	Prediction of friction factor and Nusselt number for flow through an in-line array of square pin fins under developing flow conditions	Regression	Pai and Weibel [31]	2023
	Prediction of Nu and pressure drop in microchannel heat sinks having baffle geometries	Regression Optimization	Shuqi et al. [32]	2023
	Prediction of frictional pressure gradient during two-phase flow	Regression	Mauro et al. [34]	2024
ANN-TL	Prediction of the heat transfer coefficient in a mini-channel two-phase heat exchanger having micro pin fins by using transfer learning model in new domains	Regression	Zhu et al. [35]	2024
ANN-DE	Prediction of convection and radiation heat transfer of wavy porous fins	Regression	Kumar et al. [36]	2023
DNN	Prediction of the LED junction temperature	Regression	Merenda et al. [39]	2020
	Prediction of fouling resistances of a cross-flow heat exchanger	Regression	Sundar et al. [40]	2020
	Prediction of pressure drop and heat transfer coefficient in air-cooled parallel plate-finned heat sinks	Regression Optimization	Shaeri et al. [41]	2023
DNN DRL	Controlling of conjugate heat transfer in a 3-dimensional open cavity with an immersed single heating source	Regression Control	Wang et al. [42, 43]	2024

(Continued)

TABLE 3.1 (*Continued*)
AI/ML Studies on Convective Heat Transfer

Algorithm	Task	Objective	Reference	Year
CNN	Prediction of turbulent heat transfer coefficient	Regression	Kim and Lee [44]	2020
	Prediction of effective thermal conductivity of solids	Regression Optimization	Adam et al. [45]	2024
PINN	Solving the heat transfer PDEs with convective boundary conditions	Regression	Zobeiry and Humfeld [48]	2021
	Prediction of temperature and heat flux fields in porous media	Regression	Xu et al. [49]	2023
MLPNN	Prediction of mixed and forced convection heat transfer coefficients for a radiant cooling system	Regression	Acikgoz et al. [50]	2022
	Prediction of convective, conductive, and radiative heat transfer of a porous longitudinal fin	Regression	Alhakami et al. [51]	2022
	Prediction of flow boiling heat transfer coefficient of R1234yf in a multiport mini channel tube	Regression	Agustiarini et al. [53]	2023
CS-ANN	Prediction of temperature distribution in rectangular-shaped porous fins and optimization	Regression Optimization	Waseem et al. [54]	2020
ELM	Prediction of fin temperature for rectangular and wavy fins.	Regression	Prakash et al. [55]	2023
BAFIS	Prediction of outlet velocity of water flow inside a pipe filled with copper porous media	Regression	Alotaibi et al. [58]	2023
FL	Prediction of *Nu* and *f* in conical spiral tubes	Regression	Soltanian and Beigzadeh [59]	2023
RF	Prediction of heat transfer coefficient for variable rib roughness	Regression	Kwon et al. [62]	2020
RF ANN	Prediction of pressure drop and temperature increment in lattice-structured heat sinks	Regression	Suzuki et al. [63]	2023
RF GA	Prediction of pressure drop and heated surface temperature in a cooling channel having pin-fin array	Regression Optimization	Nguyen et al. [64]	2023
GBR	Prediction of pumping power and *Nu*, and assessment of the relationship between different geometric parameters and heat transfer performance for a microchannel heat sink	Regression Optimization	Ma et al. [65]	2022
MLPNN XGBoost LightGBM	Prediction of *Nu* on micro-pin fin heat sinks	Regression	Kim et al. [66]	2022
ANN **XGBoost** RF AdaBoost	Prediction of flow condensation heat transfer coefficient in mini/micro-channel.	Regression	Zhou et al. [67]	2020

(*Continued*)

TABLE 3.1 (*Continued*)
AI/ML Studies on Convective Heat Transfer

Algorithm	Task	Objective	Reference	Year
ExtraTrees DTR AdaBoost RF GBR XGBoost k-NN Bagging	Prediction of the flow boiling heat transfer coefficient of liquid hydrogen	Regression	He et al. [70]	2022
MLR	Prediction of *Nu* for turbulent jet impingement cooling	Regression	Nagesha [71]	2023
k-NN	Prediction of the heat transfer coefficient for different geometric parameters.	Regression	Krishnayatra et al. [73]	2020
SVM **GBR** RF ANN	Prediction of *Nu* and friction factor during condensation in micro- and macro-channels	Regression	Hughes et al. [78]	2022
LS-SVM	Prediction of heat transfer and fin efficiency for different fin shapes	Regression	Shivanian et al. [80]	2023
SVM ANN GBR RF	Prediction of two-phase friction factor and Chisholm parameter in condensing and adiabatic flow in micro, mini, and macro channels.	Regression	Shourehdeli et al. [81]	2023
RF SVM ANN	Prediction of condensation frictional pressure drop and heat transfer coefficient for horizontal microchannel and macrochannel flow of different refrigerants	Regression	Hughes et al. [82]	2021
DNN MLR k-NN MLPNN DTR RF GBR	Prediction of the heat transfer coefficient and pressure drop in minichannel heat sinks.	Regression Optimization	Saeed et al. [83]	2023
GA	Enhancement of convective turbulent heat transfer in TBL	Optimization	Castellanos et al. [86]	2023
SVM GEP ANN	Prediction of flow boiling heat transfer coefficient of R134a in micro/mini-channels	Regression	Parveen et al. [90]	2020

thermohydraulic efficiency of a flow in a circular channel with corrugated spring tape inserts. Their model was able to predict the *Nu*, *f*, and thermohydraulic efficiency with 97% accuracy. Bhattacharyya et al. [25] examined turbulent flow and heat transfer through a circular tube with grooved tape inserts. They employed an ANN model for regression analysis to predict heat transfer, pressure drop, thermohydraulic efficiency, irreversibility, and exergy efficiency. Berber and Gurdal [26] used an ANN model to predict the *Nu* by using the input values of *Re*, angle of wing, and heater plate temperature in a rectangular channel with a curved winglet vortex generator. The model predicted the *Nu* with a correlation coefficient (R^2) of 0.9879. Oh and Guo [27] used ANN to predict *Nu* for forced convection heat transfer of microscale pin fin heat sinks for different geometries. They trained their model with the experimental data in the literature and reported better accuracy than the correlations in the literature. Huang et al. [28] used ML to optimize microchannel design. They improved the heat transfer rate by 5%, and decreased the pressure drop by 37% compared to their initial structure. They also reported that using the ANN model increased the prediction speed 63 times compared to traditional numerical models with a high prediction accuracy. Sikirica et al. [29] performed microchannel heat sink optimization by using different ML algorithms. They reported that neural networks can make accurate predictions, and be used to create an acceptable surrogate model. They also noted that the proposed methodology can generate acceptable solutions and requires one-fifth of computational time.

Rehman and Shatanawi [30] used an ANN model to predict the skin friction factor in a non-Newtonian stagnation point fluid flow towards two different inclined heated surfaces, where velocities ratio parameter, temperature Grashof number, Casson fluid parameter, solutal Grashof number, and magnetic field parameter are the inputs. They reported an average mean squared error (MSE) of 0.52%. Pai and Weibel [31] used ANN to predict the friction factor and Nusselt number for flow through an in-line array of square pin fins under developing flow conditions. They found that ANNs with one hidden layer having 20 nodes exhibit excellent performance for their case.

Shuqi et al. [32] performed a study on optimizing baffle geometries for maximized heat transfer efficiency in microchannel heat sinks using ANN. They used ANN to predict *Nu* and pressure drop. They reported R^2 values around 0.98, and 46% enhancement in the overall performance after the optimization is reported compared to a reference study [33]. Mauro et al. [34] used ANN to predict the frictional pressure gradient during two-phase flow and compared it with the predictions of conventional correlations. They used an extensive database with 8000 data points from 49 sources to train their ANN model with two hidden layers.

The transfer learning (TL) method is an ML technique where a model trained with a previous dataset is reused or used as the starting point for another model to predict another dataset. This approach saves time and resources and can be applied to different but related datasets. Zhu et al. [35] used ANN-TL to predict the heat transfer coefficient in a mini-channel two-phase heat exchanger having micro-pin fins. They compared the conventional ML model and the TL model. The results showed that the conventional ML model performs well, with an overall deviation of 4.11%. However, their TL model showed an overall deviation of 4.28% with 70% dataset from the new domain. This showed that TL frameworks can extend conventional ML models into different domains with reasonable prediction accuracy.

Kumar et al. [36] developed a hybrid model that combines the differential evolution (DE) algorithm with an ANN to predict the convection and radiation heat transfer from wavy porous fins. They stated that one of the drawbacks of ANN is that it converges to a local solution, instead of a global one. They proposed combining DE, developed by Storn [37], with the ANN method. The DE-ANN forecasting model showed consistent performance, while conventional ANN methods often provided overfit.

ANNs can be used by implementing different methods. Deep neural networks (DNN) are ANNs that can learn more complex data and relationships using multiple hidden layers. A significant advantage of DNNs is that they can capture the relationships in data, irrespective of the complexity or nonlinearity of the dataset, due to their higher number of hidden layers [38]. Therefore, they are preferred in engineering applications in which datasets have high complexity and nonlinearities. Merenda et al. [39] used DNN to predict the light-emitting diode (LED) junction temperatures. They reported good accuracies (±2°C) for the temperature range of interest (50–110°C). Sundar et al. [40] used DNNs to predict fouling resistances in a cross-flow heat exchanger used in waste heat recovery. They reported good accuracies, having R^2 values over 99%. Shaeri et al. [41] used DNN to predict the heat transfer coefficient and pressure drop in air-cooled parallel plate-finned heat sinks subjected to laminar flow. They conducted an optimization through DNN-based models and a greedy search algorithm.

Deep reinforcement learning (DRL) is a learning method often used with DNNs. This learning approach uses an agent to explore the environment. The agent learns from the consequences of its actions, and updates itself accordingly. Wang et al. [42] used DRL to achieve efficient forced convection control over an open cavity model with single or multiple heat sources immersed. Compared to intuitive control, they reported a reduction of 8°C due to improved cooling with the DRL agent. Wang et al. [43] used the DRL for intelligent control of conjugate heat transfer in a 3D open cavity with an immersed single heating source. They proposed a novel strategy that uses the policy learned in two-dimensions and transfers it into a 3D environment, reducing the high computational cost of DRL agent training.

Convolutional neural networks (CNN) are another type of neural network primarily used for image processing. The main idea behind the CNNs is to extract significant features from the input data by filtering the data in the connections. Kim and Lee [44] used CNN to predict the turbulent heat transfer. They reported that CNN can predict the heat flux with a correlation coefficient of 0.98, compared to direct numerical simulation (DNS). They also reported that the model shows a similar accuracy even at *Re* values three times higher than the trained domain. Adam et al. [45] used CNN to predict effective thermal conductivity. The model is inspired by the visual geometry group (VGG) networks. They trained their model using 130,000 unique binary images, and they achieved MAPE of 0.35% when the thermal conductivity of the solid was ten times higher than that of the fluid. However, they reported MAPE of 2.35% when the thermal conductivity of solid and fluid is closer to each other. They reported that the prediction time is 15 ms for a single image with 128 × 128 pixels, 3 to 5 orders of magnitude faster than the finite volume method (FVM).

Another type of ANN is the physics-informed neural network (PINN). PINNs use the governing partial differential equations (PDEs) of the physical phenomena as a loss function, along with other loss functions with the measurements, boundary conditions, and initial conditions [46]. Although PINNs can work on a coarser mesh compared to numerical solvers, speed is an issue for these algorithms, since they use gradient descent optimization [47]. Zobeiry and Humfeld [48] used PINN to solve the heat transfer PDE with convective boundary conditions. They reported that, unlike neural networks, PINNs with engineered features can capture the physics of the problem to make accurate predictions beyond the training zone. Xu et al. [49] used PINN to predict the temperature and heat flux fields in porous media. They reported only a 2.49% average relative error for effective thermal conductivity with respect to FVM. They also reported five orders of magnitude acceleration with respect to numerical simulations.

Multilayer perceptron neural networks (MLPNNs) generally refer to neural networks with one or more hidden layers with perceptron neurons. Acikgoz et al. [50] used MLPNN to predict the mixed and forced convection heat transfer coefficients of a radiant cooling system. They trained their model with 10 neurons in the hidden layer, and 35 datasets. They reported excellent agreement between the predicted and numerical results, with R values close to unity. Alhakami et al. [51] used MLPNN to predict convective, conductive, and radiative heat transfer of porous longitudinal functionally graded fins having linear, quadratic, and exponential thermal conductivity profiles. They also used the Tiki-Taka algorithm [52] and sequential quadratic programming to train the weights or the neurons in the NN architecture to minimize the loss function. They reported percentages of absolute errors and the MSE of between 10^{-4} to 10^{-5}, and 10^{-8} to 10^{-10}, respectively. Agustiarini et al. [53] used MLPNN to predict the flow boiling heat transfer coefficient of R1234yf inside a multiport mini-channel tube. The model predicted the boiling heat transfer coefficient with a mean deviation of 7.12%. They reported that this value is better than those of others, such as the superposition, asymptotic, and flow pattern models.

Cuckoo search ANN (CS-ANN) is a method that uses cuckoo search optimization to train the ANN. Waseem et al. [54] used a CS-ANN to predict the temperature distribution in rectangular-shaped porous fins and optimize the parameters.

Extreme learning machines (ELM) are also a specific type of ANN. Their distinctive feature is that they have a single hidden layer, where they initialize the input weights randomly and directly calculate the output weights. Since they don't use an iterative approach, ELMs are faster than ANNs. Prakash et al. [55] used an ELM model to predict the fin temperature for rectangular and wavy fins. They stated that ELMs have better capabilities and are less prone to overfitting than traditional neural networks.

3.1.3.2 Fuzzy Logic-Based Algorithms

Adaptive network-based fuzzy inference system (ANFIS) is a hybrid ANN/ML algorithm that combines neural network and fuzzy logic (FL) [56], when dealing with fuzzy datasets. The ANFIS algorithm is computationally costly compared to other methods since it has several fuzzy rules. Seyedashraf et al. [57] used ANFIS and MLPNN to predict the water depth and velocity. Alotaibi et al. [58] used an AI algorithm to learn and map their CFD data for water flow inside a pipe filled with copper porous media. They used a bee algorithm-based fuzzy inference system (BAFIS) and reported the highest intelligence as $R^2 = 0.97$. Soltanian and Beigzadeh [59] used

FL models to predict *Nu* and *f* with different geometric and velocity input parameters in conical spiral tubes. They used the Takagi-Sugeno Fuzzy Inference System [60], and reported good agreement between the FL prediction and numerical results with a maximum mean relative error of 6.2%.

3.1.3.3 Tree-Based Algorithms

Random forest (RF), first proposed by Tin Kam Ho in 1995 [61], is one of the most used algorithms in the literature. The RF algorithm creates several randomized decision trees and makes a decision by averaging the decisions from them. RF models are useful, especially in engineering applications, due to their simplicity and accessibility. Another advantage of RF is that it can show the feature importance, which is helpful, especially for engineering problems. Kwon et al. [62] used an RF algorithm to predict the convection heat transfer coefficients for a high-order nonlinear heat transfer problem with an R^2 higher than 0.966. Suzuki et al. [63] first analyzed the heat transfer properties of lattice-structured heat sinks by CFD/HT simulations. Then, they let their ML model learn the relationship between results and eleven structural features using RF feature importance. Using this model, they obtained the important features for pressure drop and heat transfer. They then used neural networks to predict pressure drop and temperature increment. Nguyen et al. [64] used RF algorithm to predict the pressure drop and the heated surface temperature in a cooling channel having pin fin array. Therefore, they easily obtained a large number of datasets for use in the subsequent genetic algorithm (GA) optimization. They reported an R^2 value of 0.98 for the testing dataset.

Gradient boosting (GB) methods such as extreme gradient boosting (XGBoost), light gradient boosting machine (LightGBM), or gradient boosting regression (GBR) differ from the RFs in terms of ensembling methods. Unlike classical RFs, GB methods combine the series of weak learner trees sequentially; therefore, each new tree corrects the error coming from the previous tree. The loss function calculates the error, and the negative gradient of the loss function defines the next tree. Ma et al. [65] used the GBR algorithm to assess the relationship between different geometric parameters and heat transfer performance for a microchannel heat sink. They predicted pumping power and *Nu* with R^2 values greater than 0.999. Kim et al. [66] collected 906 *Nu* data points from 15 studies on micro-pin fin heat sinks and used the collected data to train three different ML methods listed as MLPNN, XGBoost, and LightGBM. They reported superior prediction for all ML models compared to their newly proposed correlation. Zhou et al. [67] performed a study in which they used four ANN, RF, AdaBoost, and XGBoost, to predict flow condensation heat transfer coefficient in mini/micro-channel. They used 4,882 data points from 37 sources from the literature to train the algorithms. It is reported that ANN and XGBoost models showed the best predicting accuracy with MAEs of 6.8% and 9.1%, respectively.

Decision Tree Regression (DTR) uses a single tree. It splits the dataset to reduce the variance, and finally ends up dividing the dataset into several regions. Extremely Randomized Trees (ExtraTrees) [68] algorithm is another decision tree algorithm. This algorithm is similar to RF but has differences in data selection. RT uses bootstrap replicas, but ExtraTrees uses the entire original sample. Moreover, while RF chooses the optimal split, ExtraTrees chooses randomly [69]. This makes ExtraTrees faster in terms of computation.

He et al. [70] used eight different ML algorithms to predict the flow boiling heat transfer coefficient of liquid hydrogen using a dataset having 864 data points from three sources. The algorithms are listed as DTR, AdaBoost, RF, ExtraTrees, GBR, XGBoost, k-nearest neighbor (k-NN), and bootstrap aggregation (Bagging). They reported that ExtraTrees showed the best performance among the different models with an R^2 value of 0.9933.

3.1.3.4 Other Regression Algorithms

Multiple Linear Regression (MLR) is a statistical method that extends linear regression to a higher number of features. Nagesha [71] proposed eight MLR machine-learning algorithms to assess the heat transfer performance of turbulent jet impingement cooling. They reported that the ML models improved the prediction accuracy ranging from 20.4% to 66.5%, compared to the conventional regression models in estimating heat transfer performance.

k-NN, developed in 1951 by Fix and Hodges [72], is an ML algorithm used for classification and regression problems. The k is the number of data points chosen nearest to the predicted data point. The k-NN algorithm stores the training data, and uses it to predict the results using the k number of nearest data points. Krishnayatra et al. [73] used the k-NN algorithm to predict the thermal performance of fins for an axial finned-tube heat exchanger, and reported high prediction accuracies.

Gaussian process regression (GPR) [74] is another ML algorithm that can be used for regression. Unlike traditional regression methods, the Gaussian process provides a probability distribution over possible functions that fit a set of points. Also, instead of precise predictions. GPR provides a range of possible values, with a confidence level around each possible value.

Response surface methodology (RSM) is an ML regression method introduced by Box and Wilson in 1951 [75]. It is often used for optimization problems. This method usually involves conducting experiments or numerical simulations, and defining the effects of independent variables as factors on the response variable. Often, it uses two independent variables. The mathematical model defines the relation between the factors and the response variable. The regression model then optimizes the independent variables.

Support vector machine (SVM) is known mainly as a solution for binary classification problems by conducting convex optimizations [76]. However, they can also be used for regression analysis. The main idea behind the SVM is to find a hyperplane in an N-dimensional space, and classify the data in which N shows the number of features. This hyperplane can also be used for regression data as it transforms to a ε-tube for the regression [77]. Hughes et al. [78] used four ML models (SVM, RF, GBR, and ANN) to predict *Nu* and frictional pressure gradient during condensation of three zeotropic mixtures (ethane/propane, R245fa/pentane, and R410A) in micro- and macro-channels. They reported that the SVM made the best *Nu* prediction with a mean absolute percent error (MAPE) of 5%, and GB made the best friction factor prediction with a MAPE of 5.5%, among other algorithms. The Least Squares SVM (LS-SVM) algorithm, proposed by Suykens and Vandewalle [79], differs from traditional SVMs in that it directly focuses on minimizing the sum of squared errors between the data and predictions. Shivanian et al. [80] used the LS-SVM model to predict fin efficiency and heat transfer rate on convection/radiation heat transfer of rectangular, triangular, trapezoidal, and concave parabolic-shaped fins. They reported a good agreement between the numerical results and predictions.

Shourehdeli et al. [81] used four ML algorithms, namely, ANN, SVM, GBR, and RF, to predict two-phase friction factor and Chisholm parameter in condensing and adiabatic flow in micro, mini and macro channels. They collected 11,411 data points from 80 sources to train the algorithms. They reported that GBR performs better than other models, with a mean absolute relative deviation of 3.24% for the prediction of frictional pressure drop. Hughes et al. [82] used 4,000 data points to train SVM, RF, and ANN algorithms to predict the condensation frictional pressure drop and heat transfer coefficient for horizontal microchannel and macrochannel flow of refrigerants, including synthetic refrigerants, hydrocarbons, and natural refrigerants. They reported that the RF performs the best among others, with an absolute average deviation of about 4% both for pressure drop and heat transfer.

Saeed et al. [83] used different ML-based regression techniques, namely, MLR, k-NN, MLPNN, DTR, RF, GBR, and DNN, to predict the heat transfer coefficient and pressure drop in minichannel heat sinks. They reported that DNN provided the best predictions for both heat transfer coefficient and pressure drop. They also reported that the best R^2 value is obtained by a 3-layer network.

3.1.3.5 Gene-Based Algorithms

GA, developed by Holland and collaborators in the 1960s and 1970s [84, 85], is a search and optimization algorithm inspired by natural selection and genetics in nature. Hence, GAs are usually used for optimization problems in heat transfer. Castellanos et al. [86] used a GA to enhance the convective heat transfer in the turbulent boundary layer (TBL) of a flat plate. They reported a 17.5% enhancement in heat transfer with respect to the unperturbed TBL.

Gene expression programming (GEP), invented by Ferreira [87], and genetic programming (GP), developed by Koza [88], are other variations of GA. GEP also uses populations of individuals, selects them according to fitness, and introduces genetic variation using genetic operators. The three algorithms differ in the nature of the individuals. GA uses linear strings of fixed length (chromosomes), GP uses nonlinear entities of different sizes and shapes (parse trees), and GEP also uses nonlinear entities of varying size and shapes (expression trees), but these complicated entities are encoded as simple strings of constant length [89]. Parveen et al. [90] used GEP, ANN, and SVM to predict the flow boiling heat transfer coefficient of R134a in micro/mini channels. They reported SVM is the best model among other AI-based models for predicting the heat transfer coefficient of R134a with an R^2 value of 0.9749.

Different AI/ML methods used in heat transfer studies are shown in Figure 3.4. ANN is excluded here, since it includes various types of neural network methods. The most common AI/ML methods are DNN and SVM algorithms, which are slightly more intuitive than others. GA is the most common optimization tool, followed by RF. RFs are mostly preferred due to their ability to show the importance of features in datasets. This ability makes them attractive for engineering problems. Overall, it

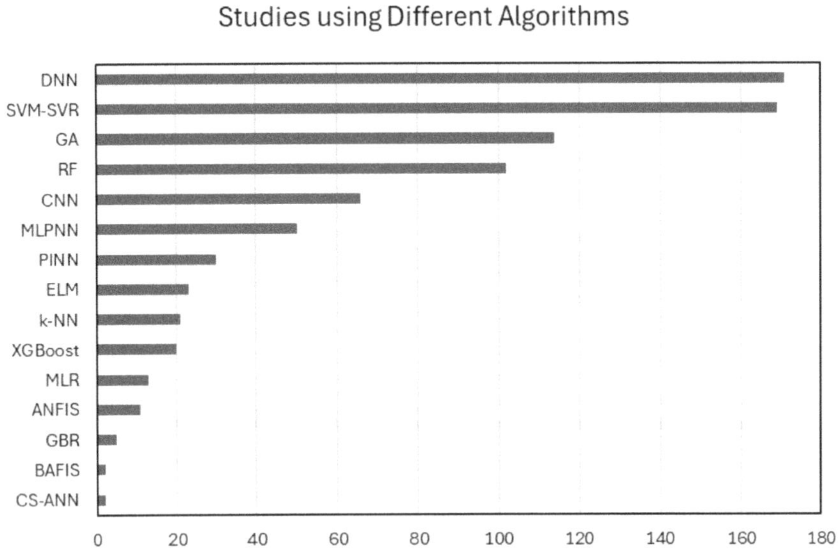

FIGURE 3.4 Different AI/ML methods used in the studies between 1993 and 2024 [20].

can be deduced that neural networks have significant dominance in the literature of AI/ML, since many of the algorithms in the list are a type of neural network. The distribution of methods in the studies on different fundamental topics is depicted in Figure 3.5. Generally, the methods are distributed similarly in different topics. However, some methods are preferred for some specific fields. GA is preferred for multi-phase transport, and conduction. DNNs, which are suitable for more complex

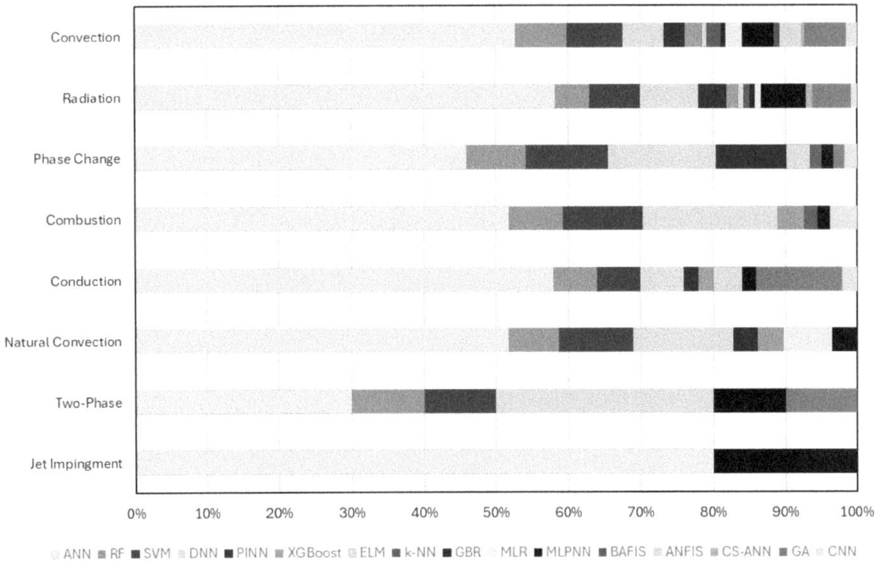

FIGURE 3.5 Different methods used in AI/ML studies between 1993 and 2024 [20].

problems, are preferred in studies on multi-phase transport and combustion. PINNs are mostly preferred in studies where phase change is present.

3.1.4 Studies on AI/ML Use in Metal Foam Heat Transfer

There are also several studies on the use of AI/ML to predict different parameters of MFs as summarized in Table 3.2.

Many studies are conducted to predict the constants representing Darcy and non-Darcy terms in MFs. Jafarizadeh et al. [91] trained DNN and SVM algorithms with CFD simulation data of MF, and used those models for the prediction of the constants representing Darcy and non-Darcy terms. They reported R^2 values of 0.9836 and 0.978 for DNN and SVM, respectively. Jafarizadeh et al. [92] performed pressure drop predictions of MFs by ML models. They used ordinary least squares (OLS), Lasso regularization regression (LRR), Ridge regularization regression (RRR), and

TABLE 3.2
AI/ML Studies on Convective Heat Transfer

Algorithm	Task	Objective	Reference	Year
DNN SVM	Prediction of the constants representing Darcy and non-Darcy terms in metal foams	Regression	Jafarizadeh et al. [91]	2023
OLS **RRR** LRR RF		Regression	Jafarizadeh et al. [92]	2023
ACOFIS ANFIS	Prediction of turbulence eddy dissipation of water flow in a heated metal foam tube	Regression	Babanezhad et al. [93]	2020
ANFIS	Prediction of axial velocity in a metal foam tube that is exposed to a constant heat flux	Regression	Babanezhad et al. [94]	2020
ANN	Prediction of the ratio of *Re* to maximum *Re*	Regression	Otaru et al. [95]	2023
SVM **ANN** *k*-NN RF XGBoost	Prediction of *Nu* and *f* in metal foam heat sinks	Regression	Tikadar and Kumar [96]	2022
ANN RF XGBoost SVM	Prediction of *Nu* and *f* for an integrated pin fin-metal foam heat sink	Regression	Tikadar and Kumar [97]	2023
MLPNN	Prediction of transient melting fraction and heat storage	Regression	Li et al. [98]	2023
ANN (LSTM-BP)	Prediction of transient PCM temperature in a multi-channel thermal storage tank with metal foam	Regression	Xiao et al. [99]	2023
SVM GA	Prediction of cell performance of a proton exchange membrane fuel cell and optimization of geometrical parameters.	Regression Optimization	Zhang et al. [100]	2020

RF. It is reported that RRR and OLS have better accuracy than the others, and give the lowest MAE and RMSE values for the coefficients for pressure drop.

Some studies focused on the use of AI/ML for the prediction of hydrodynamic characteristics of the MFs. Babanezhad et al. [93] studied the prediction of turbulence eddy dissipation of water flow in a heated MF tube using AI. They integrated the ant colony optimization-based fuzzy inference system ACOFIS with CFD/HT modeling. They reported good agreement with CFD/HT data, and faster learning compared to the adaptive network-based fuzzy inference system (ANFIS). Babanezhad et al. [94] investigated axial velocity in a MF tube that is exposed to a constant heat flux, using the ANFIS model. Otaru et al. [95] used ANN to predict the ratio of pore-diameter *Re* to maximum pore-diameter *Re* in aluminum foams. They stated that the three-layer ANN model is the best for predictions and reported that the predictions using this approach were within experimental scatter, with deviations between 0.3% and 9%.

The prediction of *Nu* and *f* in MFs is another objective for AI/ML algorithms. Tikadar and Kumar [96] predicted the thermal/hydraulic performance of MFs by using five different ML models. They reported that SVM and ANN outperformed other models in the predictions. In a subsequent study, Tikadar and Kumar [97] used RF, XGBoost, SVM, and ANN to predict the pressure drop and heat transfer of an integrated pin fin-MF heat sink. They reported that the ANN model performed the best among others in predicting the thermal-hydraulic performance of the heat sinks with a lower mean absolute percentage error.

MFs can be used with phase change materials (PCM). Therefore, some studies used AI/ML methods to predict the parameters related to PCM applications. Li et al. [98] used MLPNN to predict the transient melting fraction and heat storage of a vertical latent heat thermal energy storage. They used 204 groups of datasets and reported 2.0195% and 1.9578% mean relative deviations for melting fraction and heat storage, respectively. Xiao et al. [99] used long short-term memory back propagation (LSTM-BP) neural networks to predict liquid fraction and average temperature of PCM versus time in a multi-channel thermal storage tank filled with MF. They reported a good agreement between the predicted and simulated values.

Zhang et al. [100] used SVM to predict the cell performance of a proton exchange membrane fuel cell at various MF porosities. Then, they used GA to optimize the geometrical parameters. They obtained R^2 values of 0.99 for both training and test sets for SVM algorithm.

3.2 CONJUGATE TRANSPORT MODEL FOR METAL FOAMS

3.2.1 Model and Governing Equations

In order to illustrate results of practical interest, commercial MFs used in heat exchanger applications were analyzed in the present study. As seen in Figure 3.6, μCT scans were used to obtain the internal structure of MFs for the CFD/CHT simulations. Using the μCT data, the characteristics of the MFs have been defined, such as the pore diameter (d_p), and filament diameter (d_f). Additionally, specific surface area (a_{sf}) is calculated from the stereolithographic images of the MFs. The permeability and form coefficient are derived using the experimental pressure drop data inside the MF. Derivation details for these parameters are explained in the following section. All the geometric and derived parameters of the MFs have been listed in Table 3.3.

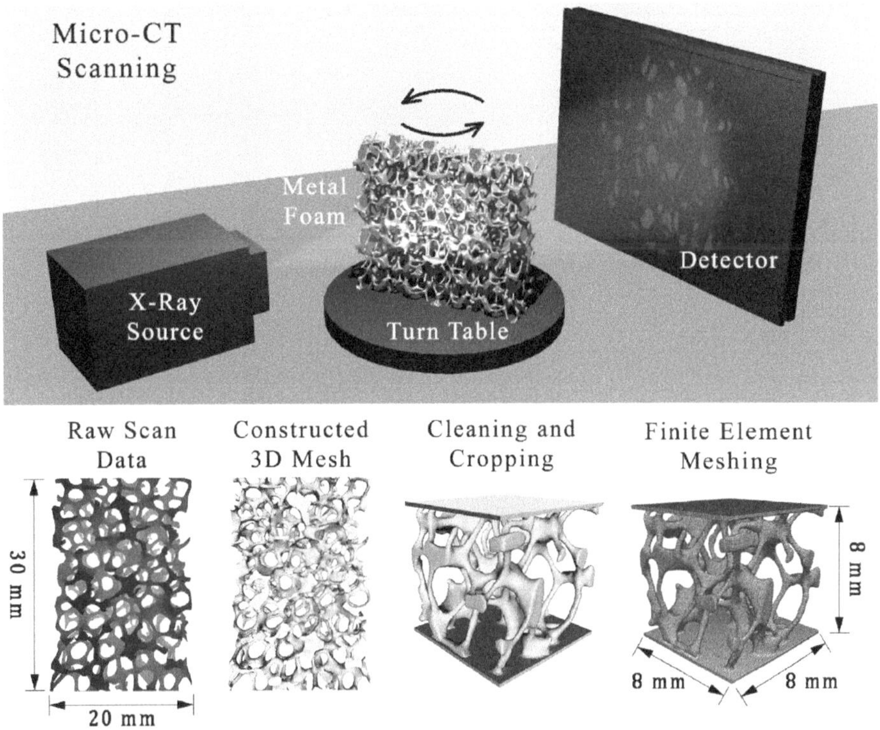

FIGURE 3.6 Preparation of μCT scans for CFD analysis.

The MFs utilized for CFD/HT simulations are cubes of side 8 mm. The computational domain is extended 12 mm upstream and 36 mm downstream to avoid any backflow effects and ensure the flow is fully developed. Inlet and outlet boundary conditions were applied from the upstream and downstream faces of the enclosure. Periodic boundary conditions were applied from both sides. Heat flux boundary condition was applied from the upper side of the heat spreader. Water is selected as the coolant, with density, viscosity, and Prandtl number as 997.78 kg/m^3, 9.772 × 10^{-4} kg/ms, and 6.587,

TABLE 3.3
Geometric Parameters of the MFs

Parameter	Symbol	Low Density	High Density
Porosity	E	0.946	0.879
Pore density	–	10	10
Permeability (m^2)	K	1.22 × 10^{-7}	0.87 × 10^{-7}
Form coefficient	F	0.090	0.079
Filament diameter (mm)	d_f	0.40	0.48
Pore diameter (mm)	d_p	3.26	3.33
Interfacial area density (m^{-1})	a_{sf}	477	732

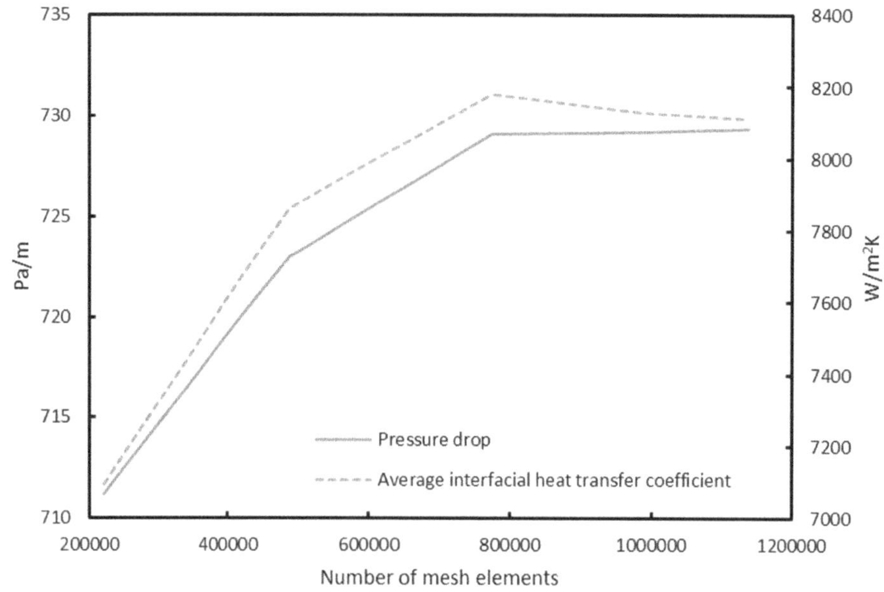

FIGURE 3.7 Mesh independence for pressure drop and heat transfer.

respectively. A mesh sizing study is conducted to ensure mesh size independence. Figure 3.7 shows the pressure drop and average heat transfer coefficients with varying numbers of mesh elements. As a compromise between accuracy and computational effort, 773.4 K mesh elements have been chosen. Numerical simulations were conducted for velocities in the 0.005 m/s–0.06 m/s range. The parameters of the numerical simulations can be seen in Table 3.4.

Convection in the fluid and coupled conduction in the solid were solved by the finite volume method with OpenFOAM with the chtMultiRegionFoam module. Laminar flow is assumed since viscous effects are dominant due to the stochastic structure of the MFs, and *Re* based on the permeability is assessed between 2 and 27, and the transition of laminar to turbulent is usually defined around 50 in the literature [101, 102]. The governing equations are:

The continuity:

$$\nabla.(\rho V) = 0 \tag{3.1}$$

TABLE 3.4
Parameters of the Numerical Simulations

Parameter	**Value**
MF dimensions	$8 \times 8 \times 8$ mm^3
Velocity range	0.005–0.06 m/s
Reynolds range (based on *H*)	40–490
Heat flux	5.65 W/cm^2

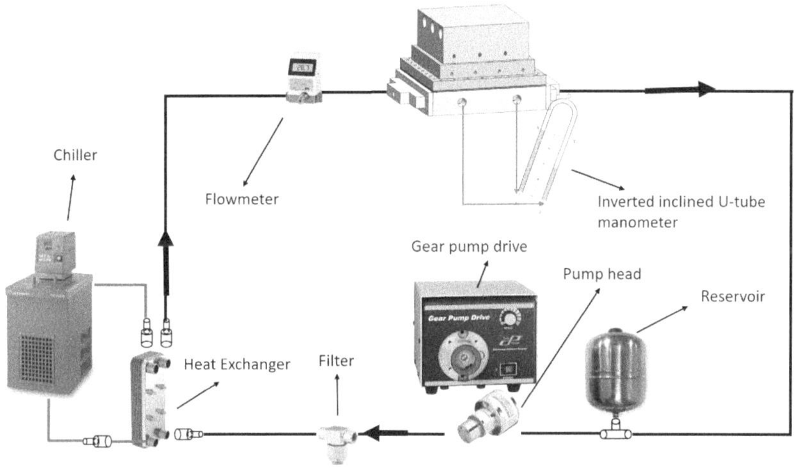

FIGURE 3.8 Flow loop for the pressure drop measurements.

The momentum transport:

$$\nabla.(\rho \boldsymbol{V}\boldsymbol{V}) = -\nabla \boldsymbol{p} + \nabla.(\mu \nabla \boldsymbol{V}) \tag{3.2}$$

The energy conversation:

$$\nabla.(\boldsymbol{V}\rho C_P T) = \nabla.(k\nabla T) \tag{3.3}$$

The numerical calculations were conducted at TUBITAK ULAKBIM, High Performance and Grid Computing Center located in Türkiye.

3.2.2 Experimental Validation Through Pressure Drop

A flow loop seen in Figure 3.8 is assembled to measure the pressure drops for the examined MFs. A Cole–Parmer gear pump with a flow rate range of 0-4550 mL/min is used for circulating the fluid. The flow rate was measured by a McMillan S-112 flowmeter located just before the test section. A plate heat exchanger connected to a chiller ensures constant temperature coolant delivery of 20°C at the inlet. The MF was cut and placed inside the inlet manifold as a flow straightener. An inclined manometer measured the pressure drop between two holes with a 60 mm distance on the side of the channel. Accordingly, pressure drop per meter in the MFs were obtained. Key parameters of the experimental setup are provided in Table 3.5.

TABLE 3.5
Measurement Parameters

Parameter	Value	Uncertainty
MF dimensions	94 × 94 × 8 mm	±0.1 mm
Velocity range	0.01–0.06 m/s	±0.0018 m/s
Re range (based on *H*)	80–490	±27

Pressure drop per meter were also obtained from the numerical solutions. A comparison of the computed pressure drop values with the experimental data can be seen in Figure 3.9, where there is generally a good agreement between 7% and 12% for low density (LD) and high density (HD), respectively.

3.2.3 Empirical Validation of Heat Transfer Coefficients

The numerical model is a representative unit cell in the MF region. Validation can be done using empirical models in the literature. Calmidi et al. [103] used the following equation for the local heat transfer coefficient:

$$h_{fs} = \frac{k_f C_T Re^{0.5} Pr^{0.37}}{d_f} \tag{3.4}$$

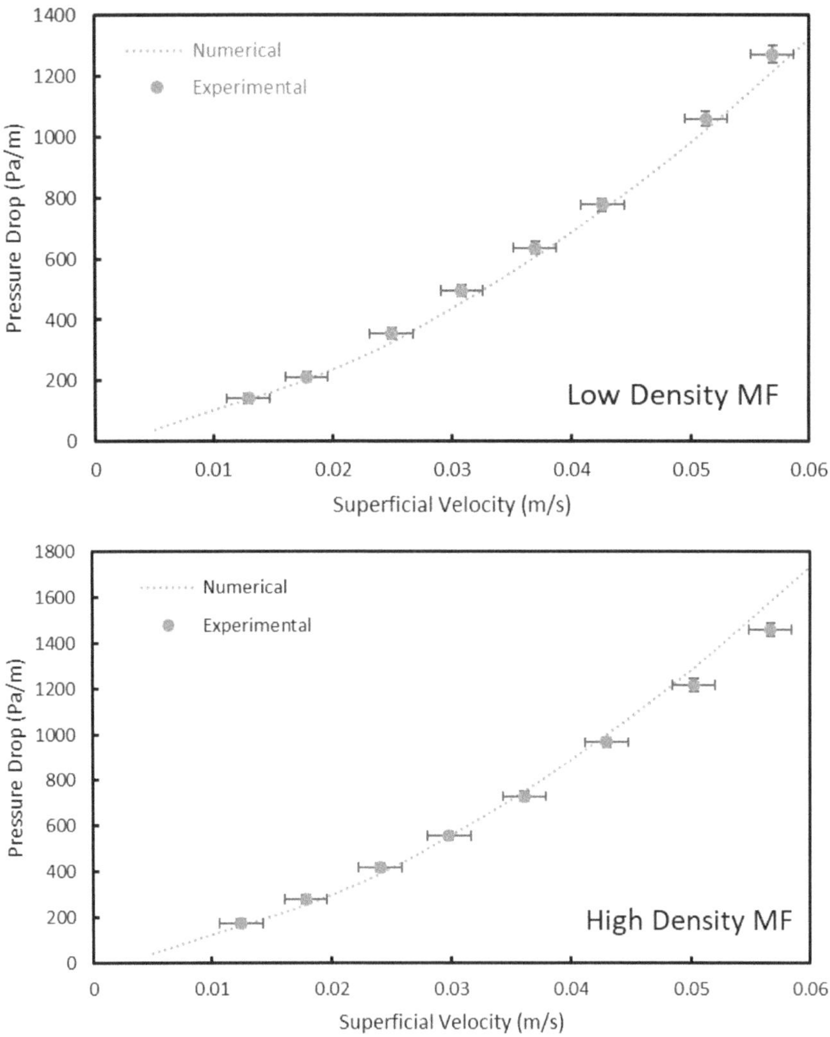

FIGURE 3.9 Validation of the numerical model by the experimental pressure drop data.

Re in Eqn. 3.4 is based on filament diameter. C_T is an empirical coefficient that varies between 0.5-0.9 in the literature [103, 104]. In this study, $C_T = 0.7$ showed the best agreement with the obtained data. Eqn. 3.4 is based on the Zukauskus correlation for the average Nusselt number, *Nu*, for flow over cylinders in forced convection [105]. Another correlation for the average *Nu* for flow over a circular cylinder for forced convection is defined by Hilpert for *Re* in the range of 4–40 [106]:

$$Nu = 0.911 Re^{0.385} Pr^{1/3} \tag{3.5}$$

The empirical correlations shown in Eqs 3.4 and 3.5 were used to validate heat transfer results, as shown in Figure 3.10. These are seen to be in good agreement with the numerical predictions.

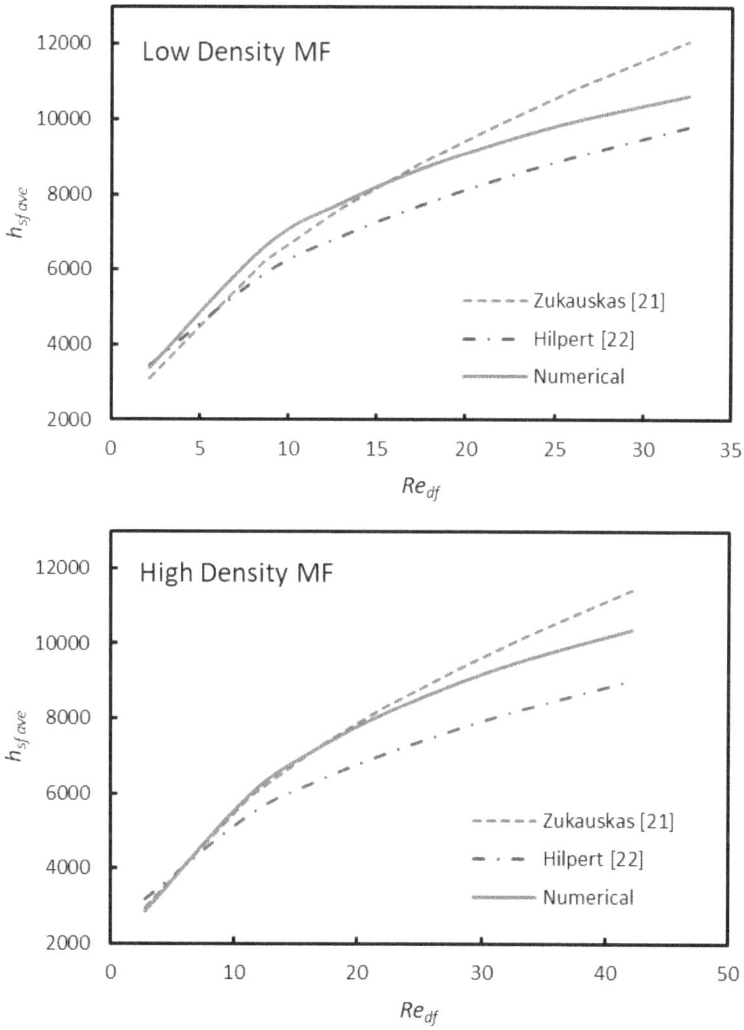

FIGURE 3.10 Agreement between the numerical results and empirical correlations in the literature.

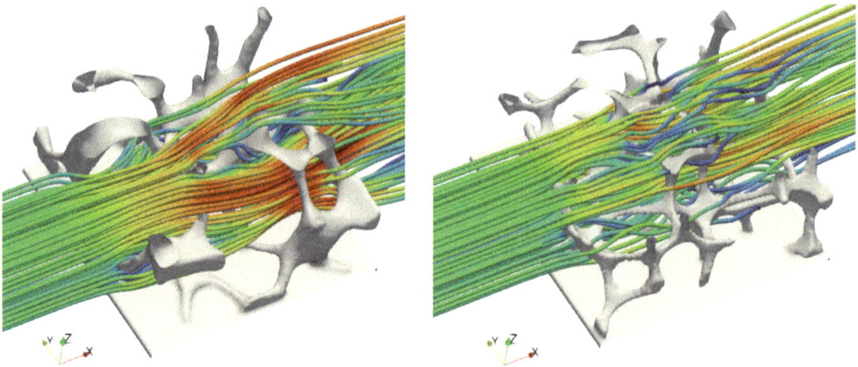

FIGURE 3.11 Streamlines for (a) high-density ($\varepsilon = 0.879$) and (b) low-density ($\varepsilon = 0.946$) MFs.

3.3 RESULTS AND DISCUSSION

3.3.1 Numerical Simulation Results

3.3.1.1 Flow Visualization

Flow patterns for MFs of 10 PPI having different porosities are depicted in Figure 3.11 for the inlet velocity of 0.02 m/s at the macro level. The flow locally accelerates in the structure up to 0.06 m/s for HD and 0.04 m/s for LD.

Since the temperature gradient is present only in the vertical direction of the fluid, mixing of fluid in this direction is desired. Figure 3.12 shows the longitudinal mixing of the fluid around the nodes and filaments. As seen for both porosities, filaments don't significantly affect the longitudinal mixing in laminar flow. However, the nodes block the flow and promote longitudinal mixing along the flow. There is a significant effect of longitudinal mixing due to the blockage

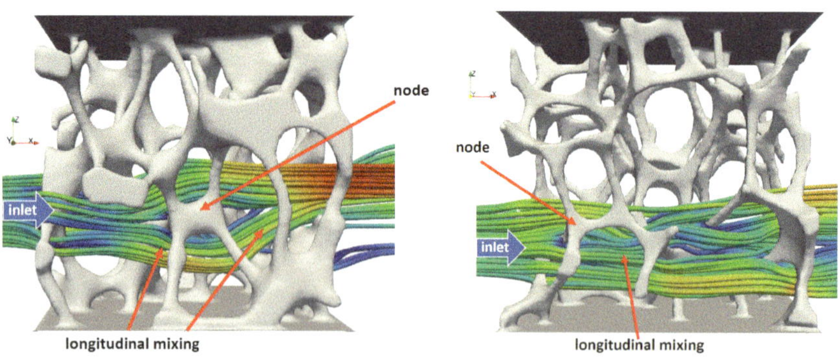

FIGURE 3.12 Longitudinal mixing within foam due to blockage of nodes in the flow field for (a) high-density ($\varepsilon = 0.879$) and (b) low-density ($\varepsilon = 0.946$) MFs.

of the nodes for the high-density MF. This phenomenon can also be seen in the low-density MF but is less pronounced. Flow mixing is usually included in the porous media equations by introducing a dispersion coefficient [107], correlated with the MF parameters [5, 103]. However, these models don't include the effect of nodes on the flow mixing. Instead, the permeability is included. Permeability is defined empirically from the measured pressure drop data and doesn't account for the form effects. Instead, the form coefficient appears in the quadratic term of the Hazen–Dupuit–Darcy equation [108, 109] and accounts for the pressure drop due to the form effects:

$$\frac{\Delta P}{\Delta x} = \frac{\mu}{K} V_{in} + C\rho V_{in}^{2} \tag{3.6}$$

Thus, the correlations for the dispersion coefficient should incorporate the impact of nodes that are directly accounted for. This is an area that could be explored further in future studies.

3.3.1.2 Heat Transfer Results

Figure 3.13 shows the temperature distribution for the two MFs. The temperature normal to the heated surface is more homogenous in the high-density MF due to the higher effective thermal conductivity. Therefore, regions with a temperature close to 20°C occupy more space in a low-density MF. This results in more unused interfacial area for heat transfer.

Figures 3.14 and 3.15 show the local heat transfer coefficients for the high-density MF. The local and average heat transfer coefficients are defined as follows:

$$h_{sf,\,local} = \frac{\dot{q}_{local}}{\left(T - T_{inlet}\right)} \tag{3.7}$$

$$h_{sf,\,ave} = \frac{\int_A h_{local} dA}{A} \tag{3.8}$$

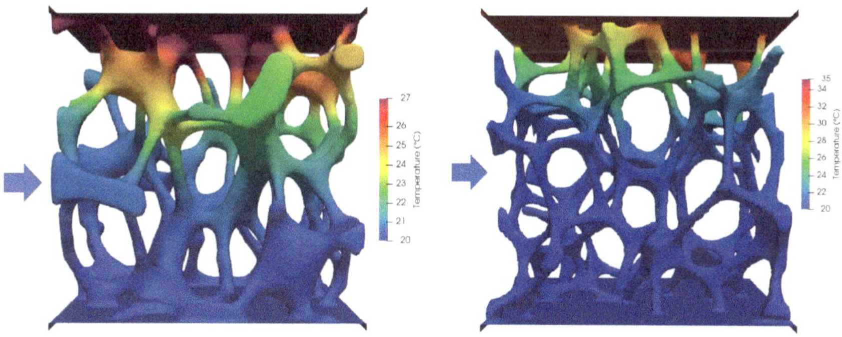

FIGURE 3.13 Temperature contours of (a) high-density ($\varepsilon = 0.879$) and (b) low-density ($\varepsilon = 0.946$) MFs.

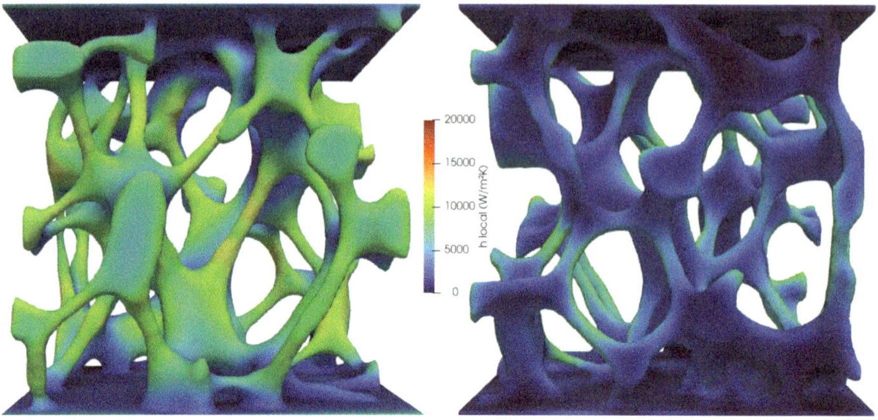

FIGURE 3.14 Front and rear view of $h_{sf,local}$ for high-density ($\varepsilon = 0.879$) MF.

Since the flow impinges on the front side, a higher $h_{sf,local}$ is observed there. The rear side of the nodes has lower $h_{sf,local}$ due to the formation of wake regions. However, the $h_{sf,local}$ on the rear side of thinner filaments is higher than that of others, since the flow separation is less pronounced. This can be seen in the side view in Figure 3.15. Also, the areas in the vicinity of the walls have lower $h_{sf,local}$ due to the lower velocities.

Figures 3.16 and 3.17 show the $h_{sf,local}$ for low-density MF. The difference in $h_{sf,local}$ between the front and rear sides of the filaments is less pronounced since thinner filaments and smaller nodes are present in a low-density MF. Since the rear side of the filaments contributes to heat transfer, the h_{ave} is slightly higher than for high-density MF.

Figure 3.18 shows the side view of the fluid and solid portions of the high-density MF. The black lines in the figure illustrate the solid–fluid interface. As can be seen, the temperatures of solid and fluid are nearly the same at the bottom. This shows the ineffective portion of the MF, where the foam structure doesn't contribute to heat transfer. This can be addressed using the appropriate channel height, or by

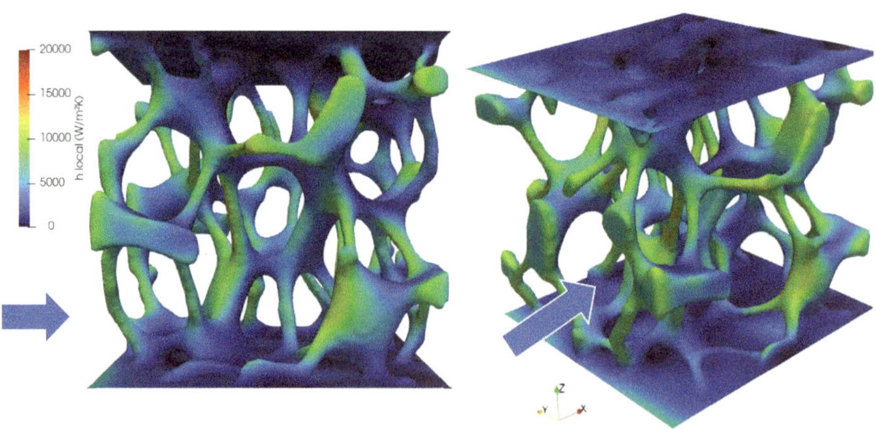

FIGURE 3.15 Side and isometric view of $h_{sf,local}$ for high-density ($\varepsilon = 0.879$) MF.

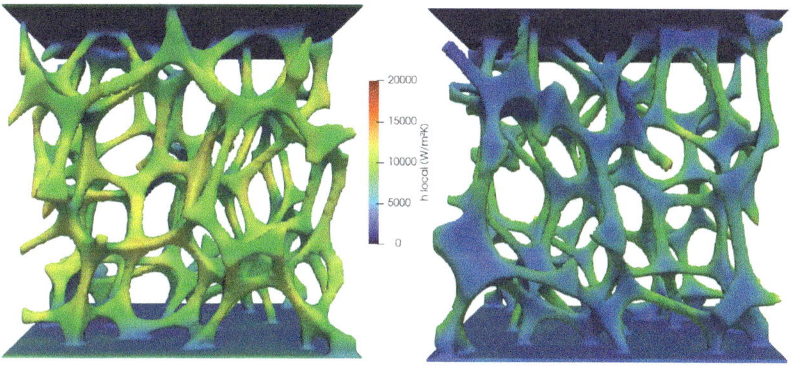

FIGURE 3.16 Front and rear view of $h_{sf,local}$ for low-density ($\varepsilon = 0.946$) MF.

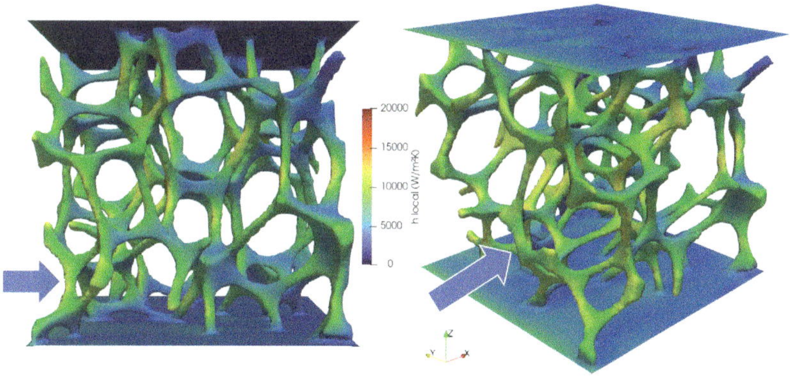

FIGURE 3.17 Side and isometric view of $h_{sf,local}$ for low-density ($\varepsilon = 0.946$) MF.

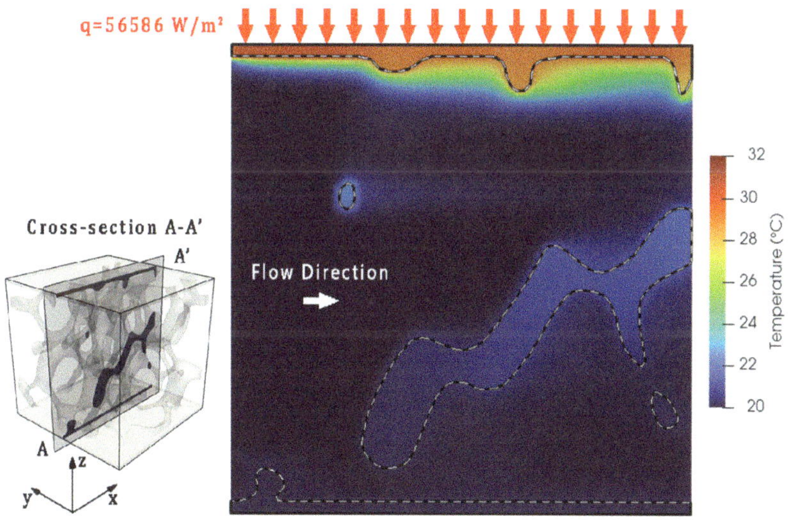

FIGURE 3.18 Temperature distribution in fluid and solid portions for high-density ($\varepsilon = 0.879$) MF (Black-white lines show the solid–fluid interface).

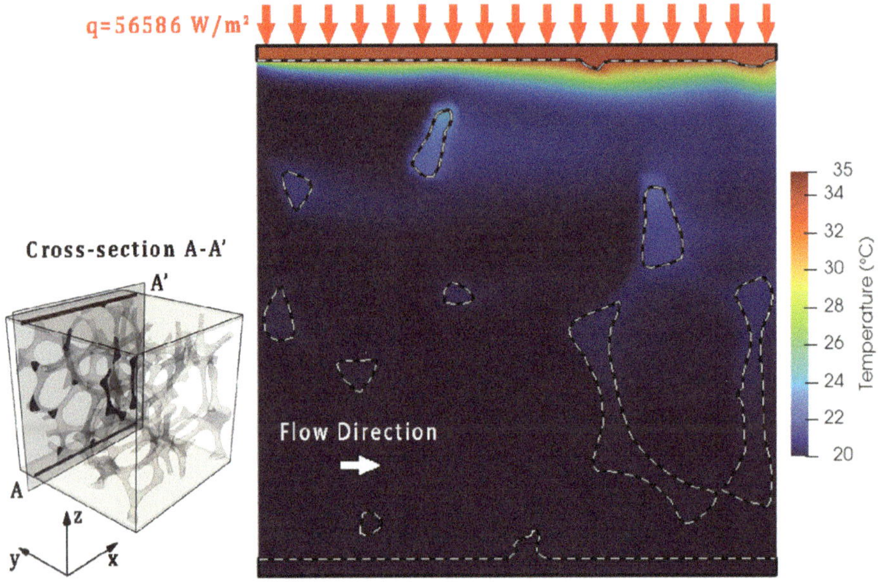

FIGURE 3.19 Temperature distribution in fluid and solid portions for low-density ($\varepsilon = 0.946$) MF (Black-white lines show the solid–fluid interface).

introducing surface enhancements such as pin fins that direct the heat to the bottom areas of the MF. Similarly, Figure 3.19 shows the temperature distribution for the fluid and solid portions of the low-density MF. For the low-density MF, the temperature difference between the fluid and solid portions is less pronounced due to the lower effective thermal conductivity.

Figures 3.20–3.23 show the heat flux on the surface of the MFs. As can be seen, the heat transfer is more pronounced at the upper part of the MFs and on the surfaces facing the flow. The rear surface of the thinner filaments also contributes to

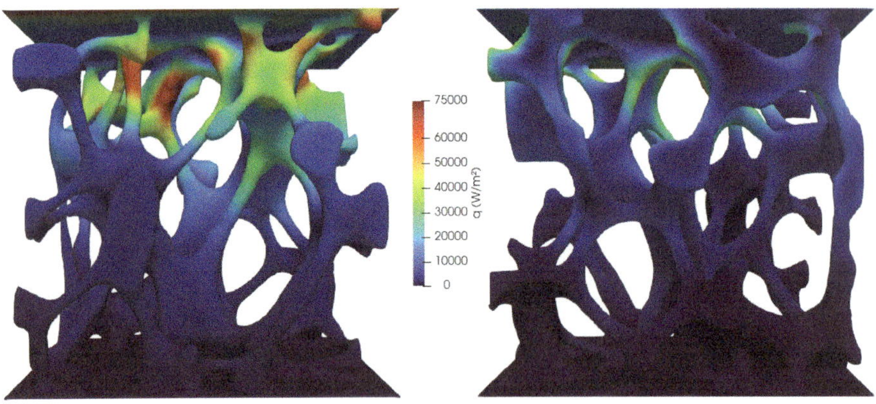

FIGURE 3.20 Front and rear view of $\dot{q}_{local}$ for high-density ($\varepsilon = 0.879$) MF.

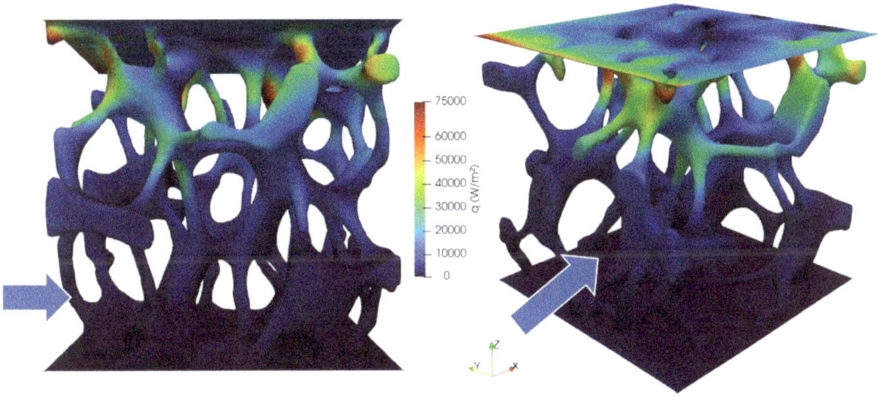

FIGURE 3.21 Side and isometric view of $\dot{q}_{local}$ for high-density ($\varepsilon = 0.879$) MF.

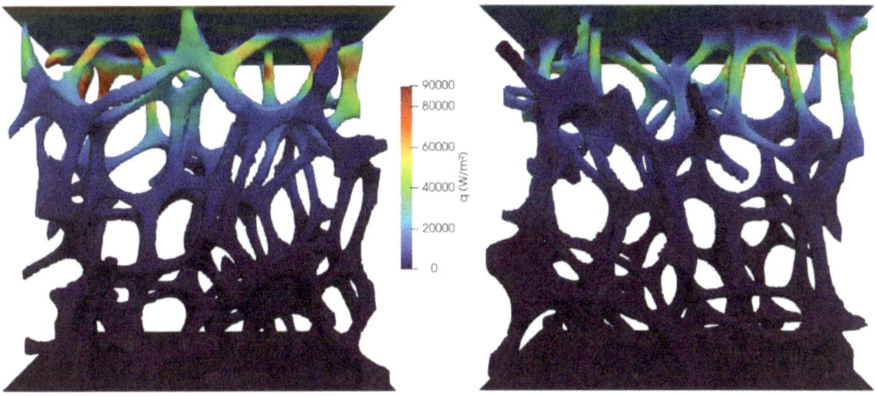

FIGURE 3.22 Front and rear view of $\dot{q}_{local}$ for low-density ($\varepsilon = 0.946$) MF.

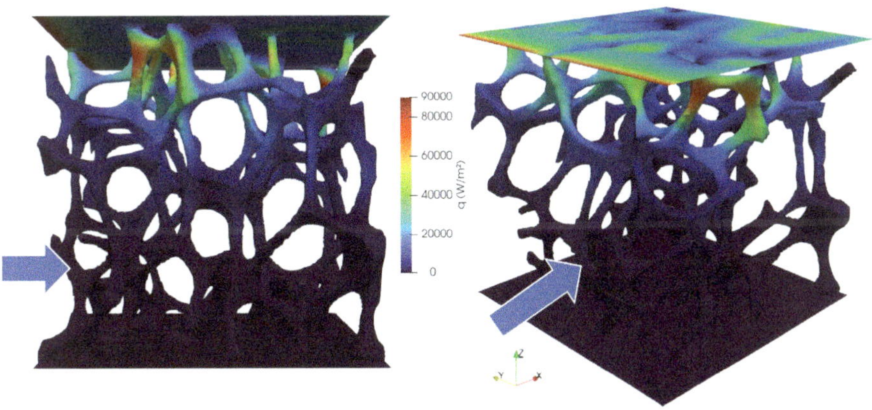

FIGURE 3.23 Side and isometric view of $\dot{q}_{local}$ for low-density ($\varepsilon = 0.946$) MF.

heat transfer, since the flow separation is less pronounced. Thus, for the low-density MF, the contribution to heat transfer of front and rear surfaces is more homogenous compared to the high-density MF.

3.3.1.3 Heat Transfer Coefficient, Wall Heat Flux, and Temperature Profiles

The variation of the average heat transfer coefficient, wall heat flux, solid temperatures, and fluid temperatures by the height at the inlet velocities of 0.02 m/s and 0.06 m/s for high and low-density MFs were obtained to reveal the height effects on the overall heat transfer in the MFs. The plane-averaged solid and fluid temperatures were calculated by slicing the domain by XY planes at height z and by integrating them along the cross-section and dividing it by the cross-section area as follows:

$$T_{solid\ pl\ ave} = \frac{\int_A TdA}{A_{solid}} \tag{3.9}$$

$$T_{fluid\ pl\ ave} = \frac{\int_A TdA}{A_{fluid}} \tag{3.10}$$

Heat transfer coefficient and heat flux quantities occur at the interface between solid and fluid. When the interface is sliced similarly, the intersection of the interface and the slice plane gives contour lines. The plane-average of the heat transfer coefficient and heat flux were calculated by integrating them along these contour lines and dividing them by the total length of the contour lines as follows:

$$h_{sf\ pl\ ave} = \frac{\int_L h_{sf,\ local} dL}{L} \tag{3.11}$$

$$\dot{q}_{pl\ ave} = \frac{\int_L q_{local} dL}{L} \tag{3.12}$$

Figures 3.24–3.27 show the variation of the averages of the mentioned parameters with height for two different MFs and velocities. $h_{sf\ pl\ ave}$ shows a similar trend for

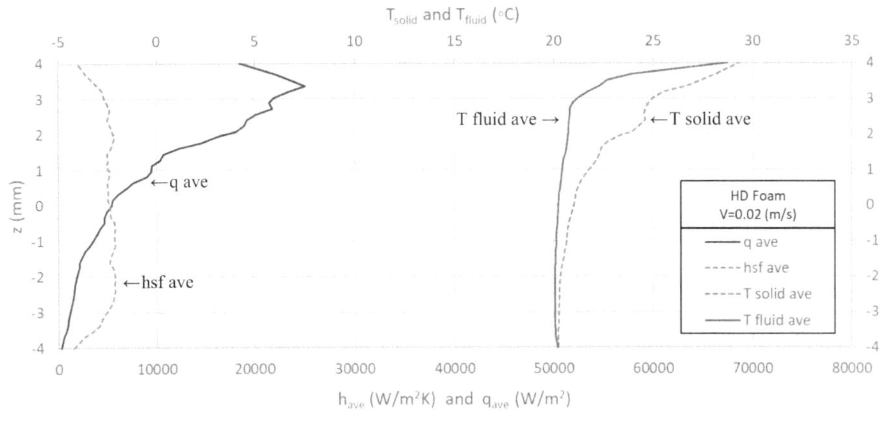

FIGURE 3.24 Variation of $h_{sf\ pl\ ave}$, $\dot{q}_{pl\ ave}$, $T_{solid\ pl\ ave}$, and $T_{fluid\ pl\ ave}$ with height for high-density ($\varepsilon = 0.879$) MF at 0.02 m/s inlet velocity.

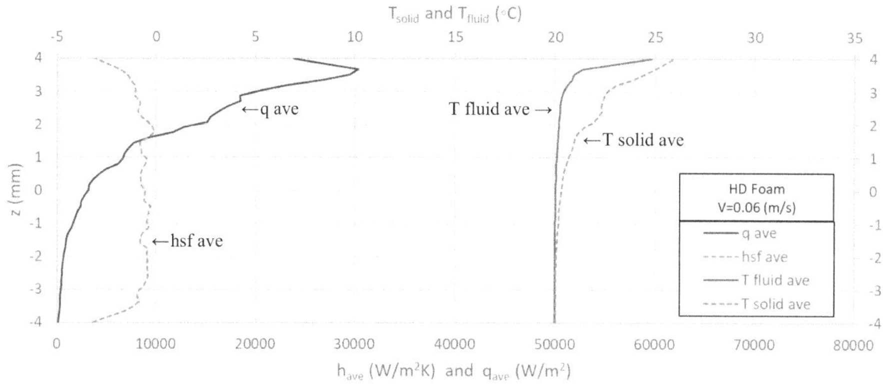

FIGURE 3.25 Variation of $h_{sf\ pl\ ave}$, $\dot{q}_{pl\ ave}$, $T_{solid\ pl\ ave}$, and $T_{fluid\ pl\ ave}$ with height for high-density ($\varepsilon = 0.879$) MF at 0.06 m/s inlet velocity.

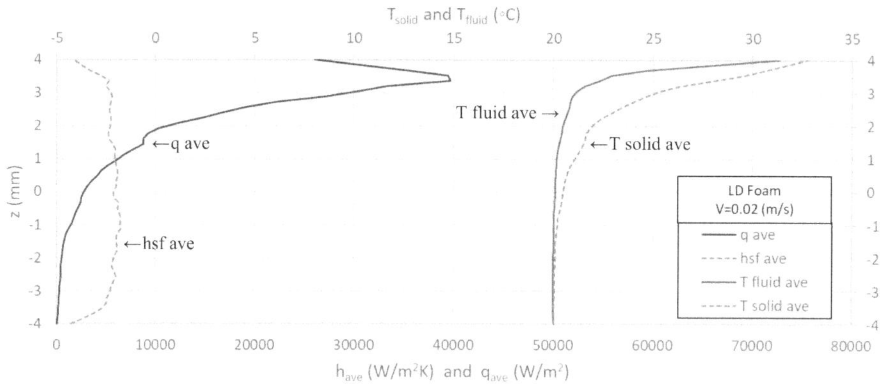

FIGURE 3.26 Variation of $h_{sf\ pl\ ave}$, $\dot{q}_{pl\ ave}$, $T_{solid\ pl\ ave}$, and $T_{fluid\ pl\ ave}$ with height for low-density ($\varepsilon = 0.946$) MF at 0.02 m/s inlet velocity.

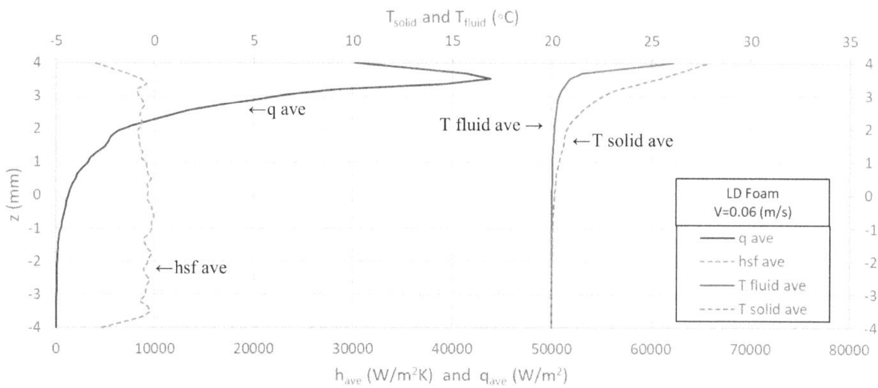

FIGURE 3.27 Variation of $h_{sf\ pl\ ave}$, $\dot{q}_{pl\ ave}$, $T_{solid\ pl\ ave}$, and $T_{fluid\ pl\ ave}$ with height for low-density ($\varepsilon = 0.946$) MF at 0.06 m/s inlet velocity.

all cases where it decreases due to the lower velocities near the wall. Moving away from the wall, $T_{fluid\ pl\ ave}$ decreases more rapidly when compared to $T_{solid\ pl\ ave}$ and that enhances heat transfer. Although the sudden decrease in the fluid temperature increases the heat transfer, the temperature difference between the solid and fluid starts to decrease after some point, resulting in the $q_{pl\ ave}$ beginning to decline after its peak point. Moving further away from the wall decreases the temperature difference as the solid and fluid temperatures come closer. Finally, temperature difference and heat flux reach zero at some point depending on the velocity and porosity, except for high-density MF at 0.02 m/s inlet velocity. $h_{sf\ pl\ ave}$ increases with velocity as expected. Although two different MFs have different $\dot{q}_{pl\ ave}$ for a given velocity, $h_{sf\ ave}$ stays at similar values. Thus, the difference in $\dot{q}_{pl\ ave}$ for two different MFs at a given velocity results from the difference in the temperature profiles of $T_{solid\ pl\ ave}$ and $T_{fluid\ pl\ ave}$.

When the velocity increases, the temperatures equalize closer to the heated surface. Thus, the effective heat transfer area reduces. When the two MFs are compared, the intersection point of solid and fluid temperatures is closer to the heated surface for low-density MF, due to the lower effective conductivity.

Since some studies have reported anisotropy for the MFs [16, 18, 19], the scanned geometries were simulated in three different directions. The results for the inlet velocity of 0.04 m/s are listed in Table 3.6 for different directions. As can be seen from the table, the maximum difference is ~4%.

3.3.2 DNN Results

3.3.2.1 DNN Modelling Parameters

Two separate models were developed for two different MFs. Each model consists of two input nodes, eight hidden layers, and four output nodes, as seen in Figure 3.28. Input nodes take velocity and height parameters as float numbers. Similarly model gives $h_{sf\ pl\ ave}$, $\dot{q}_{pl\ ave}$, $T_{fluid\ pl\ ave}$ and $T_{solid\ pl\ ave}$ parameters as float numbers. Eight hidden layers with a total of 375 nodes learn the conjugate heat transfer relationship between input and output parameters.

However, since the ML algorithms need a lot of data to be trained, the obtained data from CFD/HT simulations was initially normalized and then augmented by a B-Spline surface fitting [110]. 6,270 input and output pairs were created by data augmentation to train the model.

Dropout layers between each hidden layer and L1-L2 regularizers are used to prevent underfitting and overfitting.

TABLE 3.6
The Average Heat Transfer Coefficient of Different Directions for 0.04 m/s Inlet Velocity

Parameter	X-Direction	Y-Direction	% Diff.	Z-Direction	% Diff.
High density	8179	8372	2.3	8515	4.1
Low density	8676	8701	0.3	8518	1.8

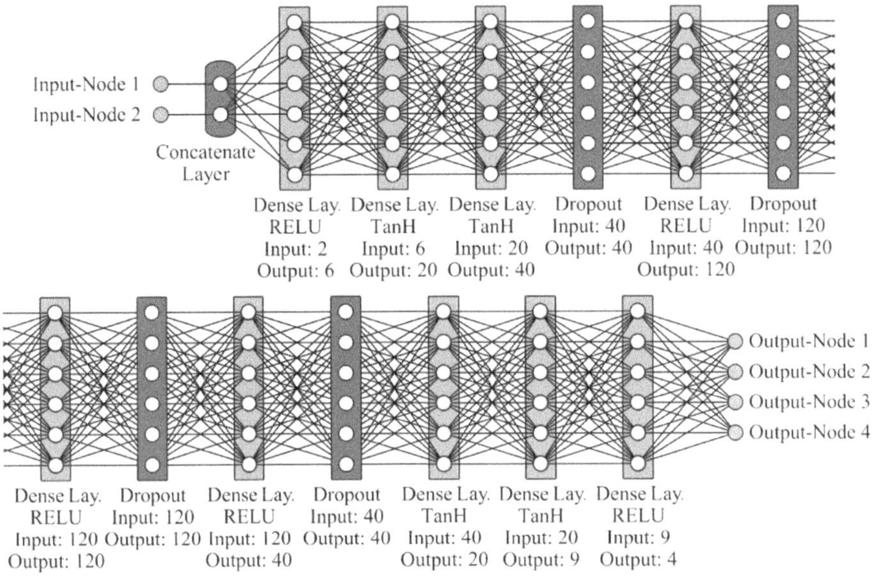

FIGURE 3.28 DNN input, output, and layer information.

3.3.2.2 DNN Training and Results

After the training of the DNN model, R^2 values of 0.996 and 0.995 have been obtained for low-density and high-density foams, respectively. Figure 3.29 shows

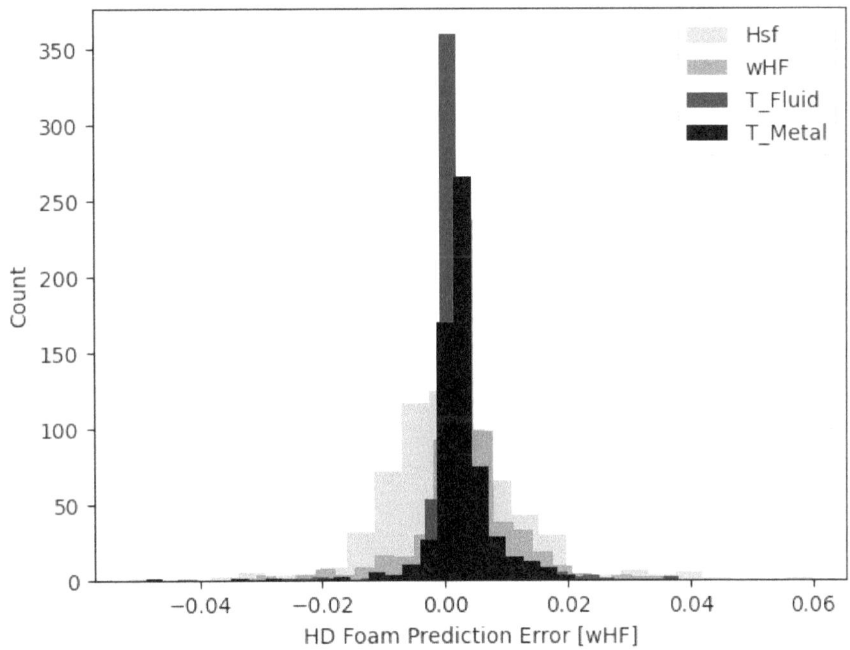

FIGURE 3.29 Error histogram of DNN algorithm for (a) high-density and (b) low-density MFs.

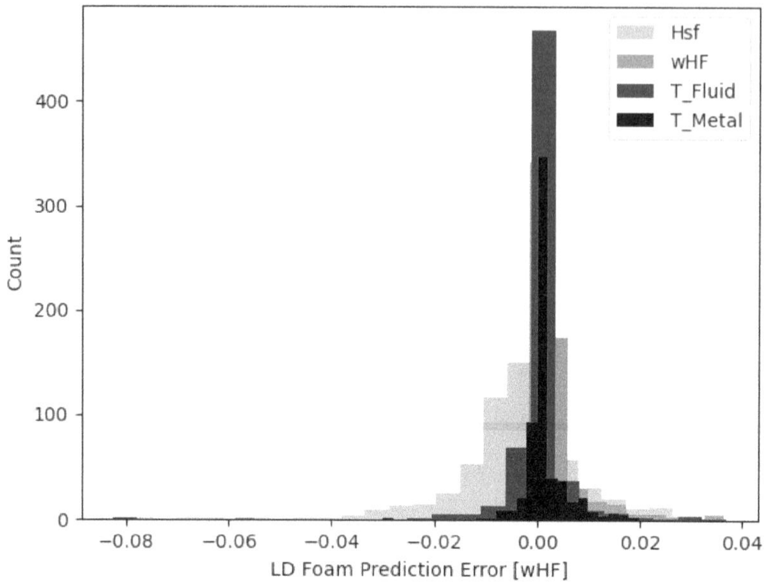

FIGURE 3.29 (***Continued***)

the error histogram for low-density and high-density foams for all parameters. This figure shows that the error sits mainly in the 2% band. In Figure 3.30, the fitting of real and predicted values for all parameters is shown, where it can be seen that the trained model does not have underfitting, overfitting, or prediction separation.

The predictions of $h_{sf\ pl\ ave}$ can be seen in Figure 3.31 for low-density and high-density MFs. A good agreement can be seen, except for some sharp oscillations of

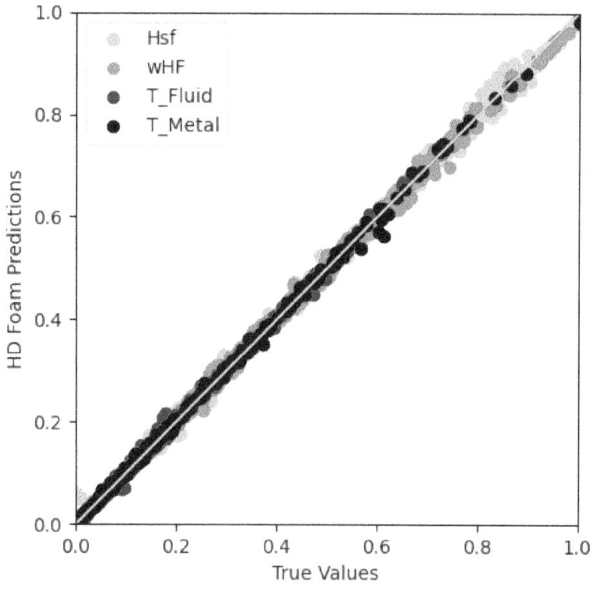

FIGURE 3.30 CFD data versus predictions of DNN for (a) high-density and (b) low-density MFs.

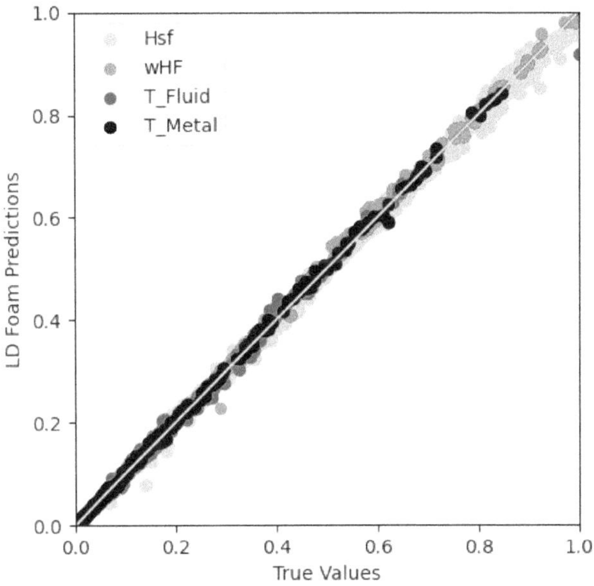

FIGURE 3.30 (*Continued*)

$h_{sf\ pl\ ave}$ towards the z-direction, which are not captured very well. The predictions of $\dot{q}_{pl\ ave}$ also shows good agreement with the CFD/HT data, as shown in Figure 3.32.

A good agreement with the CFD data and the DNN predictions for fluid and metal temperatures is also shown in Figures 3.33 and 3.34.

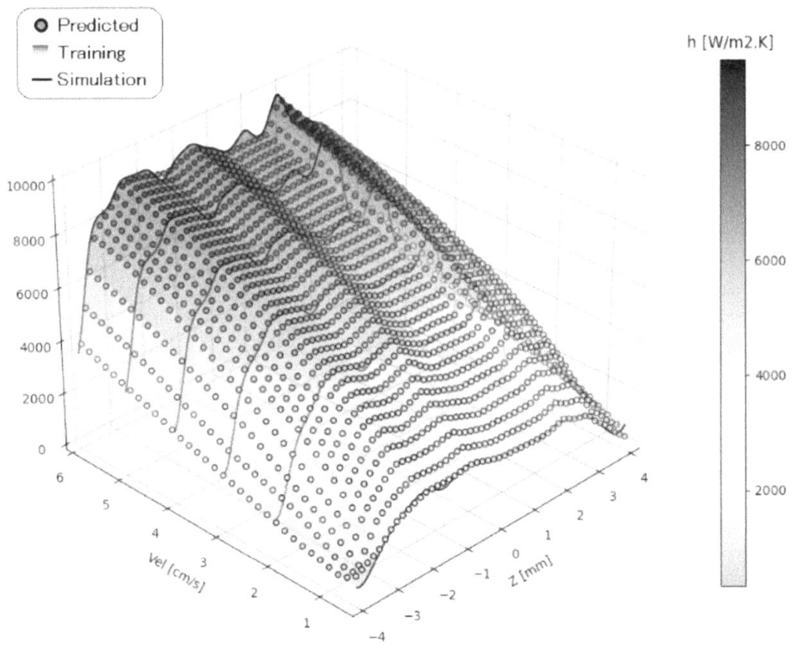

FIGURE 3.31 DNN prediction of $h_{sf\ pl\ ave}$ for (a) high-density and (b) low-density MFs.

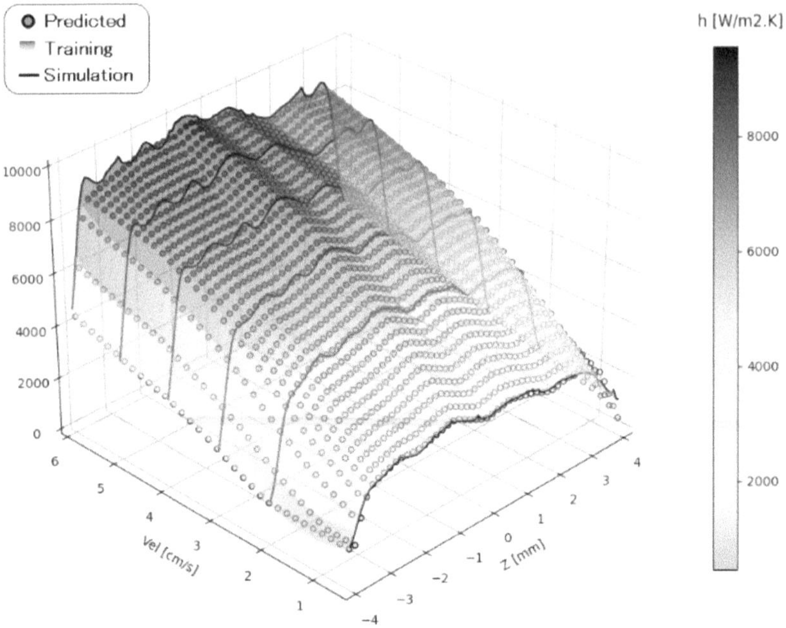

FIGURE 3.31 (***Continued***)

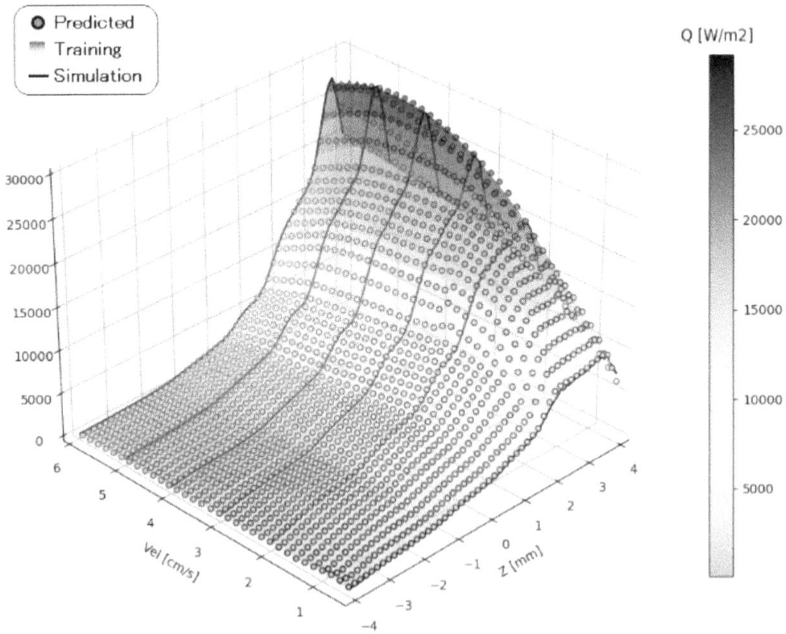

FIGURE 3.32 DNN prediction of $\dot{q}_{pl\ ave}$ for (a) high-density and (b) low-density MFs.

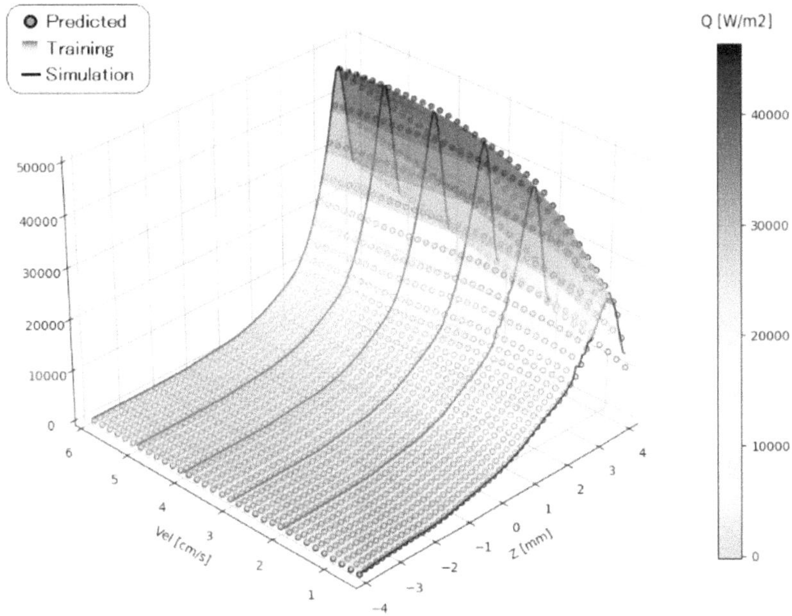

FIGURE 3.32 (*Continued*)

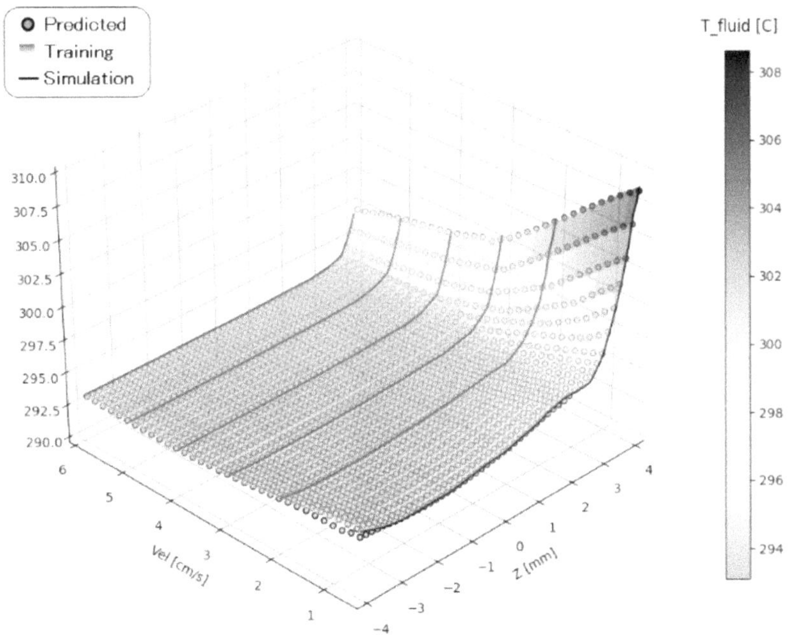

FIGURE 3.33 DNN prediction of $T_{fluid\ pl\ ave}$ for (a) high-density and (b) low-density MFs.

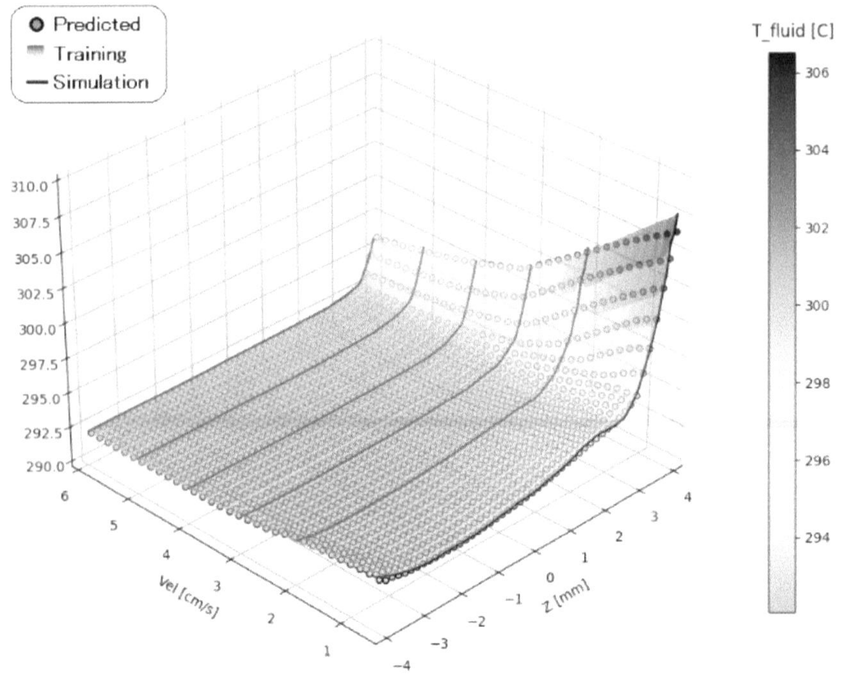

FIGURE 3.33 (*Continued*)

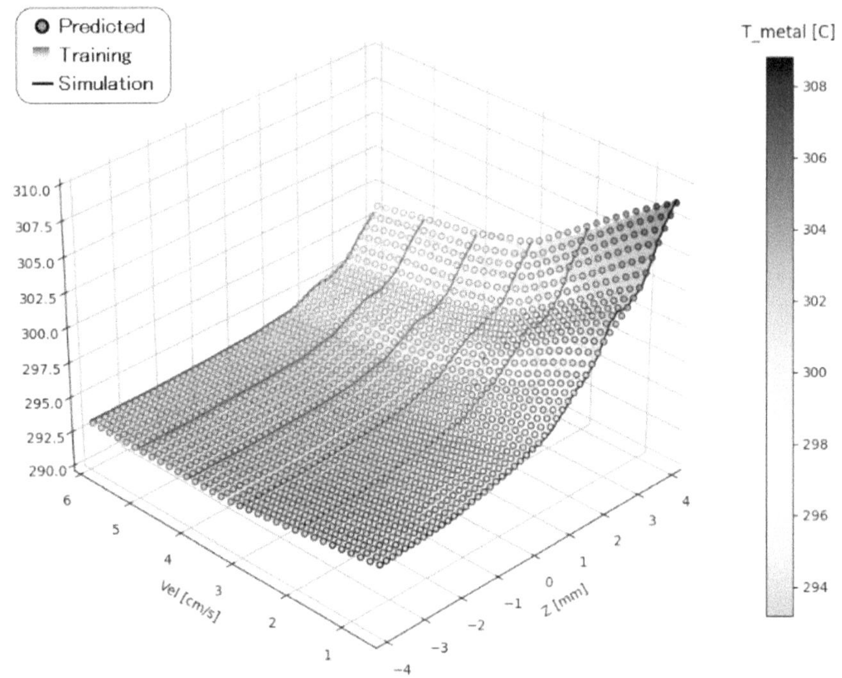

FIGURE 3.34 DNN prediction of $T_{solid\ pl\ ave}$ for (a) high-density and (b) low-density MFs.

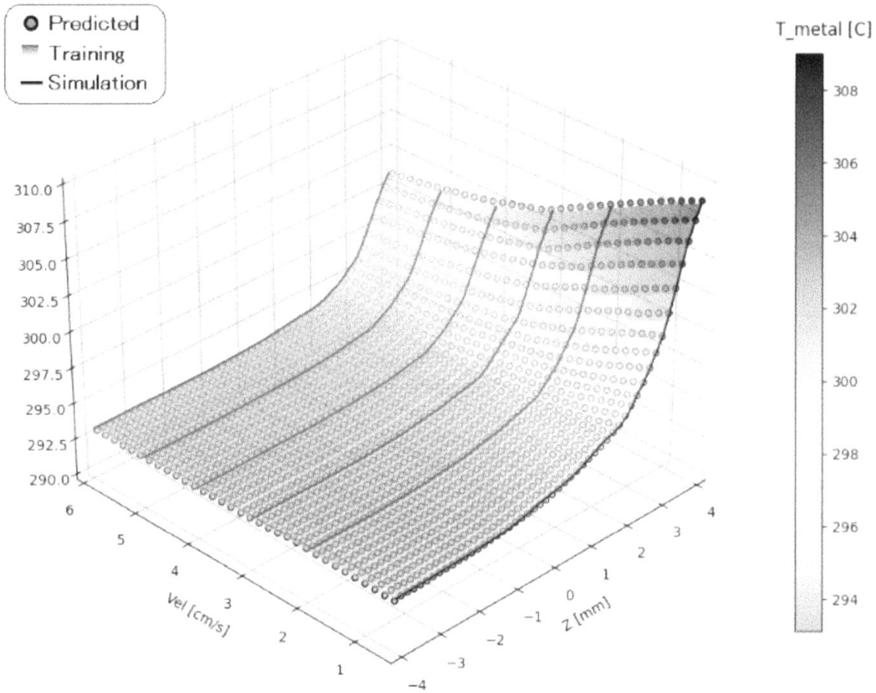

FIGURE 3.34 *(Continued)*

3.4 CONCLUSION

MFs with the same PPI but different porosity values were investigated numerically to explore the pore-level fluid flow and heat transfer characteristics, and to find ways to improve their heat transfer performance. Accordingly, the contours of temperature, interfacial heat transfer coefficients, and heat flux distributions have been obtained. The change in fluid and solid temperature, interfacial heat transfer coefficient, and heat flux with height have been shown by computing integrated values of these variables. The integrated values have been used to train a DNN algorithm to predict them without using CFD/HT simulations. The following key features have been found:

- Fluid mixing due to the nodes is an important factor for local heat transfer at the pore level, which should be included in the dispersion coefficient correlations.
- Increasing porosity decreases the effective thermal conductivity. Therefore, unused interfacial areas for heat transfer increase with porosity.
- The decrease in the heat transfer coefficient at the rear sides of the filaments is more pronounced for high-density MFs due to the thickness of the filaments. This results in a lower average interfacial heat transfer coefficient for high-density MFs.

- The temperature difference between the solid and fluid portions, which is the main factor of heat transfer, is observed to be marginal in the regions away from the heated surface. This situation must be considered in the channel design, especially for selecting the right height.
- Critically important zones for heat transfer are the solid–fluid interfaces in the upper region where the interface faces the flow, as shown in local heat flux calculations on the interfacial area.
- The bottom parts of the MF do not significantly contribute to heat transfer since the temperature differences are marginal. This situation is even more pronounced for the lower-density (higher porosity) MFs.
- Increasing the porosity reduces the effective interfacial area for the heat transfer at a given velocity.
- Anisotropy does not significantly impact heat transfer.
- 3D surface plots showed good agreement between the numerical data and the predictions obtained by using the DNN algorithm for high-density and low-density MFs, as the R^2 values are 0.996 and 0.995, respectively.
- The computational time can be drastically reduced by using DNN algorithms from several hours to minutes.

ACKNOWLEDGMENTS

The numerical calculations reported in this chapter were partially performed at TUBITAK ULAKBIM, High Performance and Grid Computing Center (TRUBA resources).

The first author was supported by the McKenney/Shiver Chair funds at the Georgia Institute of Technology, and was partially supported by the TUBITAK 2219 Grant International Postdoctoral Research Fellowship Program Grant Number 1059B192000336.

3.5 NOMENCLATURE

A	area (m^2)
a_{sf}	interfacial area density (m^{-1})
C_P	specific heat (kJ/kgK)
C_T	empirical coefficient related to the heat transfer coefficient between fluid and solid
d_f	filament diameter (m)
d_p	pore diameter (m)
f	form coefficient
H	height of the channel (m)
h_{fs}	interfacial heat transfer coefficient (W/m^2K)
K	permeability (m^2)
k	thermal conductivity (W/mK)
Nu	Nusselt number
P	pressure (Pa)
Pr	Prandtl number

Re	Reynolds number
$\dot{q}$	heat flux (W/m^2)
V	velocity (m/s)
T	temperature (°C)

Greek symbols

ε	porosity
μ	dynamic viscosity (Pa s)
ρ	density (kg/m^3)

Abbreviations

AI	artificial intelligence
ACO	ant colony optimization
ANFIS	adaptive network-based fuzzy inference system
ANN	artificial neural network
CNN	convolutional neural networks
CFD	computational fluid dynamics
DE	differential evolution
DNN	deep neural network
DNS	direct numerical simulation
DRL	deep reinforcement learning
DTR	decision tree regression
ELM	extreme learning machines
FL	fuzzy logic
GA	genetic algorithm
GB	gradient boosting
GBR	gradient boosting regression
GEP	gene expression programming
GP	genetic programming
GPR	Gaussian process regression
HT	heat transfer
k-NN	k nearest neighbor
MF	metal foam
ML	machine learning
MLR	multiple linear regression
MLPNN	multilayer perceptron neural network
MPR	multivariate polynomial regression
MSE	mean squared error
PINN	physics-informed neural network
PPI	pores per inch
RF	random forest
RMS	response surface methodology
RMSE	root mean square error
SVM	support vector machine
TBL	turbulence boundary layer
TL	transfer learning
μCT	micro-computed tomography

REFERENCES

1. Nawaz, K., et al., *Experimental studies to evaluate the use of metal foams in highly compact air-cooling heat exchangers.* 2010.
2. Ejlali, A., et al., *Application of high porosity metal foams as air-cooled heat exchangers to high heat load removal systems.* International Communications in Heat and Mass Transfer, 2009. **36**(7): p. 674–679.
3. Lu, T.J., H.A. Stone, and M.F. Ashby, *Heat transfer in open-cell metal foams. Acta Materialia*, 1998. **46**(10): p. 3619–3635.
4. Mahjoob, S., and K. Vafai, *A synthesis of fluid and thermal transport models for metal foam heat exchangers.* International Journal of Heat and Mass Transfer, 2008. **51**(15–16): p. 3701–3711.
5. Hunt, M., and C. Tien, *Effects of thermal dispersion on forced convection in fibrous media.* International Journal of Heat and Mass Transfer, 1988. **31**(2): p. 301–309.
6. Kennedy, A., *Porous metals and metal foams made from powders.* 2012: INTECH Open Access Publisher.
7. Du Plessis, P., et al., *Pressure drop prediction for flow through high porosity metallic foams.* Chemical Engineering Science, 1994. **49**(21): p. 3545–3553.
8. Calmidi, V.V., *Transport phenomena in high porosity fibrous metal foams.* 1999, University of Colorado Boulder.
9. Bhattacharya, A., V.V. Calmidi, and R.L. Mahajan, *Thermophysical properties of high porosity metal foams.* International Journal of Heat and Mass Transfer, 2002. **45**(5): p. 1017–1031.
10. Yang, K., K. Liu, and J. Wang, *Pore-Scale numerical simulation of convection heat transfer in high porosity open-cell metal foam under rotating conditions.* Applied Thermal Engineering, 2021. **195**: p. 117168.
11. Boomsma, K., D. Poulikakos, and Y. Ventikos, *Simulations of flow through open cell metal foams using an idealized periodic cell structure.* International Journal of Heat and Fluid Flow, 2003. **24**(6): p. 825–834.
12. Kopanidis, A., et al., *3D numerical simulation of flow and conjugate heat transfer through a pore scale model of high porosity open cell metal foam.* International Journal of Heat and Mass Transfer, 2010. **53**(11–12): p. 2539–2550.
13. Al-Athel, K.S., *A computational methodology for assessing the thermal behavior of metal foam heat sinks.* Applied Thermal Engineering, 2017. **111**: p. 884–893.
14. Ranut, P., E. Nobile, and L. Mancini, *High resolution X-ray microtomography-based CFD simulation for the characterization of flow permeability and effective thermal conductivity of aluminum metal foams.* Experimental Thermal and Fluid Science, 2015. **67**: p. 30–36.
15. Dixit, T., and I. Ghosh, *Simulation intricacies of open-cell metal foam fin subjected to convective flow.* Applied Thermal Engineering, 2018. **137**: p. 532–544.
16. Iasiello, M., et al., *Anisotropic convective heat transfer in open-cell metal foams: Assessment and correlations.* International Journal of Heat and Mass Transfer, 2020. **154**: p. 119682.
17. Park, S.-H., D.-H. Seo, and J.H. Jeong, *Experimental and numerical analysis of thermal flow in open-cell porous metal during Darcy-Forchheimer transition regime.* Applied Thermal Engineering, 2020. **181**: p. 116029.
18. Yu, P., et al., *Pore-scale numerical study of flow characteristics in anisotropic metal foam with actual skeleton structure.* International Communications in Heat and Mass Transfer, 2021. **126**: p. 105401.
19. Yang, T., et al., *Numerical study of flow and heat transfer in a three-dimensional metal foam considering different direction micropores in skeleton structure.* International Communications in Heat and Mass Transfer, 2022. **134**: p. 106052.

20. *Web of Science Database*, Clarivate, Editor. 2024.
21. McCullough, W.S., and W. Pitts, *A logical calculus of the ideas immanent in nervous activity*. Bulletin of Mathematical Biophysics, 1943. **5**(4): p. 115–127.
22. Pacheco-Vega, A., et al., *heat rate predictions in humid air-water heat exchangers using correlations and neural networks*. Journal of Heat Transfer, 2000. **123**(2): p. 348–354.
23. Athavale, J., Y. Joshi, and M. Yoda. *Artificial Neural Network Based Prediction of Temperature and Flow Profile in Data Centers*. in *2018 17th IEEE Intersociety Conference on Thermal and Thermomechanical Phenomena in Electronic Systems (ITherm)*. 2018.
24. Souayeh, B., et al., *Heat and fluid flow analysis and ANN-based prediction of a novel spring corrugated tape*. Sustainability, 2021. **13**(6): p. 3023.
25. Bhattacharyya, S., et al., *Turbulent flow heat transfer through a circular tube with novel hybrid grooved tape inserts: Thermohydraulic analysis and prediction by applying machine learning model*. Sustainability, 2021. **13**(6): p. 3068.
26. Berber, A., and M. Gürdal, *Estimation of forced heat convection in a rectangular channel with curved-winglet vortex generator: A machine learning approach*. Thermal Science and Engineering Progress, 2023. **37**: p. 101563.
27. Oh, Y., and Z. Guo, *Prediction of Nusselt number in microscale pin fin heat sinks using artificial neural networks*. Heat Transfer Research, 2023. **54**(1).
28. Huang, L., et al., *Machine learning assisted microchannel geometric optimization—A case study of channel designs*. Energies, 2024. **17**(1): p. 44.
29. Sikirica, A., L. Grbčić, and L. Kranjčević, *Machine learning based surrogate models for microchannel heat sink optimization*. Applied Thermal Engineering, 2023. **222**: p. 119917.
30. Rehman, K.U., and W. Shatanawi, *Non-newtonian mixed convection magnetized flow with heat generation and viscous dissipation effects: A prediction application of artificial intelligence*. Processes, 2023. **11**(4): p. 986.
31. Pai, S.S., and J.A. Weibel, *Neural net-based surrogate square pin fin correlations for thermal-fluid design optimization under developing flow conditions*. Numerical Heat Transfer, Part B: Fundamentals, 2023: p. 1–18.
32. Shuqi, Z., et al., *Artificial neural network-based optimization of baffle geometries for maximized heat transfer efficiency in microchannel heat sinks*. Case Studies in Thermal Engineering, 2023. **49**: p. 103331.
33. Liu, H.-l, et al., *An experimental and numerical investigation of heat transfer enhancement in annular microchannel heat sinks*. International Journal of Thermal Sciences, 2019. **142**: p. 106–120.
34. Mauro, A.W., R. Revellin, and L. Viscito, *Development and assessment of performance of artificial neural networks for prediction of frictional pressure gradients during two-phase flow*. International Journal of Heat and Mass Transfer, 2024. **221**: p. 125106.
35. Zhu, G., et al., *Transfer learning model to predict flow boiling heat transfer coefficient in mini channels with micro pin fins*. International Journal of Heat and Mass Transfer, 2024. **220**: p. 125020.
36. Kumar, C., et al., *Analysis of heat transfer behavior of porous wavy fin with radiation and convection by using a machine learning technique*. Symmetry, 2023. **15**(8): p. 1601.
37. Storn, R., *Differrential evolution-a simple and efficient adaptive scheme for global optimization over continuous spaces*. Technical report, International Computer Science Institute, 1995. **11**.
38. Joo, C., et al., *Machine-learning-based optimization of operating conditions of naphtha cracking furnace to maximize plant profit*, in Computer aided chemical engineering. 2023, Elsevier. p. 1397–1402.

39. Merenda, M., C. Porcaro, and F.G.D. Corte. *LED junction temperature prediction using machine learning techniques*. in *2020 IEEE 20th Mediterranean Electrotechnical Conference (MELECON)*. 2020.
40. Sundar, S., et al., *Fouling modeling and prediction approach for heat exchangers using deep learning*. International Journal of Heat and Mass Transfer, 2020. **159**: p. 120112.
41. Shaeri, M.R., et al., *Machine learning-based optimization of air-cooled heat sinks*. Thermal Science and Engineering Progress, 2022. **34**: p. 101398.
42. Wang, Y.-Z., et al., *Closed-loop forced heat convection control using deep reinforcement learning*. International Journal of Heat and Mass Transfer, 2023. **202**: p. 123655.
43. Wang, Y.-Z., et al., *Control policy transfer of deep reinforcement learning based intelligent forced heat convection control*. International Journal of Thermal Sciences, 2024. **195**: p. 108618.
44. Kim, J., and C. Lee, *Prediction of turbulent heat transfer using convolutional neural networks*. Journal of Fluid Mechanics, 2020. **882**: p. A18.
45. Adam, A., H. Fang, and X. Li, *Effective thermal conductivity estimation using a convolutional neural network and its application in topology optimization*. Energy and AI, 2024. **15**: p. 100310.
46. Cai, S., et al., *Physics-informed neural networks (PINNs) for fluid mechanics: A review*. Acta Mechanica Sinica, 2021. **37**(12): p. 1727–1738.
47. Dwivedi, V., N. Parashar, and B. Srinivasan, *Distributed learning machines for solving forward and inverse problems in partial differential equations*. Neurocomputing, 2021. **420**: p. 299–316.
48. Zobeiry, N., and K.D. Humfeld, *A physics-informed machine learning approach for solving heat transfer equation in advanced manufacturing and engineering applications*. Engineering Applications of Artificial Intelligence, 2021. **101**: p. 104232.
49. Xu, J., H. Wei, and H. Bao, *Physics-informed neural networks for studying heat transfer in porous media*. International Journal of Heat and Mass Transfer, 2023. **217**: p. 124671.
50. Acikgoz, O., et al., *Machine learning approach to predict the heat transfer coefficients pertaining to a radiant cooling system coupled with mixed and forced convection*. International Journal of Thermal Sciences, 2022. **178**: p. 107624.
51. Alhakami, H., et al., *On the computational study of a fully wetted longitudinal porous heat exchanger using a machine learning approach*. Entropy, 2022. **24**(9): p. 1280.
52. Ab. Rashid, M.F.F., *Tiki-taka algorithm: A novel metaheuristic inspired by football playing style*. Engineering Computations, 2021. **38**(1): p. 313–343.
53. Agustiarini, N., et al., *Predictive modeling for the boiling heat transfer coefficient of R1234yf inside a multiport minichannel tube*. International Journal of Heat and Mass Transfer, 2023. **211**: p. 124188.
54. Waseem, W., et al., *A study of changes in temperature profile of porous fin model using cuckoo search algorithm*. Alexandria Engineering Journal, 2020. **59**(1): p. 11–24.
55. Prakash, S.B., et al., *Investigation of the thermal analysis of a wavy fin with radiation impact: An application of extreme learning machine*. Physica Scripta, 2023. **99**(1): p. 015225.
56. Jang, J.-S., *ANFIS: Adaptive-network-based fuzzy inference system*. IEEE Transactions on Systems, Man, and Cybernetics, 1993. **23**(3): p. 665–685.
57. Seyedashraf, O., A. Rezaei, and A.A. Akhtari, *Application of computational intelligence methods for complex two-phase flow pattern recognition*. Journal of the Brazilian Society of Mechanical Sciences and Engineering, 2018. **40**(1): p. 35.
58. Alotaibi, H.F., et al., *Development of mechanistic-artificial intelligence model for simulation of numerical data of water flow in porous materials*. Engineering Applications of Artificial Intelligence, 2023. **126**: p. 106844.

59. Soltanian, S., and R. Beigzadeh, *Computational fluid dynamics and fuzzy logic for modeling conical spiral heat exchangers.* Chemical Engineering & Technology, 2023. **46**(4): p. 747–755.
60. Takagi, T., and M. Sugeno, *Fuzzy identification of systems and its applications to modeling and control.* IEEE Transactions on Systems, Man, and Cybernetics, 1985(1): p. 116–132.
61. Ho, T.K.. *Random decision forests.* in *Proceedings of 3rd international conference on document analysis and recognition.* 1995, IEEE.
62. Kwon, B., F. Ejaz, and L.K. Hwang, *Machine learning for heat transfer correlations.* International Communications in Heat and Mass Transfer, 2020. **116**: p. 104694.
63. Suzuki, A., H. Nakatani, and M. Kobashi, *Machine learning surrogate modeling toward the design of lattice-structured heat sinks fabricated by additive manufacturing.* Materials & Design, 2023. **230**: p. 111969.
64. Nguyen, N.P., et al., *Shape optimization of pin fin array in a cooling channel using genetic algorithm and machine learning.* International Journal of Heat and Mass Transfer, 2023. **202**: p. 123769.
65. Ma, Y., et al., *Research on modeling and parameter sensitivity of flow and heat transfer process in typical rectangular microchannels: From a data-driven perspective.* International Journal of Thermal Sciences, 2022. **172**: p. 107356.
66. Kim, K., et al., *A machine learning approach for predicting heat transfer characteristics in micro-pin fin heat sinks.* International Journal of Heat and Mass Transfer, 2022. **194**: p. 123087.
67. Zhou, L., et al., *Machine learning algorithms to predict flow condensation heat transfer coefficient in mini/micro-channel utilizing universal data.* International Journal of Heat and Mass Transfer, 2020. **162**: p. 120351.
68. Geurts, P., D. Ernst, and L. Wehenkel, *Extremely randomized trees.* Machine Learning, 2006. **63**: p. 3–42.
69. Aznar, P., *What is the difference between extra trees and random forest.* Quantdare, Jun, 2020.
70. He, Y., et al., *Data-driven approach to predict the flow boiling heat transfer coefficient of liquid hydrogen aviation fuel.* Fuel, 2022. **324**: p. 124778.
71. Nagesha, K., *Machine learning algorithms to study the relative roles of jet convection and natural convection in the presence of surface roughness elements on enhancement of jet impingement heat transfer.* International Journal of Thermal Sciences, 2024. **198**: p. 108847.
72. Fix, E., *Discriminatory analysis: nonparametric discrimination, consistency properties.* Vol. 1. 1985: USAF school of Aviation Medicine.
73. Krishnayatra, G., S. Tokas, and R. Kumar, *Numerical heat transfer analysis & predicting thermal performance of fins for a novel heat exchanger using machine learning.* Case Studies in Thermal Engineering, 2020. **21**: p. 100706.
74. Williams, C.K., and C.E. Rasmussen, Gaussian processes for machine learning. Vol. 2. 2006, MIT press, Cambridge, MA.
75. Box, G., and K. Wilson, *On the experimental designs for exploring response surfaces..* Ann Math Stat, 1951. **13**: p. 1–45.
76. Vapnik, V., *The support vector method of function estimation*, in *Nonlinear modeling: Advanced black-box techniques.* 1998, Springer. p. 55–85.
77. Awad, M., et al., *Support vector regression.* Efficient Learning Machines: Theories, Concepts, and Applications for Engineers and System Designers, 2015: p. 67–80.
78. Hughes, M.T., S.M. Chen, and S. Garimella, *Machine-learning-based heat transfer and pressure drop model for internal flow condensation of binary mixtures.* International Journal of Heat and Mass Transfer, 2022. **194**: p. 123109.

79. Suykens, J.A., and J. Vandewalle, *Least squares support vector machine classifiers.*. Neural Processing Letters, 1999. **9**: p. 293–300.
80. Shivanian, E., et al., *A novel learning approach for different profile shapes of convecting–radiating fins based on shifted gegenbauer lssvm.* New Mathematics and Natural Computation, 2023. **19**(01): p. 195–215.
81. Shourehdeli, S.A., and K. Mobini, *Machine learning-based models for frictional pressure drop prediction of condensing and adiabatic flow in micro, mini and macro channels utilizing universal data.* International Journal of Air-Conditioning and Refrigeration, 2023. **31**(1): p. 8.
82. Hughes, M.T., B.M. Fronk, and S. Garimella, *Universal condensation heat transfer and pressure drop model and the role of machine learning techniques to improve predictive capabilities.* International Journal of Heat and Mass Transfer, 2021. **179**: p. 121712.
83. Saeed, M., et al., *Machine learning-based optimization of a mini-channel heatsink geometry.* Arabian Journal for Science and Engineering, 2023. **48**(9): p. 12107–12124.
84. Holland, J.H., Adaptation in natural and artificial systems: an introductory analysis with applications to biology, control, and artificial intelligence. 1992, MIT press.
85. De Jong, K.A., *An analysis of the behavior of a class of genetic adaptive systems.* 1975, University of Michigan.
86. Castellanos, R., A. Ianiro, and S. Discetti, *Genetically-inspired convective heat transfer enhancement in a turbulent boundary layer.* Applied Thermal Engineering, 2023. **230**: p. 120621.
87. Ferreira, C., *Gene expression programming in problem solving*, in *Soft computing and industry: Recent applications.* 2002, Springer. p. 635–653.
88. Koza, J., *Genetic programming:* On the programming of computers by means of natural selection. 1992, MIT press, Cambridge, MA.
89. Karbasi, M., and H.M. Azamathulla, *GEP to predict characteristics of a hydraulic jump over a rough bed.* KSCE Journal of Civil Engineering, 2016. **20**: p. 3006–3011.
90. Parveen, N., S. Zaidi, and M. Danish, *Comparative analysis for the prediction of boiling heat transfer coefficient of R134a in micro/mini channels using artificial intelligence (AI)-based techniques.* International Journal of Modelling and Simulation, 2020. **40**(2): p. 114–129.
91. Jafarizadeh, A., et al., *Pressure gradient computation for foams with different geometric properties: based on ANN and SVR machine learning model and trained by CFD simulations.* 2023. **26**(9): p. 49–76.
92. Jafarizadeh, A., et al., *A new approach for predicting the pressure drop in various types of metal foams using a combination of CFD and machine learning regression models.* Transport in Porous Media, 2023. **147**(1): p. 59–91.
93. Babanezhad, M., et al., *Prediction of turbulence eddy dissipation of water flow in a heated metal foam tube.* Scientific Reports, 2020. **10**(1): p. 19280.
94. Babanezhad, M., et al., *Computational modeling of transport in porous media using an adaptive network-based fuzzy inference system.* ACS Omega, 2020. **5**(48): p. 30826–30835.
95. Otaru, A.J., Z.A. Alhulaybi, and T.A. Owoseni, *On the hydrodynamics of macroporous structures: Experimental, CFD and artificial neural network analysis.* Chemical Engineering Journal Advances, 2023. **16**: p. 100545.
96. Tikadar, A., and S. Kumar, *Investigation of thermal-hydraulic performance of metal-foam heat sink using machine learning approach.* International Journal of Heat and Mass Transfer, 2022. **199**: p. 123438.
97. Tikadar, A., and S. Kumar, *Machine learning approach to predict heat transfer and fluid flow characteristics of integrated pin fin-metal foam heat sink.* Numerical Heat Transfer, Part B: Fundamentals: p. 1–26.

98. Li, Y., et al., *Machine learning and multilayer perceptron enhanced CFD approach for improving design on latent heat storage tank.* Applied Energy, 2023. **347**: p. 121458.
99. Xiao, T., et al., *LSTM-BP neural network analysis on solid-liquid phase change in a multi-channel thermal storage tank.* Engineering Analysis With Boundary Elements, 2023. **146**: p. 226–240.
100. Zhang, G., et al., *Optimization of porous media flow field for proton exchange membrane fuel cell using a data-driven surrogate model.* Energy Conversion and Management, 2020. **226**: p. 113513.
101. Dukhan, N., Ö Bağcı, and M. Özdemir, *Metal foam hydrodynamics: Flow regimes from pre-Darcy to turbulent.* International Journal of Heat and Mass Transfer, 2014. **77**: p. 114–123.
102. Kececioglu, I., and Y. Jiang, *Flow through porous media of packed spheres saturated with water.* Journal of Fluids Engineering, 1994. **116**(1): p. 164–170.
103. Calmidi, V., and R. Mahajan, *Forced convection in high porosity metal foams.* Journal of Heat Transfer, 2000. **122**(3): p. 557–565.
104. Kemerli, U., and K. Kahveci, *Forced convection in finned metal foams: The effects of porosity and effective thermal conductivity.* International Journal of Thermal Sciences, 2021. **161**.
105. Zukauskas, A., *Convective heat transfer in cross flow.* Handbook of Single-Phase Convective Heat Transfer, 1987.
106. Hilpert, R., *Wärmeabgabe von geheizten Drähten und Rohren im Luftstrom.* Forschung auf dem Gebiet des Ingenieurwesens A, 1933. **4**: p. 215–224.
107. Hsu, C., and P. Cheng, *Thermal dispersion in a porous medium.* International Journal of Heat and Mass Transfer, 1990. **33**(8): p. 1587–1597.
108. Dupuit, J.E., *Etudes théoriques et pratiques sur le mouvement des eaux dans les canaux découverts et à travers les terrains perméables avec des considérations relatives au régime des grandes eaux, au débouché à leur donner, et à la marche des des alluvions dans les rivières à fond mobile.* 1863: Dunod, éditeur.
109. Lage, J., *The fundamental theory of flow through permeable media from darcy to turbulence.* Transport Phenomena in Porous media, 1998. **1**.
110. Liew, K.J., Ramli A., and Abd Majid A., *B-spline surface fitting on scattered points.* Applied Mathematics & Information Sciences, 2016. **10**(1): p. 273.

4 Integrating Artificial Intelligence in Nanofluid Heat Transfer

A Deep Dive into Artificial Intelligence Applications

Andaç Batur Çolak

4.1 INTRODUCTION

Nanofluids, colloidal suspensions of nanoparticles in base fluids, have garnered significant interest in various industrial and engineering sectors due to their enhanced thermophysical properties compared to conventional fluids. These properties include improved thermal conductivity, viscosity, and specific heat, making nanofluids promising candidates for applications in electronic cooling, heat exchangers, energy systems, and biomedical fields. However, accurately predicting the complex behavior of nanofluids poses challenges for traditional modeling approaches due to their dynamic nature. In recent years, artificial intelligence (AI) has emerged as a powerful alternative in modeling complex structures of nanofluids. AI techniques, such as artificial neural networks (ANN) and machine learning, provide the ability to distinguish complex patterns and relationships within large datasets, enabling high-accuracy prediction of the properties of nanofluids.

In this chapter, important thermophysical properties of nanofluids such as thermal conductivity, viscosity, and specific heat are investigated, along with applications of AI tools in predicting the heat transfer properties of their use in different applications. Through a comprehensive review of current studies, the availability of AI techniques and advances in using AI tools to model and predict the behavior of nanofluids are examined.

4.2 AI APPLICATIONS IN DETERMINING THERMOPHYSICAL PROPERTIES OF NANOFLUIDS

In the rapidly developing field of nanotechnology, the integration of AI tools facilitates the determination of nanofluids and their thermophysical properties. Nanofluids, which are colloidal suspensions of nanoparticles in base fluids, exhibit unique and improved thermal conductivity, viscosity, and specific heat properties compared to conventional fluids. Accurate prediction of these thermophysical properties is

DOI: 10.1201/9781032688121-4

crucial for a variety of industrial and engineering applications, including electronic cooling, heat exchangers, energy systems, and biomedical applications. The complex and dynamic nature of nanofluids creates challenges for traditional modeling approaches, making AI an important tool for its ability to discern complex patterns and relationships within large datasets. Leveraging the advantages of AI in predicting the thermophysical behavior of nanofluids can not only facilitate the design of more efficient and innovative technologies, but can also contribute to the ongoing research of nanomaterials for various applications in science and engineering.

4.2.1 Thermal Conductivity

Çolak [1] created five distinct ZrO_2 nanofluids using water as a base, with volumetric contents of 0.0125%, 0.025%, 0.05%, 0.1%, and 0.2%. A two-step process including the use of a magnetic stirrer and an ultrasonic homogenizer was recommended for the manufacturing of nanofluids. Their thermal conductivity was empirically tested within the temperature range of 10°C to 65°C. A multi-layer perceptron feed-forward back-propagation ANN was created using the acquired experimental data. Furthermore, a novel correlation was suggested to compute the thermal conductivity values of the ZrO_2/water nanofluid. The findings indicated that the ZrO_2/water nanofluid exhibited superior thermal conductivity in comparison to the base fluid. Moreover, the thermal conductivity showed an upward trend with rising temperature and concentration. The ANN, trained using experimental data, accurately predicted the thermal conductivity of ZrO_2/water nanofluid with an average error of −0.41%. However, the newly constructed correlation achieved an even higher level of accuracy, with an average error of −0.02%. These numbers indicate that the findings acquired from the created ANN and the correlation are in complete accord with the experimental data.

Çolak [2] examined the impact of neuron quantity on the predicting accuracy of ANN using empirical data. Six distinct ANNs were created specifically for this objective, utilizing a combined dataset of 60 experimental data points of ZrO_2/water nanofluid sourced from existing literature. An ANN was constructed with varying numbers of neurons: 5, 10, 15, 20, 25, and 30. All other parameters were held constant, and the only focus was on examining the impact of the number of neurons on prediction performance. The performance of each ANN was evaluated individually, and then their performance was assessed by comparing them with one another. The investigation revealed that the model with the most optimal predictive performance was constructed using 5 neurons, resulting in an average error rate of 0.001%. Conversely, the model with the biggest margin of error was built using 15 neurons, yielding an error rate of 0.026%. Based on the acquired data, it was determined that ANNs are typically effective predicting tools. However, it is not feasible to establish a fixed correlation to determine the optimal number of neurons for optimizing ANNs.

Öcal et al. [3] examined the thermal conductivity of a hybrid nanofluid consisting of TiO_2-$CaCO_3$ particles dispersed in water. The nanofluid was generated using a two-step technique and five different concentrations were used. The KD2 Pro gadget was utilized to conduct thermal conductivity measurements within a temperature range spanning from 10°C to 60°C. A mathematical correlation and an ANN model were created using experimental data to forecast thermal conductivity based on concentration and temperature. The multilayer perceptron model, which utilized a feed-forward back-propagation ANN with 10 neurons in its hidden layer, was chosen. The coefficient of determination (R) for the suggested novel mathematical correlation was 0.9999, whereas it was 0.99913 for the ANN model. The average error rate was computed as 0.005% for the mathematical model and −0.02% for the ANN.

Sahin et al. [4] developed two distinct ANNs to forecast the thermal conductivity and zeta potential of Fe_3O_4/water nanofluid. The experimental measurements were conducted to determine the thermal conductivity and zeta potential of the Fe_3O_4/water nanofluid at three distinct concentrations. A novel mathematical correlation is suggested to compute thermal conductivity by utilizing experimental data of temperature and concentration. The suggested model is particularly notable due to its uniqueness, as it allows for the calculation of correlations based on concentration, which is a common approach in the literature. The thermal conductivity and zeta potential values obtained by the ANN and the proposed mathematical correlation were compared to the experimental data. Furthermore, a thorough performance study was conducted by computing several performance metrics. The neural network models achieved R values exceeding 0.99, with mean-squared error values of 1.47E-05 and 1.58E-06, respectively. The mean deviation values for the thermal conductivity of the network model were estimated to be 0.03%, but for the new mathematical correlation, it was 0.05%. The study findings show that ANN models have the capability to accurately forecast the thermal conductivity and zeta potential of Fe_3O_4/water nanofluid. The newly suggested mathematical correlation exhibited greater error rates in comparison to the ANN model, despite its ability to accurately compute thermal conductivity values.

Çolak [5] did a study where they experimentally measured the thermal conductivity values of a new nanofluid. The nanofluid was synthesized by introducing yttrium oxide nanoparticles into pure water, which acted as the underlying fluid. The thermal conductivity values of the nanofluid, which is composed of yttrium oxide and water, were obtained experimentally at five different concentrations. The measurements were done within the temperature range of 10–65°C. The experimental findings revealed a 1.47% increase in the thermal conductivity of the nanofluid composed of yttrium oxide and water, compared to the original fluid. The experimental results were compared with correlations that are often employed in the literature. A multilayer neural network model and a unique mathematical correlation were developed using a dataset of 60 experimental observations to determine the link between temperature, concentration, and the thermal conductivity of yttrium oxide-water nanofluid. The network model was verified using nine datasets, and its performance was assessed using multiple performance criteria. The values obtained from the network model and mathematical correlation were compared with both the empirical findings and the results recorded in existing literature. The network model's R value was determined to be 0.99980, while the MSE value was 5.89E-07. The network model had an average deviation rate of −0.0007%, whereas the mathematical correlation resulted in a value of 0.049%. The study's findings showcased the creation of a neural network model and a unique mathematical correlation for precisely determining the thermal conductivity of yttrium oxide–water nanofluid.

The study done by Lin et al. [6] examines and predicts the thermal conductivity of a WO_3–CuO–Ag (35:40:25)/water nanofluid using an ANN and the back-propagation technique. The aim of this research is to create an advanced data analysis model capable of assessing complex correlations that influence the thermal conductivity of hybrid nanofluid, specifically in relation to temperature and solid volume fraction. This investigation is conducted at six different temperatures ranging from 25 to 50°C, and four different volume fractions ranging from 1% to 0.4%. The regression figure exhibits a striking similarity to unity, suggesting a strong connection between the projected values from the ANN and the actual values. The predicted values in the ANN were very close to 0.998, suggesting a high level of similarity between the projected and real values. The findings indicate that augmenting the volume fractions of nanoparticle and temperature has resulted in an elevation in the thermal conductivity of the examined nanofluid. Furthermore, at a high temperature level, the thermal conductivity reaches its greatest value when the volume fraction is extended to its utmost extent. Furthermore, the findings suggest that the impact of temperatures sensitivity on the increase in the thermal conductivity of WO_3–CuO–Ag (35:40:25)/water nanofluid is more substantial than that of volume fractions. The sensitivity analysis for temperature in five ANNs yielded a range of values from a maximum of 177.4501 to a minimum of 431.489098. Similarly, the sensitivity study for volume fractions resulted in a range of values from a maximum of 16.2539901 to a minimum of 37.2332646. Based on several trained ANNs, it has been shown that the optimal number of neurons in the hidden layer was eight.

The primary objective of the work done by Cui et al. [7] was to uncover the impact of influencing elements on the effective thermal conductivity of nanofluids using a combination of experimental investigation and AI. The researchers created TiO_2/water nanofluids using four different forms of TiO_2 nanoparticles: spherical, ellipsoidal, clubbed, and sheet. They then assessed the effective thermal conductivities of the samples at various temperatures ranging from 20 to 60°C, with nanoparticle concentrations ranging from 0.5 to 4.0 vol%. In addition, a total of 389 experimental datasets sourced from literature and 80 datasets obtained from the experiment were utilized to ascertain the optimal configuration of AI-driven models. The independent factors chosen were temperature, concentrations, shape factor, and thermal conductivity of nanoparticles. The dependent variable selected was the relative thermal conductivity of nanofluids. The study analyzed six models based on AI, which consisted of four types of ANNs (multi-layer perceptron, cascade feed-forward, radial basis function, and generalized regression neural network), adaptive neuro-fuzzy inference systems, and least-squares support vector machines. The study compared the performance of these models in estimation. The experimental findings demonstrated that elevating the temperature and concentrations of nanoparticles resulted in a significant enhancement in the relative thermal conductivity of nanofluids. This improvement can be attributed to the intensified Brownian motion of nanoparticles at higher temperatures and the increased frequency of effective collisions between nanoparticles at higher concentrations. In addition, nanoparticles that have a significant difference in length and width offer a direct and efficient route for heat transmission in nanofluids. This is because they seldom encounter obstacles such as antiparticle boundaries or junction points. As a result, nanofluids containing such nanoparticles exhibit a greater relative thermal conductivity. The statistical analysis validated that the optimal AI-based model for calculating the relative thermal conductivity of nanofluids is the cascade feed-forward neural network with ten hidden neurons and the Levenberg–Marquardt training procedures.

Esfe et al. [8] developed a model and conducted an investigation on an ANN and the thermal conductivity of a nanofluid that includes TiO_2 nanoparticles. The training and estimation of the ANN rely on experimental data, namely the thermal conductivity measurements of the nanofluid at various volume fractions and temperatures. The ANN architecture was determined to include of two hidden layers, each consisting of five neurons. In order to select the optimal training function, an analysis was conducted on the various training functions offered in the ANN. Based on the analysis of the highest regression coefficient and the lowest mean square error (MSE), the trainbr training function demonstrated superior performance. The 2.5% maximum error in approximating test data throughout the estimate phase demonstrates the accuracy of the developed ANN. The findings also indicate that ANN were able to accurately predict the thermal conductivity of TiO_2/water nanofluid at various temperatures, demonstrating high precision. In contrast, theoretical models such as Pak and Cho were unable to provide accurate estimations of the thermal conductivity.

Tian et al. [9] employed ANN to examine how temperature and volume percentage of nanoparticles affect the thermal conductivity of a hybrid nanofluid consisting of graphene oxide-Al_2O_3, water, and ethylene glycol. Nanofluids were synthesized using nanoparticles at volume fractions of 0.1%, 0.2%, 0.4%, 0.8%, and 1.6% within a temperature range of 25–55°C. The thermal conductivity data of the nanofluid was obtained by analyzing six distinct nanoparticle volume fractions and seven varying temperatures. Subsequently, the Perceptron feed-forward ANN was employed to generalize the data and derive a function, effectively replicating the output parameter. The results indicate that the ANN was well trained using the trainbr method, with a MSE of 1.67e-6 and a high correlation coefficient of 0.999 for thermal conductivity. Ultimately, they determined that the impact of raising the temperature of the nanofluid was smaller compared to the proportion of nanoparticles, particularly at low concentrations. The impact of this phenomenon was insignificant, and without the presence of nanoparticles, raising the temperature from 20°C to 55°C results in a 6% improvement in thermal conductivity. Nevertheless, when nanoparticles are present in large quantities, raising the temperature results in a subsequent increase in thermal conductivity. At a volume proportion of 1.6%, raising the temperature from 20°C to 55°C resulted in an increase in thermal conductivity from 0.45 to 0.54 W/m.K.

Ghazvini et al. [10] used ANN techniques to simulate the thermal conductivity ratio of a nanofluid consisting of modified copper nanoparticles and water. They achieved this by analyzing experimental data. The thermal conductivity of the nanofluids was experimentally assessed using KD2Pro, considering different fluid temperatures, nanoparticle concentrations, and diameters. In order to represent the relationship between the thermal conductivity of the nanofluid and variables such as temperature, solid volume percent, and diameter, a correlation was developed using ANNs and taking into account the experimental results. Based on statistical accuracy analysis, the optimal approach for modeling this problem was a two-layer feed-forward ANN model with 6 hidden neurons. The comparative findings indicate that the ANN model successfully predicted the favorable increase in the thermal conductivity of nanofluids.

The current study done by Esfe et al. [11] involved estimating and predicting the thermal conductivity of functionalized Graphene generated using the alkaline approach in water. This was achieved by utilizing an ANN and analyzing experimental data. The thermal conductivity of four varieties of functionalized graphene-water nanofluid was simulated at five distinct temperatures, spanning from 10 to 50°C, as the input for an ANN. The results indicated that the Relative Thermal Conductivity of nanofluids in sample 1 slightly decreased as the temperature decreased, but the other samples exhibited an increase in relative thermal conductivity as the temperature increased. Furthermore, following the training of the network and the subsequent testing of the related data, the disparity between the experimental data and the values derived from the modeling process was determined. The results demonstrated the satisfactory accuracy of the modeling and validated its findings.

In their groundbreaking study, Jamei et al. [12] introduced three novel soft computing approaches: genetic programming, model tree, and multilinear regression models. These techniques were employed to provide precise predictions of the thermal conductivity of several ethylene glycol-based hybrid nanofluids. A grand total of 275 datasets sourced from various publications were gathered and subsequently partitioned into separate testing and training groups. The outcomes derived from the suggested methodologies were evaluated against several performance indicators and empirical correlations. The performance criteria shown that the genetic programming model outperformed the model tree and multilinear regression models in predicting the relative thermal conductivity of hybrid nanofluids, based on the test dataset. The sensitivity analysis revealed that the nanoparticle volume percent exerted the greatest influence among all the input parameters in the model.

In their study, Peng et al. [13] employed an ANN method to ascertain the thermal conductivity of a mixture containing Al_2O_3–Cu and ethylene glycol, with equal volumes. To achieve this objective, a blend of Al_2O_3 and Cu nanoparticles was introduced into ethylene glycol at different percentages ranging from 0.125 to 2.0, while maintaining temperatures between 25 and 50°C. The two-step procedure was utilized to incorporate nanoparticles into the base fluid. Furthermore, the feedforward multilayer perceptron of ANN was analyzed to replicate the heat conduction coefficient of Al_2O_3–Cu nanofluid. A total of 36 measured points were obtained from the experiments. Out of these, 25 points were selected for the ANN, while the remaining eleven points were used to verify the network. The suggested technique using ANN demonstrated the ability to accurately determine the heat conduction coefficient of hybrid nanofluids, showing strong agreement with existing empirical data.

Wang et al. [14] performed studies with hybrid nanofluids consisting of 1.0 wt% Cu/Al_2O_3 in a combination of ethylene glycol and water. The experiments were conducted at temperatures ranging from 20 to 50°C, with varying ratios of the base fluid mixture ranging from 20:80 to 80:20. In order to enhance the accuracy of the anticipated thermal conductivity, a combination of a genetic algorithm and a mental evolutionary algorithm, along with a back-propagation neural network, was employed to search for people with ideal weights and thresholds. The findings indicate that the thermal conductivity exhibits a nonlinear increase when the ratio of water to ethylene glycol and temperature rise. This may be attributed to the superior thermal conductivity of water and the heightened collision frequency between molecules and nanoparticles. This suggests that the combination of back-propagation neural networks improves the accuracy of predictions, with the evolutionary algorithm-back-propagation neural network doing the best.

4.2.2 Viscosity

Çolak [15] examined the viscosity of five distinct ZrO_2/water nanofluids with concentrations of 0.0125%, 0.025%, 0.05%, 0.1%, and 0.2%. These nanofluids were generated using the two-step procedure. A multi-layer perceptron feed-forward back-propagation ANN and a novel mathematical correlation were created based on the experimental data to forecast the viscosity of ZrO_2/water nanofluid. The experimental findings demonstrated an inverse relationship between temperature and viscosity, indicating that viscosity reduces as temperature increases. In addition, the results indicated a direct relationship between concentration and viscosity, with viscosity increasing as concentration increases. The findings acquired from the ANN that was created and the newly built correlation were compared with the experimental data and examined. The findings demonstrated that the created ANN successfully forecasted the viscosity of ZrO_2/water nanofluid with an average error rate of -0.11%. The newly created mathematical model accurately determined the viscosity of ZrO_2/water nanofluid with a negligible error rate of –0.74%.

Sahin et al. [16] assessed the stability of Fe_3O_4-water magnetic nanoparticles at different mass ratios (0.1, 0.25, 0.5, 0.75, and 1) using zeta potential testing. The thermal conductivity and viscosities of magnetic nanoparticles with sufficient stability were tested within the temperature range of 20–60°C for all mass ratios. Furthermore, by utilizing experimental data, two distinct ANN models were constructed. The first model predicts thermal conductivity and viscosity based on temperature (20–60°C) and mass ratio values, while the second model predicts zeta potential based on pH and mass ratio. Ultimately, by utilizing the acquired ANN data, two novel mathematical connections were formulated to forecast thermal conductivity and viscosity. The study found that the constructed ANN model correctly predicted the thermal conductivity and viscosity of Fe_3O_4-water magnetic nanofluids using innovative mathematical correlations. The model had a MSE of 4.51E-06 and a R of 0.99968.

In this study conducted by Heidari et al. [17], a thorough literature search was conducted to gather 1490 experimental data points on the relative viscosity of several nanofluids. A feed-forward back-propagation multilayer perceptron ANN was created and evaluated using the Levenberg–Marquardt training technique to forecast nanofluid viscosity across a wide range of operational parameters. The input parameters of the model were temperature, nanoparticle size, density, volume percentage, and base fluid viscosity. The regression statistical analysis of the training and test data, together with the comparison of the projected values from the ANN with the corresponding experimental data, showed that the constructed neural network has a strong prediction capability. This was further supported by the maximum average relative deviation and the comparison with existing theoretical models.

In this investigation conducted by Dalkilic et al. [18], graphite particles were used to form a nanofluid combination with clean water as the foundation fluid. Their concentrations ranged from 0 to 2% in clean water. After preparing the stabilized nanofluid using a sonicator and ultrasonic bath, the viscosity was measured using a viscosity meter at temperatures ranging from 20°C to 60°C. The experiments were validated by comparing them with the 32 empirical correlations found in the literature. Subsequently, ANN evaluations were conducted to establish a more robust empirical correlation compared to existing research. The publication included comprehensive information on the manufacture of nanofluids, measurement of viscosity, a list of measured data, a numerical model using MATLAB software, and the effect of temperature and concentration on viscosity. The analysis indicated that viscosity correlations found in existing literature may accurately forecast the viscosity of various nanofluids, even if they were developed based on certain types and sizes of nanoparticles and their corresponding base fluids.

Gholami et al. [19] identified the key factors that affect the viscosity of nanofluids and created many AI models to forecast the viscosity of alumina nanoparticles in different base fluids. Subsequently, the predicted accuracy of the produced models was assessed by comparing them with the existing empirical correlations, and the most optimal model was chosen. Correlation matrix analyses verified that the viscosity of nanofluids is primarily influenced by the reduced pressure, the reciprocal of the reduced temperature, the acentric factor of the base fluids, and the diameter and volume concentration of the nanoparticles in the base fluids. Different statistical criteria were employed to numerically assess the correctness of different models. The results indicated that the multi-layer perceptron neural network exhibited superior prediction accuracy compared to other intelligence and empirical models. As a consequence, it was deemed the most effective solution for the given job.

The major objective of the study done by Vakili et al. [20] was to forecast the viscosity of graphene nanoplatelets nanofluid using a multi-layered perceptron ANN and genetic algorithm. In order to get the experimental outcomes, a nanofluid comprising of graphene nanoplatelets and deionized water at temperatures ranging from 20 to 60°C and concentrations of 0.025, 0.05, 0.075, and 0.1 wt% was employed. Moreover, genetic algorithms were employed in ANNs to enhance the learning process. Put simply, varying weights were selected for the connections between neurons. Furthermore, the fixation on bias was rooted on advancements made in the genetic algorithm. Conversely, many indices were employed to evaluate the precision of the model proposed, which provides predictions for nanofluid viscosity. Regarding the comparison, the findings demonstrated that the proposed model, which integrated evolutionary algorithms and ANNs, was congruent with the experimental study.

Esfe et al. [21] sought to forecast the viscosity of a hybrid nanofluid consisting of 10% MWCNT and 90% Al_2O_3 in a lubricant with a viscosity grade of 5W50. They employed ANNs for this purpose. An ANN model was developed to simulate temperature variations from 5 to 55°C. The model considered volumetric fractions ranging from 0.05% to 1% and was trained using 174 experimental data points. The ANN was evaluated using three input variables and one output variable, which represents viscosity. The findings demonstrated the precision and dependability of the suggested model based on the evaluation criteria of R-squared and MSE. Based on the findings from both one factor and two factor analyses, it can be concluded that temperature change had the most significant impact on viscosity when compared to shear rate and solid volume fractions. A correlation was suggested to forecast the viscosity of the hybrid nanofluid, which has a 5W50 fluid as its base fluid, based on temperature, solid volume percentage, and shear rate.

Esfe et al. [22] employed ANNs to simulate the viscosity of an aqueous nanofluid containing TiO_2, based on experimental data. ANNs accurately predicted the changes in dynamic viscosity based on temperature and nanoparticle mass fraction. A network employing a single hidden layer consisting of four neurons was utilized. The regression coefficient achieved a value of 0.9998 in this modeling, indicating the exceptional accuracy of neural networks despite their simplistic structure. Furthermore, a correlation between the mass fraction and temperature was introduced to forecast the viscosity of this nanofluid. This correlation provides an estimation of the viscosity of TiO_2-water nanofluid across a broad range of nanoparticle mass fractions, with a maximum error of 0.5%.

Toghraie et al. [23] examined the dynamic viscosity of Ag/Ethylene glycol nanofluid at temperatures between 25 and 55°C, using different volume fractions of nanoparticles ranging from 0.2% to 2%. The dataset comprised 42 samples. Initially, an ANN was created with the purpose of forecasting the dynamic viscosity of this nanofluid. Subsequently, the outcomes of the ANN were juxtaposed with those of the correlation. A novel technique was introduced to determine the optimal architecture of an ANN. The most effective ANN model was then employed to accurately forecast the dynamic viscosity of a silver/ethylene glycol nanofluid. The study revealed that the ANN had a high level of accuracy in predicting the viscosity of Ag/Ethylene glycol nanofluid, surpassing the performance of the correlation approach. In the correlation approach, the MSE was 0.0012, the sum of squared errors (SSE) was 0.0512, and the maximum error value was 0.0858.

Li et al. [24] sought to determine the stability of Al_2O_3-ethylene glycol nanofluids by analyzing the particle size distribution and velocity ratio. The thermal conductivity and viscosity were determined using ultrasonic techniques for different time durations, mass percentages (ranging from 0 to 2.0 wt%), and temperature ranges (from 25 to 60°C). Furthermore, many criteria were introduced to assess the thermal efficiency of convective heat transmission. Novel correlations and optimum ANN models were proposed based on various sets of experimental data. The findings indicated that Al_2O_3-EG nanofluids, produced using a 60-minute ultrasonication process, exhibit highly promising characteristics. Furthermore, both the experiment and ANN models shown strong correlations in predicting these two values. Nevertheless, the ANN model exhibited more precision. The results were anticipated to have use in further investigations on nanofluid stability, particularly due to their provision of appropriate selection criteria derived from heat transfer behavior, which may be applied in practical scenarios.

Ali et al. [25] examined the viscosity characteristics of TiO_2 nanotubes that were distributed in a nanofluid consisting of ethylene glycol and water. They studied the impact of several process factors, including the mass concentration of nanotubes, temperature, and shear rate. The findings demonstrated a significant rise in viscosity of 30% at 55°C when the mass concentration of nanotubes was increased from 0 to 1%. Similarly, a 22% increase in viscosity was seen at 25°C. The second half of the study employed a multivariable correlation and ANN to forecast the viscosity at different temperatures and shear rates, utilizing the available experimental data. Statistical analyses were conducted to examine the precision of both empirical correlation and ANN modeling. The results indicate that the accuracy of ANN prediction is significantly higher than that of empirical correlations.

Singh et al. [26] introduced a novel single feed-forward multilayer perceptron neural network that utilizes a data transformation technique. They also proposed mathematical models for predicting the density and dynamic viscosity of Al_2O_3/distilled water, MWCNT/distilled water, and graphene nanoparticle/distilled water nanofluids. The density and dynamic viscosity of nanofluids were empirically determined within the temperature range of 30–80°C. The network received nanofluids, volume concentration, and temperature as input, and produced density and dynamic viscosity as output. A statistical study was done to forecast the accuracy of both models. The investigations demonstrated that the density and viscosity of the nanofluid increase as the concentration of nanoparticles increases, and drop as the temperature rises. An increase of 41.59% in dynamic viscosity and 5.06% in density was observed for the Al_2O_3/distilled water nanofluid. The results indicate that the ANN model has been effectively trained using 12 perceptrons in the hidden layer using the Levenberg–Marquardt method. A comparison was made between the findings generated via ANN and the mathematical model, and the experimental data. Both models had a maximum error of less than 0.2% in density measurement and less than 1% in viscosity measurement.

Esfe et al. [27] examined the viscosity of MWCNT-TiO_2/SAE40 nanofluid by employing an ANN. The ANN was developed at temperature ranges ranging from 25 to 50°C and with varying solid volume percentages of nanoparticles. A total of 174 data points from laboratory experiments were utilized. The work involved the utilization of the MLP technique and the Levenberg–Marquardt training algorithm for ANN modeling. The most suitable configuration chosen for the MWCNT-TiO_2/SAE40 nanofluid involves an ANN with two hidden layers. The best structure of this network consists of 10 neurons in the first hidden layer and 4 neurons in the second hidden layer. The correctness of the suggested model is confirmed by the R and MSE coefficients, which were determined to be 0.9999507 and 0.004599708, respectively, in the final step. The margin of deviation (MoD) in the grid dataset ranges from –3% to +3%. According to research that examined the association between ANN and computational data in comparison to laboratory data, the findings of this comparison indicate that the ANN is a more precise and dependable method for calculating viscosity. The focal focus of this research is to develop and forecast the viscosity of MWCNT-TiO_2/SAE40 nanofluid using ANN as a cost-effective and time-efficient alternative to repeated and time-consuming tests.

4.2.3 Specific Heat

Çolak et al. [28] developed an ANN model to predict the specific heat of Cu-Al_2O_3/water hybrid nanofluid. The model was based on temperature and volume concentration. The specific heat values of the Cu-Al_2O_3/water hybrid nanofluid, generated with a concentration of five volumes, were experimentally measured within the temperature range of 20°C to 65°C. The dataset was divided into three main segments, consisting of 901 samples for training, 257 samples for testing, and 129 samples for validation. Upon comparing the model's predictions with experimental results, it was determined that the model accurately predicts specific heat with an R-value of 0.99994 and an average relative error of around 5.84E-9. Furthermore, a mathematical correlation was formulated to calculate the specific heat of the Cu-Al_2O_3/water hybrid nanofluid. The results obtained from the mathematical correlation exhibited a strong connection with all the experimental values, with an average variance of −0.005%. The findings indicated that the mathematical correlation created was a suitable model for accurately calculating the specific heat of the Cu-Al_2O_3/water hybrid nanofluid.

Çolak [29] conducted an experiment to determine the specific heat values of a nanofluid made of yttrium oxide and water. The nanofluid was created with Y_2O_3 nanoparticles and its volumetric content was varied in five different levels. The specific heat values were evaluated using the DTA technique. A multilayer perceptron was created using the experimental data, employing a feed-forward back-propagation ANN with 15 neurons in its hidden layer. Out of the total 60 experimental data, 42 were utilized for the training phase, 12 for the validation phase, and 6 for the test phase. Furthermore, a novel mathematical correlation has been suggested to compute the specific heat values of yttrium oxide-water nanofluid. The ANN accurately predicted the specific heat values of yttrium oxide-water nanofluid with a negligible average error of −0.0007%. The average error rate of the suggested new correlation was determined to be −0.011%.

Çolak [30] examined the impact of several techniques utilized in the training of ANNs on the predictive accuracy of ANNs. For this investigation, three distinct ANN models were created utilizing the Levenberg–Marquardt, Bayesian regularization, and scaled conjugate gradient training techniques, which are often employed in academic research. The training of ANNs utilized the experimentally recorded specific heat values obtained using the differential thermal analysis method for ZrO_2/water nanofluid produced in five distinct volumetric concentrations. The multilayer perceptron feed-forward back-propagation ANN model utilized temperature and volumetric concentration as input parameters. The model consisted of 15 neurons in the hidden layer and predicted particular heat values in the output layer. The findings demonstrated that ANNs were very suitable for forecasting the thermophysical characteristics of nanofluids. Nevertheless, it was discovered that the ANN created using the Bayesian regularization training approach exhibits the most superior predictive capability, with an average deviation margin of just 0.00009%. The ANN trained using the scaled conjugate gradient training approach had the lowest prediction performance, with an average error of –0.0032%.

In their study, Deymi et al. [31] employed sophisticated machine learning methods to forecast the specific heat capacity of nanofluids. The target variable was assessed as a function of the mean nanoparticle size, particle volume fraction, temperature, specific heat of nanoparticles, and specific heat of the base-fluids. 2084 experimental data pertaining to 10 distinct types of nanoparticles and 6 different types of base-fluids were gathered from the literature to forecast the specific heat values of various nanofluids. The results showed that the random forest model had the best accuracy among the others, with an average absolute percent relative error of 0.1928% and an R^2 value of 0.9995. In addition, a sensitivity analysis was conducted to establish the connections between the input and output parameters for the optimal model. The Leverage statistical approach was utilized to identify the legitimate data points, outliers, and data points that are beyond the expected range.

Alade et al. [32] demonstrated the modeling of the specific heat capacity of CuO/water nanofluids using support vector regression and ANN. The models provided were derived from experimental data on the specific heat capacity of CuO nanoparticles, the volume fractions of CuO nanoparticles, and the fluid temperature. The CuO nanoparticles were evaluated at volume fractions ranging from 0.4% to 2%, and the temperature range spans from 293 K to 338 K. The findings indicated that the support vector regression model has marginally superior accuracy in comparison to the ANN model. Nevertheless, both the support vector regression and ANN models exhibited superior predictive capabilities for the specific heat capacity of CuO/water nanofluids in comparison to the current theoretical models. The findings of this work demonstrated that machine learning models offer a more precise assessment of the specific heat capacity of CuO/water nanofluids.

Seawram et al. [33] created an ANN prediction model to forecast specific heat capacity. They utilized feedforward and cascade forward propagation networks together with the Levenberg–Marquardt learning method. The optimal topology of the ANN model was chosen to forecast the specific heat capacity of a hybrid nanofluid. The R values obtained for training, validation, and testing were 0.9919, 0.9473, and 0.9673, respectively. Subsequently, the specific heat capacity value obtained from the ANN model was utilized to assess and examine the rates at which heat is transferred. The hybrid nanofluid consisting of Al_2O_3 and CuO suspended in water demonstrated superior heat dissipation properties when used as a heat exchanger medium.

4.3 AI APPLICATIONS IN DETERMINING FLOW CHARACTERISTICS OF NANOFLUIDS

Shafiq et al. [34] developed a new intelligent numerical computer solver using ANN and the Levenberg–Marquard algorithm. This solver was used to analyze heat generation/absorption and radiation in an unsteady electrically conducting Williamson liquid flow along a porous stretched surface. The investigation of the heat phenomenon involved considering a convective boundary condition, as well as including both velocity and thermal slip phenomena. The initial nonlinear coupled partial differential equations that describe the fluidic model were converted into a corresponding nonlinear ordinary differential equations system by integrating suitable transformations. A dataset for the proposed ANN was created by systematically varying the relevant parameters of the fluidic model using the Galerkin weighted residual approach. An ANN was constructed to forecast the values. The prediction accuracy of the ANN model was evaluated by comparing the outcomes generated by the ANN model, which utilized the Levenberg-Marquard algorithm as the training procedure, with the desired target values. The current investigation has revealed that the ANN technique is very accurate, efficient, and easily usable for modeling the slip flow of Williamson fluid towards a stretched plate with heat production or absorption.

Shafiq et al. [35] demonstrated the phenomenon of magnetohydrodynamic squeezing flow, where fluid is forced through a non-Darcian medium towards a stretched permeable surface. An examination was conducted on the thermal and mass transfer processes under convective circumstances and nonlinear stratification. The heat transmission was improved by including the radiation and viscous dissipation processes. The nonlinear simplified equations were assessed using a numerical Runge–Kutta fourth-order method through the shooting process. Graphs depicting crucial metrics were shown to illustrate the variability in the pertinent domains. Numerical calculations were performed to determine the Sherwood number, Nusselt number, and skin friction coefficient for different parameters. Subsequently, three distinct ANNs were created using the collected data. The findings gained demonstrate that ANNs have the capability to accurately forecast and optimize various tasks.

In the current study, Shafiq et al. [36] investigated the slip and Darcy–Forchheimer phenomena in a Powell–Eyring nanofluid model. The study focused on bioconvective implementations on a stretched surface, using ANNs. Throughout the investigation, the unique influences of activation energy, convective boundary condition, and thermal radiation phenomena were taken into account. The regulating terms were devised in accordance with essential principles. The answer was obtained using numerical simulations utilizing a fourth-order Runge–Kutta approach with a shooting operation, and then an ANN was applied. A dataset was generated to encompass different flow situations, and an ANN model was constructed to forecast values for skin friction coefficient, local Sherwood number, local motile density of microorganisms, and local Nusselt number. The study's findings demonstrate that the created ANN models had the capability to make very accurate predictions, with an average error rate not surpassing 0.53%.

Shafiq et al. [37] examined the forced convective heat and mass transfer of a nanofluid by applying the Buongiorno model. They studied the movement of the nanofluid through a thin needle using the Runge–Kutta fourth-order technique with a firing approach. The study focused on examining the impact of thermo-diffusion and diffusion-thermo effects on the flow. Specifically, the Dufour and Soret effects were explored. In addition, the mass transport phenomena were analyzed using activation energy. The flow model was derived as a system of partial differential equations using the boundary-layer technique. These equations were then changed through suitable transformations to be linked with nonlinear ordinary differential equations. The Runge–Kutta method was the predominant approach for getting numerical solutions to differential equations. It is capable of assessing numerical solutions of higher order and delivering very accurate results. Hence, by employing the Runge–Kutta fourth-order technique together with a shooting approach, a dataset was generated for various flow scenarios of the intricate and inclusive nanofluid model (Boungiorno's model), which encompasses Brownian motion and thermophoresis. An ANN model was constructed using the given dataset to forecast the values of skin friction coefficient, Sherwood number, and Nusselt number. Seventy percent of the data utilized in ANN models, generated with varying amounts of datasets, were allocated for training, validation, and testing purposes. The findings demonstrated that ANN models have the capability to accurately forecast the values of skin friction coefficient, Sherwood number, and Nusselt number.

Shafiq et al. [38] developed a theoretical bioconvective model within the current framework to study the flow of thermally developed thixotropic nanoparticles. The model incorporates various flow features, including thermophoresis, Brownian motion, convective condition, radiation features, and ANN models. The nonlinear complex equations were computed numerically using the Runge–Kutta fourth-order shooting method. The Sherwood number, microorganism motile density, skin friction coefficient, and Nusselt number were computed using various parameters. Based on the results, four distinct ANNs were created. The R values for the constructed neural network models were found to be more than 0.99. The findings demonstrated that ANNs yield highly accurate results when analyzing the flow of thermally generated thixotropic nanoparticles. Theoretical results have uncovered applications in industry, engineering, and thermal processes related to heat transport.

Shafiq et al. [39] investigated the synergistic impact of activation energy and binary chemical reactant on a steady magnetohydrodynamic mixed convective third-grade nanofluid flow. The study focused on a radially radiative stretching plate and employed an AI methodology. A heat transfer study was performed, taking into account heat generation, Joule heating, and the Soret and Dufour effects. The initial nonlinear coupled partial differential equations representing the fluid model were transformed into a system of equivalent nonlinear ordinary differential equations by applying suitable transformations. Three distinct ANN models were suggested to forecast the skin friction, Nusselt number, and Sherwood number values of the fluid model using the Shooting Runge–Kutta Fehlberg 4 approach. This was accomplished by employing a dataset that encompassed diverse values of the pertinent parameters. It is worth mentioning that the average deviation values for each output parameter stayed below 5%. The results demonstrate that ANNs are a precise engineering tool for accurately estimating the combined impacts of activation energy and binary chemical reaction in a fixed magnetohydrodynamic mixed convective third-grade nanofluid flow with a radial radiative stretched plate.

Shafiq et al. [40] studied the behavior of a bio-convective Walter's B nanofluid with thermophoresis and Brownian diffusion. The study focused on a cylindrical disk and utilized ANNs for analysis. Furthermore, the analysis accounted for the thermal conductivity, radiation, and motile density of microorganisms. The Buongiorno model was employed to examine the characteristics of nanofluids in mobile microorganisms. A dimensionless system of a differential system was achieved by employing suitable similarity variables. The system of nonlinear equations was solved numerically using the fourth-order Runge–Kutta shooting method. The impact of flow parameters on the velocity field, temperature distribution, species volumetric concentration, and microorganism fields were also examined. Two separate ANN models were created utilizing numerical data, and their predictive accuracy was extensively evaluated. Based on the error histograms, it was observed that the training phase of the ANN model had minimal error. Based on the measured performance parameters, both ANN models demonstrated a high level of accuracy in their predictions.

Shafiq et al. [41] examined the Darcy–Forchheimer flow paradigm, a valuable framework utilized in industries like petroleum engineering where the impact of high flow velocities is prevalent, using an AI methodology. The study focused on analyzing the Darcy–Forchheimer flow of Ree–Eyring fluid on a permeable stretching surface. The investigation included the examination of heat and mass transfer mechanisms, taking into account factors such as chemical processes, heat generation/absorption, and activation energy. The Cattaneo–Christov heat flow model was employed to examine the characteristics of heat transmission. Three distinct ANN models were created to forecast the values of Nusselt number, Sherwood number, and skin friction coefficient within the context of improving Darcy–Forchheimer flow of Ree-Eyring fluid. The ANN model that was created demonstrated exceptional accuracy in predicting values for Nusselt number, Sherwood number, and skin friction coefficient. The study's findings demonstrate that ANNs are a very suitable method for modeling the flow of Darcy–Forchheimer Ree–Eyring fluid towards a permeable stretch layer, considering activation energy and a convective boundary condition.

In their study, Çolak et al. [42] utilized an AI approach to estimate the significant components of the process, which included the number of tubes, the cleanliness factor, and the total cost. This estimation was based on the utilization of 339 data points. An investigation was conducted to determine how the outputs were affected by the Reynolds number, thermal conductivity, specific heat, viscosity, and total fin surface efficiency of the individual input components. According to the solution with six inputs, the total tube number, cleanliness factor, and total cost analysis were calculated with variations of –0.66%, 0.001%, and 0.12%, respectively. These deviations were recorded as a consequence of the solution.

The Sisko fluid model was initially proposed by Shafiq et al. [43] as a means of assessing lubricating greases' high shear rates. They found that our model accurately predicted shear thickening and thinning, and for some fluids, they were able to just choose three of its constants. The nanomaterial flow of the Darcy–Forchheimer Sisko fluid model was examined in this study using the Levenberg-Marquardt approach with backpropagated neural networks. The development of the nanofluid model took into account the effects of thermophoresis and Brownian motion. A nonlinear ordinary differential system that is similar to the original one was obtained by transforming the fluidic model's nonlinear coupled partial differential system. By adjusting the relevant variables using the Galerkin weighted residual method, a dataset for the suggested multilayer perceptron ANN was generated for various fluid model situations. A multilayer perceptron, a type of ANN, has been developed for the purpose of making predictions about these values.

In their study, Shafiq et al. [44] improved the thermal performance of base solvents by adding small-sized metal solids to the base fluids. They used a machine learning technique to assemble numerical data on the thermal application behavior of nanofluids based on graphene oxide water/ethylene glycol in the presence of electromagnetohydrodynamic and Darcy–Forchheimer medium. A convective boundary condition and heat radiation were also utilized in the investigation of nanofluid flow. The system of equations was calculated using the Runge–Kutta fourth-order shooting technique. After running simulations with different values for the skin friction coefficient and Nusselt parameter, two separate ANNs were built from the results. A big Biot number was useful for estimating the fluid temperature. The created ANNs had an R-value greater than 0.99. In addition, extremely modest numbers were used to compute the deviation rate. Skin friction coefficient and Nusselt number may be reliably predicted using the suggested models of ANNs, according to the results.

The Darcy–Forchheimer Williamson nanofluid model was tested by Shafiq et al. [45] on the flow of nanomaterials using the Levenberg-Marquardt method in conjunction with backpropagated neural networks. The nanofluid model was enhanced by include thermophoresis and Brownian motion. The required transformations were applied to this system, transforming it into a similar nonlinear ordinary differential system. The suggested multilayer perceptron ANN's dataset was created by using a Runge–Kutta fourth-order shooting technique to change the relevant variables. In order to predict the values of the multiple-layer perceptron, it constructed an ANN. The parameter with the smallest mean deviation and the one with the largest mean deviation were both found.

Shafiq et al. [46] sought to utilize an ANN to analyze the velocity of heat, mass, and motile microbe migration in the convective flow of magnetohydrodynamic second-grade nanofluid towards a vertical surface. The suspension of nanoparticles was successfully stabilized by the process of bioconvection, helped by microorganisms. Both the characteristics of the nanoparticles and the forces of buoyancy had an impact on this process. The model takes into account not just thermophoretic dynamics and Brownian motion, but also includes radiation and Newtonian heating effects. The nonlinear equation systems were derived using suitable transformations. The simplified equations, which were nonlinear, were subjected to numerical computations using the fourth-order Runge–Kutta shooting technique. The Sherwood number, Nusselt number, and coefficient of motile microbe density were calculated using different parameters, and three separate ANNs were constructed based on these results.

4.4 USE OF AI IN NANOFLUID APPLICATIONS IN HEAT EXCHANGER SYSTEMS

Çolak et al. [47] created three distinct ANN models to forecast the radiative, convective, and total heat transfer coefficients on the floor surface of a radiant floor heating system in an actual-sized room, using three different working fluid scenarios. The working fluids used were pure water, multiwall carbon nanotube with a volume fraction of 0.7% and 0.07%, and aluminum oxide with a volume fraction of 1.26%. The input temperatures ranged from 30°C to 60°C, and the mass flow rates were 0.056 kg/s, 0.09 kg/s, and 0.125 kg/s. The study involved developing multilayer perceptron networks using three different training algorithms: Levenberg–Marquardt, Bayesian regularization, and scaled conjugate gradient. The networks were trained with varying numbers of neurons. After conducting detailed computational numerical analyses, it was found that the Levenberg–Marquardt algorithm had the highest prediction performance, achieving an accuracy of 99%. Therefore, this algorithm was selected as the optimal choice.

Kamsuwan et al. [48] conducted a sophisticated computation that combines AI with easily accessible computational software, resulting in an advanced calculation system. A response surface approach, based on statistical mathematics, was employed to establish a correlation between the calculation model and the data. This method was utilized to determine the design parameters that optimize the performance of the system under different operating situations. The optimized findings demonstrated that the integration of a polymer-based microchannel heat exchanger with nanofluid showed great potential as an innovative solution. The use of TiO_2/water resulted in a noteworthy enhancement in heat transfer, leading to a 12% augmentation in the total heat transfer coefficient as compared to the use of water alone. In addition, the use of CuO/water nanofluid resulted in a performance index of 1.03. Furthermore, a horizontal parallel connection of 66 polymer-based microchannel heat exchangers demonstrated that they can achieve the same heat transfer performance as metal-based microchannel heat exchangers when using TiO_2/water nanofluid. This suggests a balance between enhancing heat transfer and minimizing energy consumption.

Sundar et al. [49] conducted an experimental analysis on the efficacy and quantity of transfer units of water-dispersed Fe_3O_4–SiO_2 hybrid fluids passing through a plate heat exchanger. The Fe_3O_4–SiO_2/water hybrid nanofluids in the plate heat exchanger maintained laminar flow conditions. The experimental investigations were conducted using particle concentrations varying from 0.2% to 1.0% and flow speeds ranging from 0.05 kg/s to 0.1166 kg/s. The efficacy and quantity of transfer units produced from these tests were anticipated using the scaled conjugate gradient approach, which is based on ANNs. The Fe_3O_4–SiO_2 nanoparticles were synthesized using chemical reaction and in-situ growth techniques, and subsequently analyzed using x-ray diffraction and vibrating sample magnetometer apparatus. Hybrid nanofluids, which were diluted with water, were created and utilized for the experimental investigations. The results demonstrated that when the volume loadings and Reynolds number increased, both the efficacy and number of transfer units were improved. The efficiency and quantity of transfer units were enhanced by 13.23% and 24.51% respectively, while using a hybrid nanofluid at a volume concentration of 1.0% and a Reynolds number of 571.89, compared to the values obtained with water.

Zakeri [50] employed a water-based nanofluid containing graphene oxide in a heat exchanger with a twin pipe configuration, where the flow of fluids occurs in opposite directions. The inner conduit transported the heated substance. The cold fluid was circulated via the outer pipe, and the tests were conducted separately using deionized water-based fluid and graphene oxide nanofluid. The findings indicated that augmenting the flow velocity and concentration of nanoparticles resulted in an enhancement in the heat transfer coefficient. The impact of the intake hot fluid temperature on this coefficient was found to be insignificant in comparison to the concentration and flow rate. In addition, radial basis function neural networks and multilayer perceptron neural network models were developed specifically for the purpose of predicting the heat transfer coefficient. The radial basis function neural network model demonstrated superior accuracy compared to the multi-layer perceptron neural network model, but both models successfully predicted the coefficient. The results further demonstrated that the nanofluid exhibited superior performance compared to the base fluid in relation to the heat transfer coefficient.

In order to mitigate the substantial expenses and time required for examination, Kamsuwan et al. [51] opted for a novel simulation technique that is both efficient and cost-effective. This approach was employed to examine the performance of a microchannel heat exchanger using nanofluids. In order to examine the heat recovery process at low temperatures, namely in the range of 100–200°C, prediction models for nanofluid properties were created using an ANN. Subsequently, the prognostic models were included and merged into computational fluid dynamics with the purpose of developing a microchannel heat exchanger. The utilization of nanofluids has been seen to enhance the thermal conductivity of this heat exchanger, hence improving its heat transfer efficiency. The appropriate nanofluid kinds and concentrations were chosen based on their thermal-hydraulic efficiency. Among the tested nanofluids, the fluid containing 3% weight of TiO_2 in water had the most promising performance in terms of thermal-hydraulic ratio, with a value of 1.03. This nanofluid shows great potential for use in the given setting.

Kamsuwan et al. [52] presented a new approach that combines ANNs with traditional process modeling to study heat transfer in a heat exchanger utilizing nanofluid. Furthermore, an analysis was conducted to examine the responsiveness of factors in the design and evaluation of heat exchangers. The construction of the ANN involved the utilization of 2723 datasets including a wide range of nanofluid kinds and their respective attributes. The novel ANN nanofluid predictive model exhibited superior accuracy compared to alternative numerical techniques, as evidenced by the simulation results. The result had a maximum inaccuracy of just 4.1%. The approach replicated the traditional simulation to enhance its flexibility and performance, while maintaining acceptable levels of inaccuracy. Ultimately, the study utilized both ANN and traditional process modeling techniques to thoroughly analyze the impact of nanofluid augmentation on heat exchangers. The parameter sensitivity test was conducted on a plate heat exchanger, focusing on the impact of increasing the heat transfer coefficient by approximately 7%, while keeping the pressure drop constant and evaluating the performance efficiency coefficient.

Through the use of a three-dimensional shell and tube heat exchanger, Fuxi et al. [53] studied the effects of hybrid nanofluids and of helical coil pitch. Phase Coupled SIMPLE is the algorithm that was utilized in this investigation, and Eulerian was the approach that was utilized. The current research demonstrates that in order to achieve a positive gain in efficiency, it was possible to combine the utilization of hybrid nanofluids with the modification of the geometry of shell and tube heat exchangers. Finally, the findings were utilized in the process of training an ANN. The optimal feed-forward network was obtained in this respect in order to anticipate the effectiveness of the material. This was done in order to identify the optimal number of neurons in the hidden layer.

Said et al. [54] employed multi-walled carbon nanotubes (MWCNT)/water nanofluids to enhance the efficiency of a shell and tube heat exchanger, while simultaneously decreasing energy consumption and total expenses. When the volume percentage was increased to 0.3%, there was a significant enhancement in the heat transfer coefficient, with an approximate improvement of 31.08% compared to the original fluid. Tests were performed on a shell and tube heat exchanger, and the findings indicate that employing nanofluid with a volume percentage of 0.3% enhances the efficiency of the heat exchanger by 5.49% in comparison to the base fluid. A high level of concordance was achieved between the experimental and analytical findings. In addition, a numerical model was created using the ANSYS commercial software to investigate the incorporation of semicircular baffles with nanofluid. Numerical data indicates that the addition of MWCNT/water nanofluid at a volume percentage of 0.3%, along with semicircular baffles, increased the overall efficiency of the shell and tube heat exchanger by 15.4%. The temperature variation data obtained from the experiment was used to develop an AI-driven prognostic model. An ANN of the multilayer perceptron type was used to model and predict the thermal behavior of MWCNT nanofluids on the tube side and water on the shell side. The tube side model had exceptional R and R^2 values of 0.998 and 0.996, respectively, while the shell side model demonstrated R and R^2 values of 0.994 and 0.988, respectively. These values indicate the presence of a strong and reliable predictive model.

In their study, Jery et al. [55] performed a simulation of a geothermal heat exchanger to examine the most effective diameter and nanoparticle concentration for reducing entropy formation. The findings indicated a positive correlation between the increase in thermal resistance of the inner wall and the rise in outlet temperature. The study extensively examined the impact of thermal resistance and diameter, and the findings were utilized to choose the option with the lowest entropy generation. The findings indicate that the diameter at which entropy is reduced expands as the heat resistance of the inner wall rises. The optimal scenario entailed a significant level of thermal resistance in the outside walls, while exhibiting little resistance in the inside walls. Employing inner walls with strong heat resistance might mitigate the detrimental consequences of entropy formation in situations with large diameters. Furthermore, the researchers examined the impact of including nanoparticles, which resulted in a nearly 10% rise in the average Nusselt number and a corresponding drop in entropy production. Proposed were ANN models to forecast the Nusselt number and entropy production, utilizing numerical outcomes. The models successfully attained mean absolute error (MAE) below 3% and R^2 above 0.95.

4.5 USE OF AI IN NANOFLUID APPLICATIONS IN RENEWABLE ENERGY SYSTEMS

Chilambarasan et al. [56] examined how the combination of an absorber tube with internal grooves and a nanofluid based on Al_2O_3-WEG may enhance the performance of solar flat plate collectors. A comparison was made between the results obtained by using Al_2O_3-WEG-based nanofluid and ordinary working fluid. The solar flat plate collector models 1 and 2 achieved the best efficiency while operating with mass flow rates of 0.036 kg/s and 0.2% volume. The efficiency improvement of model 2 compared to model 1 of plain fluid was 46.37%, 54.13%, and 33.83% at mass flow rates of 0.024, 0.036, and 0.048 kg/s, respectively. Model 2 achieved a 54.1% higher efficiency compared to Model 1, when operating at a mass flow rate of 0.036 kg/s with a volume concentration of 0.2%. The ANN models accurately forecasted the real-time output values of the solar flat plate collector experiment.

Jakeer et al. [57] investigated the heat transfer characteristics of a Sutterby hybrid nanofluid flow with magnetohydrodynamics, using a non-uniform heat source/sink and linear thermal radiation across a non-Darcy curved permeable surface. The current study presented a new intelligent numerical computing solution that utilizes an MLP feed-forward back-propagation ANN with the Levenberg-Marquardt method. Data were collected for testing, certifying, and training the ANN model. Graphs depicted how various physical elements influenced variables such as pressure, velocity, and temperature. The thermal energy was enhanced more by the SiO_2–Au hybrid nanofluid compared to the SiO_2–TiO_2 hybrid nanofluid throughout the investigation. Higher values of the internal heat generation/absorption parameter led to an increase in temperature.

Rawat et al. [58] examined the heat transfer of a ternary hybrid nanofluid flowing inside a receiver tube in a parabolic trough solar collector. The flow inside the receiver tube was mathematically approximated using spinning parallel plates. The investigation examined the impact of the Hall effect, the Cattaneo–Christov model, suction/injection, and thermal radiation (linear and quadratic). Conventional methods for analyzing parameters might struggle to produce meaningful results when dealing with sophisticated models and numerical techniques. This study assessed the effectiveness of soft computing approaches in predicting the behavior of an issue with several interconnected factors. The simulated data was utilized to train the ANN and particle swarm optimization technique. An ANN and particle swarm optimization technique were used to accurately predict the Nusselt number values at the lower and top plates. The algorithms' prediction ability was assessed using mean squared error, average forecasting error rate, and coefficient of correlation. The examination demonstrated the benefits of utilizing soft computing approaches for accurately studying complicated flow models.

Alnaqi et al. [59] studied the impact of utilizing an off-center finned absorber tube on the performance of a two-fluid parabolic trough solar collector containing a nanofluid with various nanoparticle shapes and sizes. The energy efficiency of the two-fluid parabolic trough solar collector was determined to be greater than that of the single-fluid parabolic trough solar collector. Furthermore, the results showed that the energy efficiency of the parabolic trough solar collector improves when an off-center finned absorber tube is used, by increasing the Reynolds number and reducing the solid volume portion. The highest energy efficiency observed in the study was 58.9%. It was achieved by the two-fluid parabolic trough solar collector with an off-center finned absorber tube filled with nanofluid containing spherical nanoparticles with a diameter of 40 nm at a solid volume fraction of 1% and a Reynolds number of 5000. An ANN was used to evaluate how sensitive energy efficiency is to changes in the Reynolds number.

Ibrahim et al. [60] simulated an 80-W photovoltaic technology with its cooling system. The cooling system comprises a copper pipe, with a copper plate of 1 mm thickness positioned beneath the panel. Copper/water-ethylene nanofluids were circulated at various flow rates via the cooling system while being subjected to varying heat flux levels at different times of the day. A neural network was used to predict the thermal characteristics of a solar system that generates heat. An optimization was conducted on the variables of pipe diameter, heat flux, and nanofluid flow rate, as well as the output parameters of maximum temperature and pressure drop to enhance the analysis. The simulation results demonstrated the exceptional capability of the neural network in predicting the output variables. The optimization findings indicated that the pipe with the smallest diameter and lowest heat flux should be used to achieve the highest fluid flow and reduce the maximum temperature. The highest unwanted panel temperature was recorded at the opposite location. The lowest pressure drop was seen when the panel included the pipe with the greatest diameter, and the fluid flowed through it at the slowest speed.

Delfani et al. [61] used an ANN based system to forecast the thermal efficiency of a nanofluid-based direct absorption solar collector. Nine collector prototypes with varying geometries were tested under different conditions in the experimental phase of the project to analyze the impact of collector depth and length on thermal performance. The input parameters of the network chosen to estimate the collector efficiency and Nusselt number are collector depth, length, working fluid flowrate, concentration, and decreased temperature differential. The Nusselt number of the collector significantly increased due to the collector depth and nanofluid flow rate. The network performed best in predicting collector efficiency when the nanofluid concentration was set at 1000 ppm. The Nusselt number prediction had the highest performance while using a collector length of 300 mm. The agreement between the actual and anticipated findings demonstrated the ANN's strong capability in forecasting the thermal performance of direct absorption solar collectors.

Tomy et al. [62] utilized an ANN to simulate the efficiency of a flat plate solar collector using silver/water nanofluid. The solar radiation heat flux ranged from 900 W/m^2 to 1000 W/m^2. The Reynolds number ranged from 5,000 to 25,000. Analyzed the impact of radiation heat flux, mass flow rate, and input temperature on the heat transfer coefficient and thermal efficiency. The ANN findings were compared with actual data. In addition, the ANN was utilized to forecast the thermal efficiency of a flat plate solar collector up to a Reynolds number of 100,000. The study found that the ANN findings closely matched the experimental results, with a difference of less than ±2%.

Moghadam et al. [63] utilized entropy generation minimization, quadratic optimization method, and ANN to determine the ideal operating conditions for turbulent flow of Al_2O_3-EG/W nanofluid within the absorber tube of a parabolic trough solar collector. A three-input ANN was used to estimate the appropriate volume fraction. The procedure aimed to optimize nanoparticle concentration, nanoparticle diameter, nanofluid average flow temperature, and Reynolds number. The study found that the entropy production rate reduces when the volume percentage falls, particle diameter increases, and average flow temperature increases. Introducing nanoparticles to the base fluid elevates frictional entropy production and reduces thermal entropy production. It enhanced heat transfer efficiency but led to a rise in viscous irreversibility. It was noted that there is a definite ideal volume fraction for each particle size and average flow temperature, regardless of the number. An ideal volume fraction existed for all numbers with a constant particle size and mean flow temperature.

Fuxi et al. [64] studied the numerical simulation and ANN modeling of turbulent flow in a pipe with two spring turbulator samples of various sizes and a segmental cross-section. A spring turbulator was expected to enhance heat transfer in a TiO_2–Cu–Water hybrid nanofluid using a single-phase model, feed-forward ANN, and fitting approach. The study analyzed the impact of Reynolds number, size, and volume fraction on Nusselt number, pressure drop, performance evaluation coefficient, solar collector efficiency, and the field synergy principle in comparison to a basic pipe using the finite volume approach. The findings indicated that enlarging the spring turbulator scale enhanced the contact area between the working fluid and the spring turbulator. The enhanced flow turbulence resulted in improved mixing of the nanofluid within the solar collector absorber pipe. The outputs and fitting results of the ANN were compared, revealing that the ANN could properly anticipate the targets.

4.6 CONCLUSION

The use of AI techniques can lead to increased efficiency and innovation in multiple fields, offering a promising option to overcome challenges arising from the complex and dynamic nature of nanofluids. In this comprehensive research, the use of AI to understand various aspects of nanofluidic heat transfer was analyzed. Within the scope of the analysis, different applications were taken into account, including thermophysical properties of nanofluids, flow properties, heat exchanger systems, and renewable energy applications. In the study, firstly, the use of AI in determining the thermophysical properties of nanofluids was examined, focusing on thermophysical properties such as thermal conductivity, viscosity, and specific heat. By applying ANNs and other AI-enabled models, remarkable accuracy has been achieved in predicting the thermophysical properties of various nanofluid compositions. Subsequently, AI applications in determining the flow properties of nanofluids were investigated and the potential of AI in modeling complex fluid dynamics events was demonstrated. From analyzing heat generation in unsteady fluid flows to estimating surface friction coefficients and Nusselt numbers, AI-based approaches have demonstrated robustness and efficiency in capturing complex flow behaviors. The versatility of ANNs has been exemplified in the optimized design and operation of nanofluidic systems in various engineering applications. Integration of AI into heat exchanger systems and renewable energy applications plays an important role in increasing energy efficiency and performance optimization. As a result, the use of AI tools in applications of nanofluidic heat transfer can offer an innovative approach to overcoming complex challenges and unlocking new opportunities in various science and engineering disciplines.

4.7 NOMENCLATURE

AI	artificial intelligence
ANN	artificial neural network
MAE	mean absolute errors

MLP	multilayer perceptron
MSE	mean-squared error
MWCNT	multi-walled carbon nanotubes
R	coefficient of determination
SSE	sum of squared errors
WEG	water/ethylene glycol mixture

REFERENCES

1. A.B. Çolak, Experimental study for thermal conductivity of water-based zirconium oxide nanofluid: Developing optimal artificial neural network and proposing new correlation, International Journal of Energy Research 45(2) (2020) 2912–2930.
2. A.B. Çolak, A novel comparative investigation of the effect of the number of neurons on the predictive performance of the artificial neural network: An experimental study on the thermal conductivity of ZrO_2 nanofluid, International Journal of Energy Research 45(13) (2021) 18944–18956.
3. S. Öcal, M. Gökçek, A.B. Çolak, M. Korkanç, A comprehensive and comparative experimental analysis on thermal conductivity of TiO_2-$CaCO_3$/Water hybrid nanofluid: Proposing new correlation and artificial neural network optimization, Heat Transfer Research, Heat Transfer Research 52(17) (2021) 55–79.
4. F. Sahin, O. Genc, M. Gökcek, A.B. Çolak, An experimental and new study on thermal conductivity and zeta potential of Fe_3O_4/water nanofluid: Machine learning modeling and proposing a new correlation, Powder Technology 420 (2023) 118388.
5. A.B. Çolak, Advancing Nanofluid Technology: Experimental Thermal Conductivity Measurements and Accurate Predictive Models for Yttrium Oxide-Water Nanofluid Using Neural Networks and Mathematical Correlations, Numerical Heat Transfer, Part B: Fundamentals (2024) https://doi.org/10.1080/10407790.2023.2300691
6. C. Lin, J. Zhou, Q. Lu, M.K. Khabaz, A.K. Andani, M. Al-Yasiri, G. Pan, Thermal conductivity prediction of WO_3-CuO-Ag (35:40:25)/water hybrid ternary nanofluid with Artificial Neural Network and back-propagation algorithm, Materials Today Communications 36 (2023) 106807.
7. W. Cui, Z. Cao, X. Li, L. Lu, T. Ma, Q. Wang, Experimental investigation and artificial intelligent estimation of thermal conductivity of nanofluids with different nanoparticles shapes, Powder Technology 398 (2022) 117078.
8. M.H. Esfe, S. Esfandeh, D. Toghraie, Investigation of different training function efficiency in modeling thermal conductivity of TiO2/Water nanofluid using artificial neural network, Colloids and Surfaces A: Physicochemical and Engineering Aspects 653 (2022) 129811.
9. S. Tian, N.I. Arshad, D. Toghraie, S.A. Eftekhari, M. Hekmatifar, Using perceptron feed-forward Artificial Neural Network (ANN) for predicting the thermal conductivity of graphene oxide-Al2O3/water-ethylene glycol hybrid nanofluid, Case Studies in Thermal Engineering 26 (2021) 101055.
10. M. Ghazvini, H. Maddah, R. Peymanfar, M.H. Ahmadi, R. Kumar, Experimental evaluation and artificial neural network modeling of thermal conductivity of water based nanofluid containing magnetic copper nanoparticles, Physica A: Statistical Mechanics and its Applications 551 (2020) 124127.
11. M.H. Esfe, D. Toghraie, Cascade forward Artificial Neural Network to estimate thermal conductivity of functionalized graphene-water nanofluids, Case Studies in Thermal Engineering 26 (2021) 101194.
12. M. Jamei, R. Pourrajab, I. Ahmadianfar, A. Noghrehabadi, Accurate prediction of thermal conductivity of ethylene glycol-based hybrid nanofluids using artificial intelligence

techniques, International Communications in Heat and Mass Transfer 116 (2020) 104624.

13. Y. Peng, A. Parsian, H. Khodadadi, M. Akbari, K. Ghani, M. Goodarzi, Q.V. Bach, Develop optimal network topology of artificial neural network (AONN) to predict the hybrid nanofluids thermal conductivity according to the empirical data of Al2O3 – Cu nanoparticles dispersed in ethylene glycol, Physica A: Statistical Mechanics and its Applications 549 (2020) 124015.
14. J. Wang, Y. Zhai, P. Yao, M. Ma, H. Wang, Established prediction models of thermal conductivity of hybrid nanofluids based on artificial neural network (ANN) models in waste heat system, International Communications in Heat and Mass Transfer 110 (2020) 104444.
15. A.B. Çolak, A novel comparative analysis between the experimental and numeric methods on viscosity of zirconium oxide nanofluid: Developing optimal artificial neural network and new mathematical model, Powder Technology 381 (2021) 338–351.
16. F. Sahin, O. Genc, M. Gökcek, A.B. Çolak, From experimental data to predictions: Artificial intelligence supported new mathematical approaches for estimating thermal conductivity, viscosity and zeta potential in Fe_3O_4-water magnetic nanofluids, Powder Technology 430 (2023) 118974.
17. E. Heidari, M.A. Sobati, S. Movahedirad, Accurate prediction of nanofluid viscosity using a multilayer perceptron artificial neural network (MLP-ANN), Chemometrics and Intelligent Laboratory Systems 155 (2016) 73–85.
18. A.S. Dalkilic, A. Çebi, A. Celen, O. Yıldız, O. Acikgoz, C. Jumpholkul, M. Bayrak, K. Surana, S. Wongwises, Prediction of graphite nanofluids' dynamic viscosity by means of artificial neural networks, International Communications in Heat and Mass Transfer 73 (2016) 33–42.
19. E. Gholami, B. Vaferi, M.A. Ariana, Prediction of viscosity of several alumina-based nanofluids using various artificial intelligence paradigms: Comparison with experimental data and empirical correlations, Powder Technology 323 (2018) 495–506.
20. M. Vakili, S. Khosrojerdi, P. Aghajannezhad, M. Yahyaei, A hybrid artificial neural network-genetic algorithm modeling approach for viscosity estimation of graphene nanoplatelets nanofluid using experimental data, International Communications in Heat and Mass Transfer 82 (2017) 40–48.
21. M.H. Esfe, M. Reiszadeh, S. Esfandeh, M. Afrand, Optimization of MWCNTs (10%) – Al_2O_3 (90%)/5W50 nanofluid viscosity using experimental data and artificial neural network, Physica A: Statistical Mechanics and its Applications 512 (2018) 731–744.
22. M.H. Esfe, M.R.H. Ahangar, M. Rejvani, D. Toghraie, M.H. Hajmohammad, Designing an artificial neural network to predict dynamic viscosity of aqueous nanofluid of TiO_2 using experimental data, International Communications in Heat and Mass Transfer 75 (2016) 192–196.
23. D. Toghraie, N. Sina, N.A. Jolfaei, M. Hajian, M. Afrand, Designing an Artificial Neural Network (ANN) to predict the viscosity of Silver/Ethylene glycol nanofluid at different temperatures and volume fraction of nanoparticles, Physica A: Statistical Mechanics and its Applications 534 (2019) 122142.
24. L. Li, Y. Zhai, Y. Jin, J. Wang, H. Wang, M. Ma, Stability, thermal performance and artificial neural network modeling of viscosity and thermal conductivity of Al_2O_3-ethylene glycol nanofluids, Powder Technology 363 (2020) 360–368.
25. A. Ali, S.U. Ilyas, S. Garg, M. Alsaady, K. Maqsood, R. Nasir, A. Abdulrahman, M. Zulfiqar, A.B. Mahfouz, A. Ahmed, S. Ridha, Dynamic viscosity of Titania nanotubes dispersions in ethylene glycol/water-based nanofluids: Experimental evaluation and predictions from empirical correlation and artificial neural network, International Communications in Heat and Mass Transfer Volume 118, November 2020, 104882.

26. S. Singh, S.K. Ghosh, A unique artificial intelligence approach and mathematical model to accurately evaluate viscosity and density of several nanofluids from experimental data, Colloids and Surfaces A: Physicochemical and Engineering Aspects 640 (2022) 128389.
27. M.H. Esfe, D. Toghraie, F. Amoozadkhalili, Increasing the accuracy of estimating the viscosity of SAE40-based nanofluid containing MWCNT-TiO_2 nanoparticles using a creative method in optimizing the structure of the artificial neural network, Tribology International 179 (2023) 108161.
28. A.B. Çolak, O. Yıldız, M. Bayrak, B.S. Tezekici, Experimental study for predicting the specific heat of water based Cu-Al_2O_3 hybrid nanofluid using artificial neural network and proposing new correlation, International Journal of Energy Research 44(9) (2020) 7198–7215.
29. A.B. Çolak, Developing optimal artificial neural network (ANN) to predict the specific heat of water-based yttrium oxide (Y_2O_3) nanofluid according to the experimental data and proposing new correlation, Heat Transfer Research 51(17) (2020) 1565–1586.
30. A.B. Çolak, Experimental analysis with specific heat of water-based zirconium oxide nanofluid on the effect of training algorithm on predictive performance of artificial neural network, Heat Transfer Research 52(7) (2021) 67–93.
31. O. Deymi, F. Hadavimoghaddam, S. Atashrouz, A. Abedi, A. Hemmati-Sarapardeh, A. Mohaddespour, Employing ensemble learning techniques for modeling nanofluids' specific heat capacity, International Communications in Heat and Mass Transfer 143 (2023) 106684.
32. I.O. Alade, M.A.A. Rahman, Z. Abbas, Y. Yaakob, T.A. Saleh, Application of support vector regression and artificial neural network for prediction of specific heat capacity of aqueous nanofluids of copper oxide, Solar Energy 197 (2020) 485–490.
33. S. Seawram, P. Nimmanterdwong, T. Sema, R. Piemjaiswang, B. Chalermsinsuwan, Specific heat capacity prediction of hybrid nanofluid using artificial neural network and its heat transfer application, Energy Reports 8 (2022) 8–15.
34. A. Shafiq, A.B. Çolak, T.N. Sindhu, Q.M. Al-Mdallal, T. Abdeljawad, Estimation of unsteady hydromagnetic Williamson fuid flow in a radiative surface through numerical and artificial neural network modeling, Scientific Reports 11, (2021) 14509.
35. A. Shafiq, A.B. Çolak, T.N. Sindhu, T. Muhammad, Optimization of Darcy-Forchheimer squeezing flow in nonlinear stratified fluid under convective conditions with artificial neural network, Heat Transfer Research 53(3) (2022) 67–89.
36. A.B. Çolak, A. Shafiq, T.N. Sindhu, Modeling of Darcy-Forchheimer bioconvective Powell Eyring nanofuid with artifcial neural network, Chinese Journal of Physics, 77 (2022) 2435–2453.
37. A. Shafiq, A.B. Çolak, T.N. Sindhu, Modeling of Soret and Dufour's convective heat transfer in nanofluid flow through a moving needle with artificial neural network, Arabian Journal for Science and Engineering 48 (2023) 2807–2820.
38. A. Shafiq, A.B. Çolak, T.N. Sindhu, Significance of bioconvective flow of MHD thixotropic nanofluid passing through a vertical surface by machine learning algorithm, Chinese Journal of Physics 80 (2022) 427–444.
39. A. Shafiq, A.B. Çolak, T.N. Sindhu, Analyzing activation energy and binary chemical reaction effects with artificial intelligence approach in axisymmetric flow of third grade nanofluid subject to Soret and Dufour effects, Heat Transfer Research 54(3) (2023) 75–94.
40. A. Shafiq, A.B. Çolak, T.N. Sindhu, Optimization of bioconvective magnetized Walter's B nanofluid flow towards a cylindrical disk with artificial neural network, Lubricants 10(9) (2022) 209.
41. A. Shafiq, A.B. Çolak, T.N. Sindhu, Optimization of the Numerical Treatment of the Darcy-Forchheimer Flow of Ree-Eyring Fluid with Chemical Reaction by Using

Artificial Neural Networks, International Journal for Numerical Methods in Fluids 95(1) (2023) 176–192.

42. A.B. Çolak, H. Mercan, Ö. Açıkgöz, A.S. Dalkılıç, S. Wongwises, Prediction of Nanofluid flows' optimum velocity in finned tube-in-tube heat exchangers using artificial neural network, Kerntechnik 88(1) (2023) 100–113.
43. A. Shafiq, A.B. Çolak, T.N. Sindhu, Construction of neural network based intelligent computing for treatment of Darcy-Forchheimer Sisko nanofluid flow with Rosseland's radiative process, Heat Transfer Research 54 (9) (2023) 77–98.
44. A. Shafiq, A.B. Çolak, T.N. Sindhu, Significance of EMHD Graphene Oxide (GO) Water Ethylene glycol nanofluid flow in a Darcy-Forchheimer Medium by machine learning algorithm, The European Physical Journal Plus 138 (2023) 213.
45. A. Shafiq, A.B. Çolak, T.N. Sindhu, Modeling of Darcy-Forchheimer magnetohydrodynamic Williamson nanofluid flow towards nonlinear radiative stretching surface using artificial neural network, International Journal for Numerical Methods in Fluids 95(9) (2023) 1502–1520.
46. A. Shafiq, A.B. Çolak, T.N. Sindhu, Development of an intelligent computing system using neural networks for modeling bioconvection flow of second-grade nanofluid with gyrotactic microorganisms, Numerical Heat Transfer, Part B: Fundamentals (2023).
47. A.B. Çolak, Y. Karakoyun, O. Açıkgöz, Z. Yumurtacı, A.S. Dalkılıç, A numerical study aimed at finding optimal artificial neural network model covering experimentally obtained heat transfer characteristics of hydronic underfloor radiant heating systems running various nanofluids, Heat Transfer Research 53 (5) (2022) 51–71.
48. C. Kamsuwan, X. Wang, L.P. Seng, C.K. Xian, R. Piemjaiswang, P. Piumsomboon, K. Manatura, M. Kaewbumrung, Y. Pratumwal, S. Otarawanna, B. Chalermsinsuwan, Enhancing performance of polymer-based microchannel heat exchanger with nanofluid: A computational fluid dynamics-artificial neural network approach, South African Journal of Chemical Engineering 46 (2023) 361–375.
49. L.S. Sundar, K.V.V.C. Mouli, Effectiveness and number of transfer units of plate heat exchanger with Fe_3O_4–SiO_2/Water hybrid nanofluids: Experimental and artificial neural network predictions, Case Studies in Thermal Engineering 53 (2024) 103949.
50. F. Zakeri, M.R.S. Emami, Experimental and numerical investigation of heat transfer and flow of water-based graphene oxide nanofluid in a double pipe heat exchanger using different artificial neural network models, International Communications in Heat and Mass Transfer 148 (2023) 107002.
51. C. Kamsuwan, X. Wang, L.P. Seng, C.K. Xian, R. Piemjaiswang, P. Piumsomboon, Y. Pratumwal, S. Otarawanna, B. Chalermsinsuwan, Simulation of nanofluid microchannel heat exchanger using computational fluid dynamics integrated with artificial neural network, Energy Reports 9 (2023) 239–247.
52. C. Kamsuwan, X. Wang, P. Piumsomboon, Y. Pratumwal, S. Otarawanna, B. Chalermsinsuwan, Artificial neural network prediction models for nanofluid properties and their applications with heat exchanger design and rating simulation, International Journal of Thermal Sciences 184 (2023) 107995.
53. S. Fuxi, N. Sina, A. Ahmadi, E.H. Malekshah, M.Z. Mahmoud, H.Ş. Aybar, Effect of different pitches on the 3D helically coiled shell and tube heat exchanger filled with a hybrid nanofluid: Numerical study and artificial neural network modeling, Engineering Analysis with Boundary Elements 143 (2022) 755–768.
54. Z. Said, S. Rahman, P. Sharma, A.A. Hachicha, S. Issa, Performance characterization of a solar-powered shell and tube heat exchanger utilizing MWCNTs/water-based nanofluids: An experimental, numerical, and artificial intelligence approach, Applied Thermal Engineering 212 (2022) 118633.
55. A.E. Jery, A.K. Khudhair, S.Q. Abbas, A.M. Abed, K.M. Khedher, Numerical simulation and artificial neural network prediction of hydrodynamic and heat transfer in a

geothermal heat exchanger to obtain the optimal diameter of tubes with the lowest entropy using water and Al_2O_3/water nanofluid, Geothermics 107 (2023) 102605.
56. L. Chilambarasan, V. Thangarasu, P. Ramasamy, Solar flat plate collector's heat transfer enhancement using grooved tube configuration with alumina nanofluids: Prediction of outcomes through artificial neural network modeling, Energy 289 (2024) 129953.
57. S. Jakeer, M.L. Rupa, S.R.R. Reddy, A.M. Rashad, Artificial neural network model of non-Darcy MHD Sutterby hybrid nanofluid flow over a curved permeable surface: Solar energy applications, Propulsion and Power Research 12 (2023) 410–427.
58. S.K. Rawat, M. Yaseen, M. Pant, C.S. Ujarari, D.K. Joshi, S. Chaube, A.S. Negi, M. Kumar, Designing soft computing algorithms to study heat transfer simulation of ternary hybrid nanofluid flow between parallel plates in a parabolic trough solar collector: Case of artificial neural network and particle swarm optimization, International Communications in Heat and Mass Transfer 148 (2023) 107011.
59. A.A. Alnaqi, J. Alsarraf, A.A.A.A. Al-Rashed, Effect of off-center finned absorber tube and nanoparticle shape on the performance of two-fluid parabolic solar collector containing nanofluid: An application of artificial neural network, Sustainable Energy Technologies and Assessments 48 (2021) 101668.
60. M. Ibrahim, T. Saeed, M.A. El-Shorbagy, T.A. Nofal, N. Aamir, Implementation of the artificial neural network to predict the effectiveness of the solar system using Cu/water-ethylene nanofluid to save energy, Engineering Analysis with Boundary Elements 138 (2022) 30–41.
61. S. Delfani, M. Esmaeili, M. Karami, Application of artificial neural network for performance prediction of a nanofluid-based direct absorption solar collector, Sustainable Energy Technologies and Assessments 36 (2019) 100559.
62. A.M. Tomy, N. Ahammed, M.S.P. Subathra, L.G. Asirvatham, Analysing the Performance of a Flat Plate Solar Collector with Silver/Water Nanofluid Using Artificial Neural Network, Procedia Computer Science 93 (2016) 33–40.
63. A.E. Moghadam, B.M. Gharyehsafa, M.F. Gord, Using artificial neural network and quadratic algorithm for minimizing entropy generation of Al_2O_3-EG/W nanofluid flow inside parabolic trough solar collector, Renewable Energy 129 (2018) 473–485.
64. S. Fuxi, N. Sina, S.M. Sajadi, M.Z. Mahmoud, A. Abdelrahman, H.Ş. Aybar, Artificial neural network modeling to examine spring turbulators influence on parabolic solar collector effectiveness with hybrid nanofluids, Engineering Analysis with Boundary Elements 143 (2022) 442–456.

5 Developing an Artificial Neural Network Algorithm for Heat and Mass Transfer Assessment in Ternary Hybrid Nanofluid Flow

Shilpa B. and Naveen Kumar R.

5.1 INTRODUCTION

The metallurgical, chemical and petroleum engineering industries are just a few major industrial examples that highlight the essence of fluid's magnetohydrodynamic (MHD) flow over a deforming structure. The grasp of the physical repercussions that occur in MHD is closely linked to understanding MHD. According to Lenz's law, a conductor moving through a magnetic field (MF) induces an electric current, which produces its MF. The lines of the MF will be eliminated from the conductor because the induced MF seeks to erase the initial and external supported field. Meanwhile, the induced field increases the applied field when the MF moves the conductor out of the field. Overall, this procedure gives the impression that the conductor and the lines of force are being pulled. The repercussion of MHD on the nanofluid flow (NFF) via a stretchable sheet was examined by Hamid et al. [1]. Yusuf et al. [2] inspected the stream of MHD liquid past an inclined wall in the presence of porous media. In the presence of an MHD, the performance of a rough slider bearing coated with fluid was elaborated by Cyriac et al. [3]. The MHD liquid stream across a vertical channel with the Dufour and Soret effects was investigated by Shilpa et al. [4]. The flow of nanoliquid via a vertical surface with MHD heat transport was debriefed by Kumar et al. [5].

Air purifiers, heat exchangers, nuclear reactor safety, building ventilation systems, and powdered coal burners are just a few technical applications that depend on thermophoretic particle deposition (TPD). Different kinds of particles behave differently when subjected to temperature gradients, which is how the phenomenon of thermophoresis originated. Thermophoresis increases the velocity of minuscule particles placed in a gas that isn't isothermal, increasing their accumulation speed towards lower temperatures. Larger particles are unaffected by this method. TPD causes small particles to settle on a cold surface because the flow of molecules of gas

DOI: 10.1201/9781032688121-5

in the colder zone is usually slower than that of particles arriving from the warmer region. Consequently, high-velocity particles collide with one another throughout this process. The influence of TPD on the liquid stream across a stretchable surface was delineated by Gowda et al. [6]. The stream of Oldroyd-B liquid through the moving surface with the TPD was scrutinized by Wang et al. [7]. Prasannakumara and Gowda [8] probed the heat transport analysis of the liquid flow across a gyrating disk with TPD impact. Madhukesh et al. [9] examined the flow of nanoliquid via a Riga surface with the influence of TPD. The impact of Soret and Dufour on the liquid stream past a stretchy cylinder with TPD was explored by Yasir et al. [10].

Heat transfer and flow processes fall under the broad category of thermal radiation (TR), which has several uses in industry and science. This behavior is frequently seen in several industries, such as spacecraft, power production, nuclear reactor cooling, aerospace technology, hypersonic flights, gas turbines, combustion applications, high-temperature operations, etc. The relationships between densities and the linear variations in the temperature difference between the wall and remote sites were established via many investigations. In recent years, the study has focused on linear TR, with the linear Rosseland approximation being used to analyze the effect of radiative flow at the surface. However, certain thermal systems still show very high-temperature differences, which significantly impact fluid flow characteristics due to the nonlinear fluctuation of the density relation with temperature. The thermal analysis in a dovetail fin with the influence of TR was delineated by Nimmy et al. [11]. The heat transport attributes of the wavy extended surface, along with the impact of TR, were explored by Prakash et al. [12]. Wang et al. [13] inspected the flow of nanoliquid across the disk and cone with the impact of TR. The TR effect on the flow of liquid across a Riga plate was scrutinized by Nadeem et al. [14]. An ideal investigation of TR's impact on the convective flow of fluid contained in carbon nanotubes was carried out by Yu and Wang [15].

A new class of fluids called nanofluids (NFs) emerged due to the development of metal particles. NFs are composed of small materials with a dimension of 1 to 100 nanometers. These fluids have substantially higher thermal conductivities than the base liquid. NFs have been shown to possess superior thermal and physical properties compared to ordinary fluids. In nanotechnology innovation, a unique class of heat-transmission fluids known as hybrid NFs with improved thermal conductivity has been developed, resulting in global technological improvements. Distributing two or more nanoparticles smaller than 100 nm results in a hybrid NF, an improvement over a regular NF. The researchers discovered a new NF called ternary nanofluid (T-NF) when they mixed three different kinds of nanoparticles with the base fluid. In response to the growing demand for cooling agents with superior thermal features at the industrial level, researchers have altered existing NFs to boost their thermal qualities, leading to the discovery of T-NF. To enhance the thermal characteristics of ternary nanoparticles that are now available on the market, more experimental research has been conducted to suspend three different types of solid nanoparticles in liquids. Using computational modeling, Khan et al. [16] investigated T-NF flow via a stretchable sphere. Sarada et al. [17] examined the flow of T-NF through a curved surface with an impact on heat generation using the heat flux model. Nagapavani et al. [18] discussed the flow of T-NF over a gyrating stretchable cylinder in its state of torsional motion. Using numerical analysis, Sharma et al. [19] explored the T-NF steam via a microchannel. Karthik et al. [20] probed the consequence of radiation on the T-NF stream via a wedge.

The artificial neural network (ANN) is a common machine learning method, which can extract related attributes from input with higher dimensions. A neural network is an algorithm that simulates the composition and functions of the human brain. There are several advantages to applying ANNs to solve complicated problems. An ANN's autonomous learning characteristic enables it to optimize across many cycles and extract features from data without prior knowledge or understanding. ANNs may be used for a wide range of variables and can quickly estimate the statistics supplied during the training phase since they are universal approximators. ANN is trained using the Levenberg–Marquardt (LM) optimization technique for complex problems. The LM methodology is a supervised training strategy that uses any feedforward neural network. The LM technique has been utilized for curve fitting because it combines the decreasing amount of error obtained by gradient descent with the higher convergence of the Gauss–Newton technique at the minima. The LM algorithm (LMA) works on the principle of integrated process training within and outside a complexly curved region. The LMA uses the steepest descent method initially to construct a quadratic approach. By converting the LMA, the Gauss–Newton method is achieved, leading to a considerable convergence acceleration. To train the network, both the input and output levels are populated with the inputs and corresponding outputs. The initial step in creating an NN-BLMA model is to choose its input and output data efficiently. The impact of heat source on the liquid stream across a surface using NN-BLMA was studied by Raja et al. [21]. Using the ANN, Srilatha et al. [22] inspected the stream of NF via a porous disk. The NN-BLMA was employed by Kumar et al. [23] to study the thermal transport attributes of the wavy profiled extended surface. Goud et al. [24] explored thermal transmission in a trapezoidal fin in permeable media using NN-BLMA. Sowmya and Kumar examined the influence of heat generation on thermal dispersal in a moving extended surface using NN-BLMA [25]. Recently, Shilpa et al. [26–28] and Rani et al. [29] have adopted LM backpropagating algorithm to analyze the heat and mass transfer of NF, nanolubricant in various channels under the influence of various physical phenomena.

The Soviet researcher Boris Galerkin is the inspiration for the name Galerkin techniques in numerical analysis. By applying linear constraints established by finite sets of basic functions, they transform a continuous operator issue – such as a differential equation – into a discrete problem, often in a weak formulation. The Maxwell liquid flow via a vertical annulus with a heat source using the Galerkin finite element approach was deliberated by Shilpa and Leela [30]. Rekha et al. [31] investigated how the solute transport in an unsaturated zone varies exponentially with time using the finite element approach. Pasha et al. [32] studied the statistical evaluation of viscous hybridized NFF using Galerkin's finite element approach. Bejawada et al. [33] addressed the flow of NF across an inclined plate using the Galerkin finite element technique. Using the Galerkin finite element scheme,

Bouslimi et al. [34] scrutinized the flow of Sutterby NF on a slippery surface. Recently, Shilpa et al. [35–37] have applied the CFD technique to analyze the heat and mass transfer of NF and nanolubricant in distinct regions, the authors illustrated and presented the results graphically.

In addition, this study evaluates how well the soft computing approach predicts the behavior of a problem with several interdependent parameters. Because parametric studies are difficult by nature and require various procedures, traditional methods frequently struggle to yield meaningful results. This chapter investigates the possibilities of machine learning techniques to forecast the performance of a flow model with several interconnected parameters to solve the previously discussed problem. The study uses numerical findings to instruct an ANN, focusing on finding methods that provide reasonably accurate predictions. The ANN algorithm can now reliably predict the desired parameters across a wide range owing to this training. Furthermore, the Nusselt and Sherwood number values on the shrinking/stretching convergent channel are precisely determined using the ANN method. The fields that use NFF via a channel, such as medicine, architectural design systems, nuclear waste treatment, conveyance processes, etc., will benefit from the present findings. The items below outline the uniqueness, purpose, and primary research questions addressed in this study:

a. A stretching/shrinking convergent channel is subjected to the mathematical flow model of THNF (Al_2O_3–Cu–CuO/water).
b. Examining the effects of TPD and heat radiation on THNF flow.
c. Using the existing modeling, analyze the heat and mass transfer rate of THNF flow across a channel and determine the conditions under which THNF exhibits the maximum heat and mass transfer rates.
d. Creating an algorithm for machine learning (ANN) to forecast the rates of mass and heat transport at the surface.

5.2 MATHEMATICAL FORMULATION

Consider a laminar, viscous, incompressible flow of (Al_2O_3–Cu–CuO/water) T-NF in a stretchable/shrinkable convergent/divergent channel as shown in Figure 5.1. The main presumptions of this heat and mass transport model are

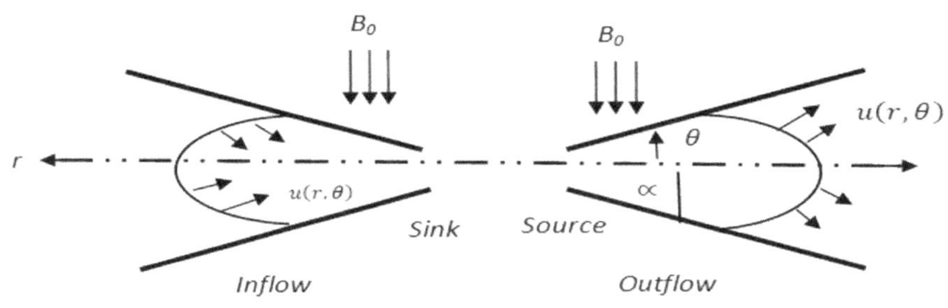

FIGURE 5.1 Physical configuration.

that the velocity of the fluid layer in adjacent walls is equal to the velocity of the wall and that variably shaped nanoparticles are equally disseminated in the base fluid. It is supposed for the walls to be expanding or contracting at a rate s. It is possible to write the velocity at the wall as $u_r = U_w = s / r$. Furthermore, the angle α is greater than zero for the divergent channel and less than zero ($\alpha < 0$) for the converging channel. Our study also considers the impact of buoyancy forces and thermal equilibrium between the base fluid and the nanoparticles. To study flow, a transverse uniform MF with strength $B_0 = B_0\ (r)$ is applied across the channel. The shape factors and thermophysical properties of nanofluid are mentioned in Tables 5.1 and 5.2. Below is a list of the physical equations that describe how T-NFs inside stretchable convergent/divergent channels when they are affected by TR, TPD, and MF properties:

$$\frac{1}{r}\frac{\partial(ur)}{\partial r} = 0 \tag{5.1}$$

$$u\frac{\partial u}{\partial r} = -\frac{1}{\rho_{thnf}}\frac{\partial p}{\partial r} + \upsilon_{thnf}\left[\frac{\partial^2 u}{\partial r^2} + \frac{1}{r}\frac{\partial u}{\partial r} + \frac{1}{r^2}\frac{\partial^2 u}{\partial \theta^2} - \frac{u}{r^2}\right] - \frac{\sigma_{thnf} B_0^2 u}{\rho_{thnf}} \tag{5.2}$$

$$-\frac{1}{\rho_{thnf}}\frac{1}{r}\frac{\partial p}{\partial \theta} + \frac{2}{r^2}\upsilon_{thnf}\frac{\partial u}{\partial \theta} = 0 \tag{5.3}$$

TABLE 5.1
Shape Factors of Distinct Nanoparticles

Shape of Nanoparticles	Shape Factor Values
Bricks	3.7
Cylinder	4.9
Platelets	5.7
Blades	8.6

TABLE 5.2
Thermophysical Values of Nanoparticles and the Base Fluid

Components	ρ	c_p	k	σ
CuO	6500	540	18	6.9×10^{-2}
H_2O	997.1	4180	0.6071	5.5×10^{-6}
Cu	8933	385	400	59.6×10^{6}
Al_2O_3	3970	765	40	35×10^{6}

$$u\frac{\partial T}{\partial r}=\frac{k_{thnf}}{(\rho c_p)_{thnf}}\left[\frac{\partial^2 T}{\partial r^2}+\frac{1}{r}\frac{\partial T}{\partial r}+\frac{1}{r^2}\frac{\partial^2 T}{\partial \theta^2}\right]-\frac{1}{(\rho c_p)_{thnf}}\frac{1}{r}\frac{\partial[rq_{r,rad}]}{\partial r}$$
$$-\frac{1}{(\rho c_p)_{thnf}}\frac{1}{r^2}\frac{\partial[rq_{\theta,rad}]}{\partial \theta} \tag{5.4}$$

$$u\frac{\partial C}{\partial r}=D_B\left[\frac{\partial^2 C}{\partial r^2}+\frac{1}{r}\frac{\partial C}{\partial r}+\frac{1}{r^2}\frac{\partial^2 C}{\partial \theta^2}\right]-\frac{\partial[CV_T]}{\partial \theta} \tag{5.5}$$

Using Roseland approximation for radiation

$$q_{r,rad}=-\frac{4\sigma^*}{3k^*}\frac{\partial T^4}{\partial r} \text{ and } q_{\theta,rad}=-\frac{4\sigma^*}{3k^*}\frac{\partial T^4}{\partial \theta} \tag{5.6}$$

The thermophoretic deposition velocity in θ - direction is

$$V_T=-k\frac{V}{T_r}\frac{dT_\theta}{d\theta} \tag{5.7}$$

Due to the symmetry assumption at the channel centerline

$$\theta=0,\ \frac{\partial C}{\partial \theta}=\frac{\partial T}{\partial \theta}=\frac{\partial u}{\partial \theta}=0, u=\frac{u_c}{r} \tag{5.8}$$

Because of the plates stretching/shrinking convergent/divergent wall conditions,

$$\theta=\pm\alpha,\ u=\frac{S}{r},\ T=\frac{T_w}{r^2},\ C=\frac{C_w}{r^2} \tag{5.9}$$

$$\text{The radial velocity is } u(r,\theta)=\frac{f(\theta)}{r} \tag{5.10}$$

The stretching pressure can be expressed as follows once the azimuthal velocity equation is integrated:

$$\frac{p}{\rho}=\frac{2V}{r^2}f(\theta)+g \tag{5.11}$$

Similarity variables

$$\left.F(\eta)=\frac{f(\theta)}{u_c},\ \eta=\frac{\theta}{\alpha},\ \Theta(\eta)=r^2\frac{T}{T_w},\ \Phi(\eta)=r^2\frac{C}{C_w}.\right\} \tag{5.12}$$

Non-dimensional quantities

$$\mathrm{Re}=\frac{u_c\alpha}{\upsilon_f},\ M=\frac{B_0^2\sigma_f}{\mu_f},\ \mathrm{Pr}=\frac{\upsilon_f(\rho c_p)_f}{k_f},\ N=\frac{16\sigma^* T_\infty^3}{3k^* k_f},\ Sc=\frac{\upsilon_f}{D_B}. \tag{5.13}$$

Reduced equations

$$F'''+FF'\alpha\,\mathrm{Re}\left(\frac{\rho_{thnf}}{\rho_f}\right)\left(\frac{\mu_f}{\mu_{thnf}}\right)+4\alpha^2F'-M\alpha^2F'\left(\frac{\sigma_{thnf}}{\sigma_f}\right)\left(\frac{\mu_f}{\mu_{thnf}}\right)=0 \tag{5.14}$$

$$\left[1+N\left(\frac{k_f}{k_{thnf}}\right)\right]\Theta''+\left[1+N\left(\frac{k_f}{k_{thnf}}\right)\right]4\Theta\alpha^2+2F\Theta\alpha\,\mathrm{Re}\,\mathrm{Pr}\frac{(\rho c_p)_{thnf}}{(\rho c_p)_f}\frac{k_f}{k_{thnf}}=0 \quad (5.15)$$

$$\Phi''+4\alpha^2\Phi+2F\,\mathrm{Re}\,Sc\,\alpha\,\Phi+K\alpha^2\left(\frac{\rho_f\mu_{thnf}}{\rho_{thnf}\mu_f}\right)Sc[\Phi\Theta''+\Theta\Phi']=0 \quad (5.16)$$

Boundary conditions are

$$F(0)=1, F'(0)=0, F(1)=\lambda,\ \Theta'(0)=0, \Theta(1)=1, \Phi'(0)=0, \Phi(1)=1, \quad (5.17)$$

Thermophysical properties of T-NF

$$\mu_{(Al_2O_3-CuO-Cu)H_2O}=\frac{\mu_{H_2O}}{(1-\phi_{Al_2O_3})^{25/10}(1-\phi_{CuO})^{25/10}(1-\phi_{Cu})^{25/10}} \quad (5.18)$$

$$\begin{aligned}\rho_{(Al_2O_3-CuO-Cu)H_2O}&=(1-\phi_{Cu})\left[(1-\phi_{CuO})\left\{(1-\phi_{Al_2O_3})\rho_{H_2O}+\phi_{Al_2O_3}\rho_{Al_2O_3}\right\}\right.\\&\quad\left.+\phi_{CuO}\rho_{CuO}\right]+\phi_{Cu}\rho_{Cu}\end{aligned} \quad (5.19)$$

$$\begin{aligned}&(\rho c_p)_{(Al_2O_3-CuO-Cu)H_2O}\\&=(1-\phi_{Cu})\left[(1-\phi_{CuO})\left\{\begin{array}{l}(1-\phi_{Al_2O_3})(\rho c_p)_{H_2O}+\\ \phi_{Al_2O_3}(\rho c_p)_{Al_2O_3}\end{array}\right\}+\phi_{CuO}(\rho c_p)_{CuO}\right]+\phi_{Cu}(\rho c_p)_{Cu}\end{aligned} \quad (5.20)$$

$$\frac{k_{(Al_2O_3-CuO-Cu)H_2O}}{k_{(Al_2O_3-CuO)H_2O}}=\frac{k_{Cu}+(n-1)k_{(Al_2O_3-CuO)H_2O}-(n-1)\phi_{Cu}\left(k_{(Al_2O_3-CuO)H_2O}-k_{Cu}\right)}{k_{Cu}+(n-1)k_{(Al_2O_3-CuO)H_2O}+\phi_{Cu}\left(k_{(Al_2O_3-CuO)H_2O}-k_{Cu}\right)} \quad (5.21)$$

$$\frac{k_{(Al_2O_3-CuO)H_2O}}{k_{nf}}=\frac{k_{CuO}+(n-1)k_{nf}-(n-1)\phi_{CuO}(k_{nf}-k_{Cu})}{k_{CuO}+(n-1)k_{nf}+\phi_{CuO}(k_{nf}-k_{Cu})} \quad (5.22)$$

$$\frac{k_{nf}}{k_f}=\frac{k_{Al_2O_3}+(n-1)k_{H_2O}-(n-1)\phi_{Al_2O_3}\left(k_{H_2O}-k_{Al_2O_3}\right)}{k_{Al_2O_3}+(n-1)k_{H_2O}+\phi_{Al_2O_3}\left(k_{H_2O}-k_{Al_2O_3}\right)} \quad (5.23)$$

$$\frac{\sigma_{(Al_2O_3-CuO-Cu)H_2O}}{\sigma_{(Al_2O_3-CuO)H_2O}}=\frac{\sigma_{Cu}+2\sigma_{(Al_2O_3-CuO)H_2O}-2\phi_{Cu}\left(\sigma_{(Al_2O_3-CuO)H_2O}-\sigma_{Cu}\right)}{\sigma_{Cu}+2\sigma_{(Al_2O_3-CuO)H_2O}+\phi_{Cu}\left(\sigma_{(Al_2O_3-CuO)H_2O}-\sigma_{Cu}\right)}$$

$$\begin{aligned}\text{where }\frac{\sigma_{(Al_2O_3-CuO)H_2O}}{\sigma_{nf}}&=\frac{\sigma_{CuO}+2\sigma_{nf}-2\phi_{CuO}(\sigma_{nf}-\sigma_{CuO})}{\sigma_{CuO}+2\sigma_{nf}+\phi_{CuO}(\sigma_{nf}-\sigma_{CuO})}\ \frac{\sigma_{nf}}{\sigma_{H_2O}}\\&=\frac{\sigma_{Al_2O_3}+2\sigma_{H_2O}-2\phi_{Al_2O_3}(\sigma_{H_2O}-\sigma_{Al_2O_3})}{\sigma_{Al_2O_3}+2\sigma_{H_2O}+\phi_{Al_2O_3}(\sigma_{H_2O}-\sigma_{Al_2O_3})}.\end{aligned} \quad (5.24)$$

Skin friction, Nusselt, and Sherwood numbers:

$$\tau = \frac{\mu_{thnf}}{\mu_f} f''(0), \tag{5.25}$$

$$Nu = -\left(\frac{k_{thnf}}{k_f} + N \right) \Theta'(0), \tag{5.26}$$

$$Sh = -\Phi'(0). \tag{5.27}$$

5.3 NUMERICAL METHOD

High nonlinearities in the developed thermal and solutal transport model prevent it from computing its closed-form solutions. As a result, residual techniques are helpful for these models, which are highly accurate and computationally efficient. Consequently, the GFEM handles the current model. The main components of GFEM implementation are the choice of residue functions, trial solutions, and weight functions that take the form of linearly autonomous functions. After that, a set of integral equations was produced, and from that integral model, indeterminate coefficients could be correctly computed. Ultimately, when those known coefficients were introduced, there was remarkable precision in the model's solution and the trial solution. Figure 5.2 provides the whole GFEM implementation procedure for the current T-NF model in convergent/divergent channels under the physical effects of TR and TPD and with distinct shape factors of NF.

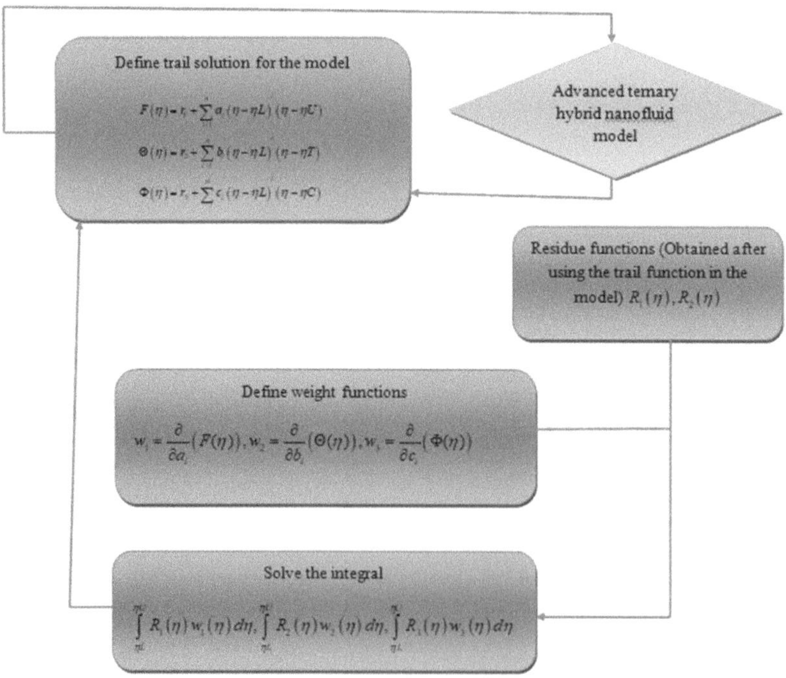

FIGURE 5.2 GFEM implementation.

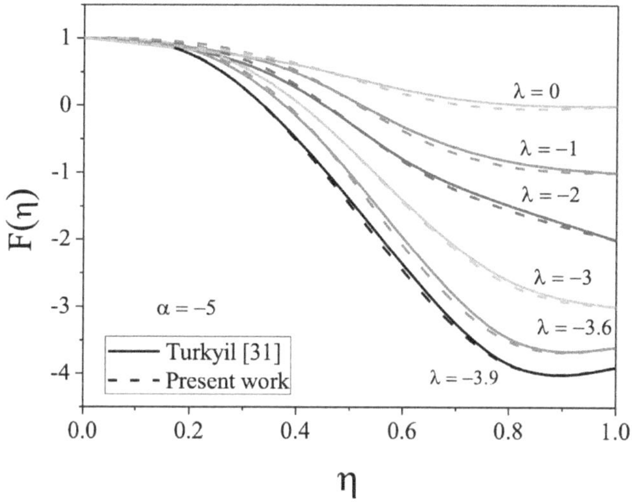

FIGURE 5.3 Comparison of present work with existing results.

5.4 VALIDATION

The results acquired by GFEM are compared with the results of Turkyilmazoglu [38] under a limiting case. A good agreement between both results can be noticed in Figure 5.3.

5.5 RESULTS AND DISCUSSION

Figure 5.4 represents the effect of *Re* on the velocity profile in both expanding/shrinking convergent channels. Re illustrates the significance of the inertial-to-viscous force ratio for a flow. Physically, an increase in Re results in a rise in the inertial force inside the current flow system, generating a velocity increment.

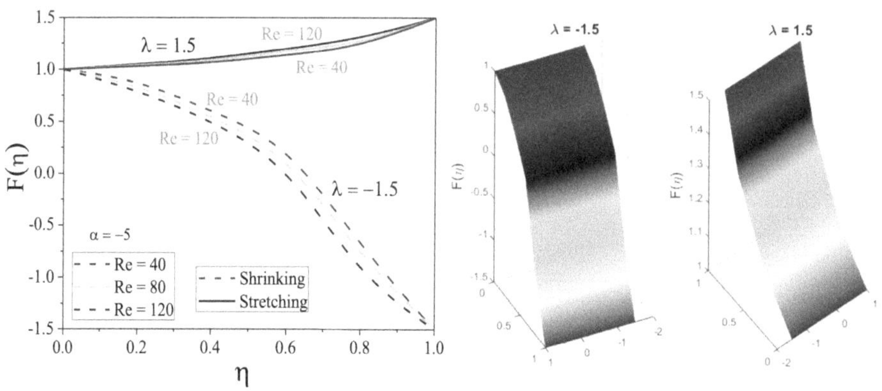

FIGURE 5.4 Influence of Re on velocity profile in shrinkable and stretchable convergent channel.

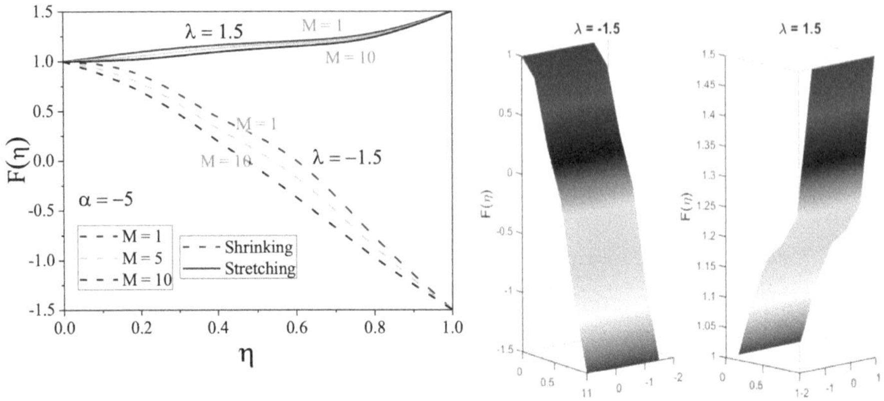

FIGURE 5.5 Influence of M on velocity profile in shrinking and stretching convergent channel.

The velocity profile gets intensified by increasing the value of the Reynolds number in the stretching channel, whereas the velocity profile gets diminished for the increased Reynolds number in the shrinking channel. However, when Re enhances, the forced convection weakens, resulting in a decline in the transport in T-NF. Figure 5.5 represents the effect of M on the velocity profile in a shrinking/stretching convergent channel. The Hartmann number reduces the velocity of the T-NF in both stretching and shrinking channels. The MF can greatly affect a fluid's temperature and velocity, especially if the fluid is electrically conducting, like liquid metal or plasma. It can potentially cause variations in the temperature and flow velocity by exerting a Lorentz force on the charged particles inside the fluid. When an MF is present, the fluid may take on a distinctive flow pattern that includes vortices and eddies that point in the same direction as the MF. This may modify the fluid's flow structure and velocity profile. Figure 5.6 represents the

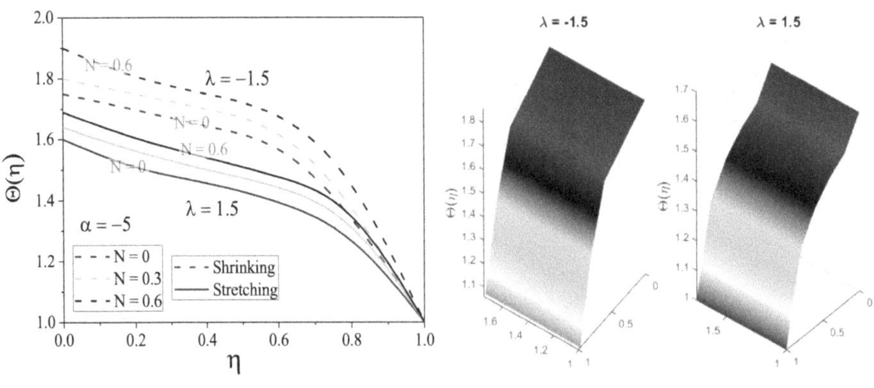

FIGURE 5.6 Influence of N on temperature profile in shrinking and stretching convergent channel.

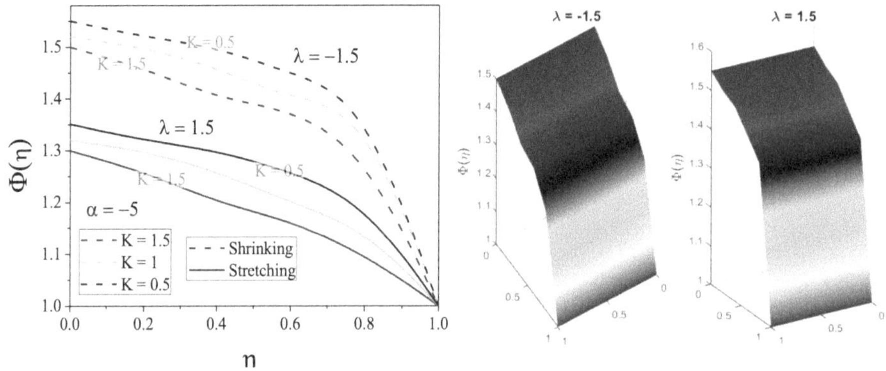

FIGURE 5.7 Influence of *K* on concentration profile in shrinking and stretching convergent channel.

influence of radiation parameters on the ternary hybrid NF temperature in shrinking/stretching convergent channels. The radiation parameter defines the proportionate impact of conductive heat transmission on radiation's heat transfer. Both stretching/shrinking convergent channels' boundary layers warm up in response to an increase in the radiation parameter. Figure 5.7 illustrates the effect of TPD parameter influence on the ternary hybrid NF concentration profile. As the thermophoretic parameter augments, the concentration of the fluid diminishes since when there is a significant thermal variance, the concentration decreases as more particles are drawn to the surface by the thermophoretic force. The Schmidt number impact on ternary hybrid NF concentration profile can be seen in Figure 5.8. *Sc* is defined as the ratio of viscosity and mass diffusivity. So, concentration intensifies with the rise of *Sc*.

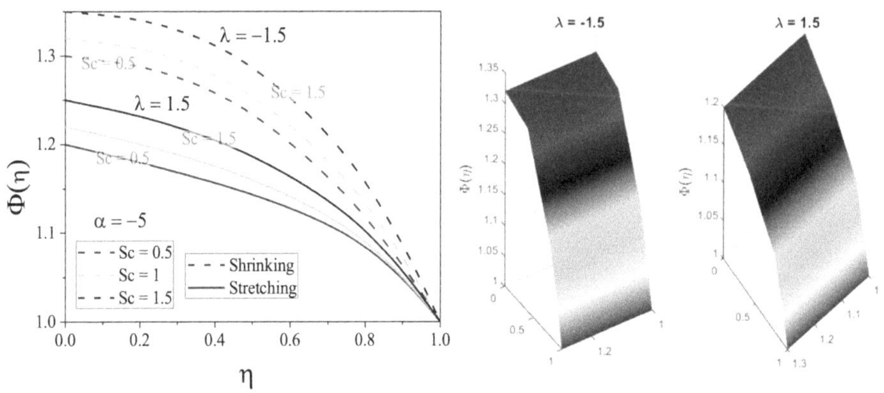

FIGURE 5.8 Influence of *Sc* on concentration profile in shrinking and stretching convergent channel.

5.6 ARTIFICIAL NEURAL NETWORK MODEL ALGORITHM

Biological neural networks influenced the computational prototypes known as ANNs. They are made up of artificial neurons that are coupled. Because ANNs can accurately simulate complicated and nonlinear functions, they have become increasingly popular recently. The authors can predict the Nusselt and Sherwood number in the present flow model using a model built on an ANN. The multilayer perceptron (MLP) network is one of the most effective models. The input, hidden, and output layers make up the three layers of MLP networks, which are feedforward NN (see Figure 5.9). Neurons in the input layer gather information, process it, and transfer it to the hidden layer, which then processes it even more before passing it to the output layer. Each layer is fully networked, and each network has been given a unique weight (see Figure 5.9). Both the hidden and output layers use activation functions to turn on the neurons. They can be added if more hidden layers are needed to handle more difficult tasks. The connection weight is applied to the input signal (X_i) that the jth neuron in the hidden layer receives, and the result is added up. The resultant sum is then supplied into an activation function to excite the neurons. The same data processing pattern is followed from the hidden layer to the output layer. The following is an expression for the jth neuron's output:

$$Y_j = f\left(\sum_{i=1}^{n} W_{ij}^{n} X_i + b_j^{n}\right) \tag{5.28}$$

where "W_{ji}" stands for the weight of the connection from the ith neuron in the previous layer and "f" stands for the activation function in Eqn. (5.28). Reducing errors between the generated and required output is the main goal of the ANN's training procedure.

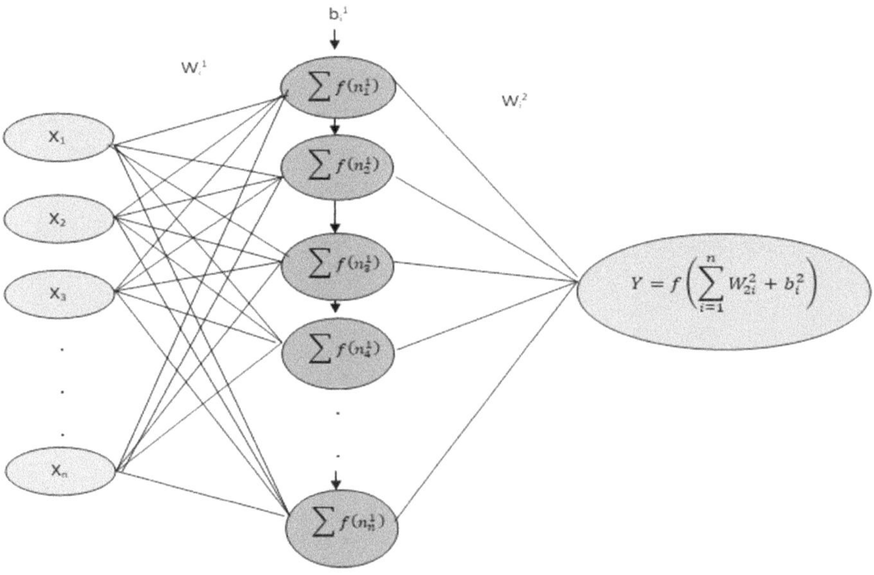

FIGURE 5.9 ANN model schematic diagram.

Various activation functions, such as exponential, logistic, and hyperbolic functions, can be applied when expressed by the variable "f". Neural network training frequently uses the error propagation algorithm. The weights between network layers are changed during this supervised learning process. The learning error rate is ascertained by comparing the NN outcomes with the predicted results from the training dataset. It is significant to consider that the MLP network's configuration greatly impacts its performance. Generally, the best network setup is found by repeatedly testing various options to reduce the error rate. This procedure comprises carrying out tests with various parameter values, such as selecting the number of neurons in the middle layer and the activation functions used in the output and hidden layers. The model's error value is computed for various parameter values during these tests. The primary structural arrangement of the ANN model put out in the present work is seen in Figure 5.9. The neural network architecture for heat and mass transport is presented in Figure 5.10. This work presents an ANN prediction model with four parameters in the first layer, one resultant parameter (Nusselt number) in the output layer, and six parameters in the input and one output parameter (Sherwood number). An essential component of an ANN model's performance is its neuron count. The correct number of neurons to utilize in ANNs is not a widely accepted strategy. Therefore, several models with varied numbers of layers were used in the design phase of both MLP networks. Numerous neurons have been tested, and the accuracy of their predictions has been assessed. The ideal number of neurons to produce the accurate prediction performance has been determined based on the analysis data. The number of neurons in the hidden layer is one of the input parameters used by the authors. The following section contains the remaining specifics of the ANN model created for this investigation.

Other metrics used to evaluate the effectiveness of ANN training status include the correlation factor and regression diagram between the target values and the output data. Regression analysis measures the relationship between the two datasets (R). R is a number ranging from 0 to 1. Cap R equals 1 indicates a reasonable overlap, while cap R equals 0 indicates no relationship between the two data sets. With the aid of Figures 5.11(a) and 5.12(a), the authors have demonstrated the alignment between the actual outcomes of the ANN and the intended values. The desired values are shown on one horizontal axis in

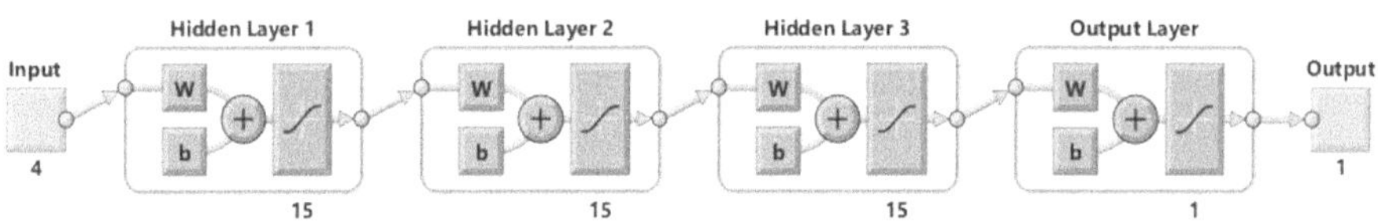

(a)

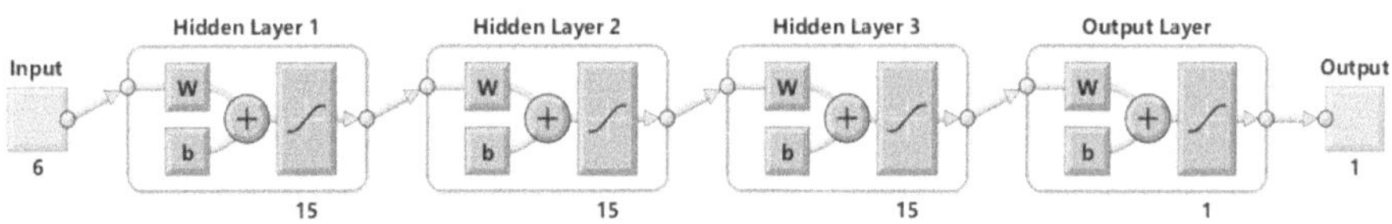

(b)

FIGURE 5.10 Neural network architecture for (a) heat transport and (b) mass transport.

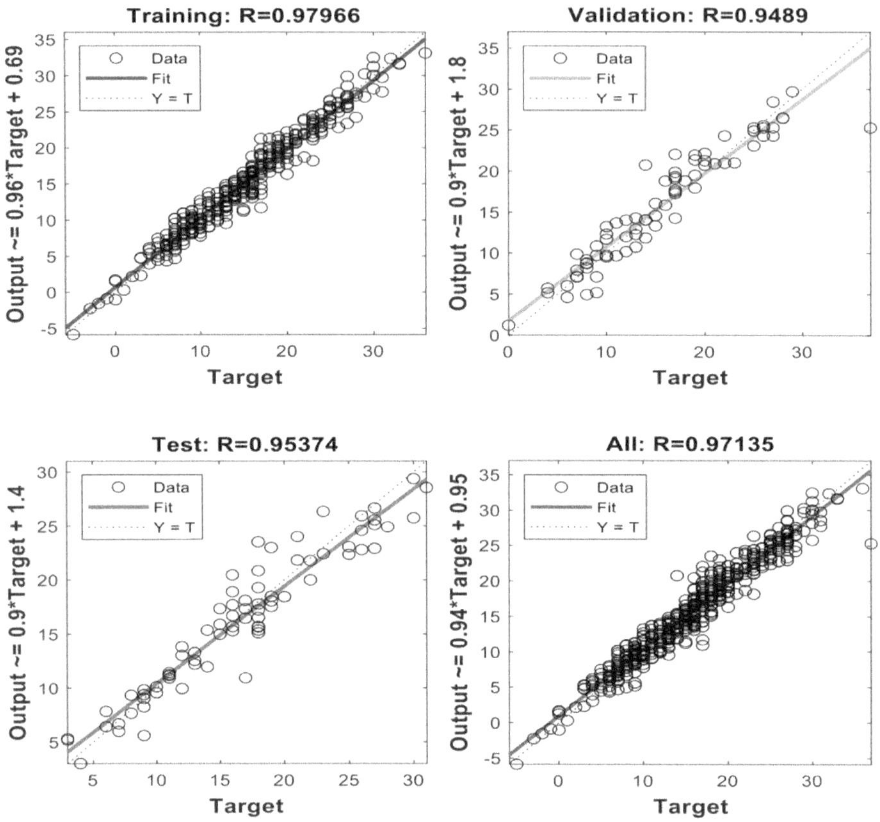

(a)

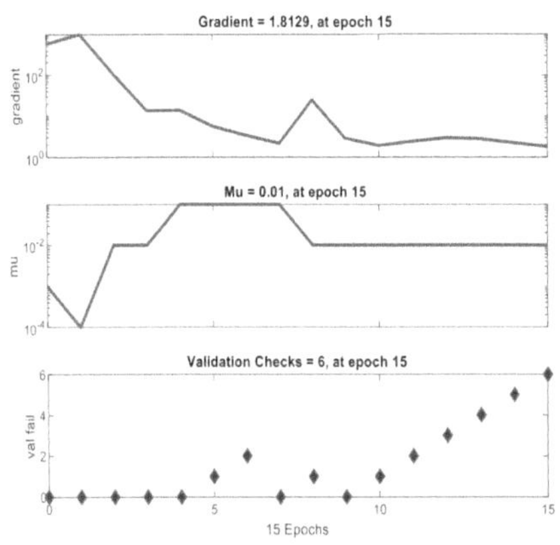

(b)

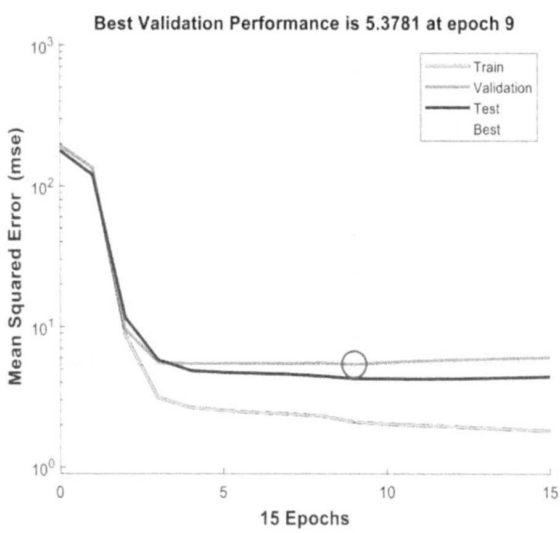

(c)

FIGURE 5.11 Heat transfer rate (a) regression plot, (b) training state performance plot, (c) validation performance plot, and (d) error histogram.

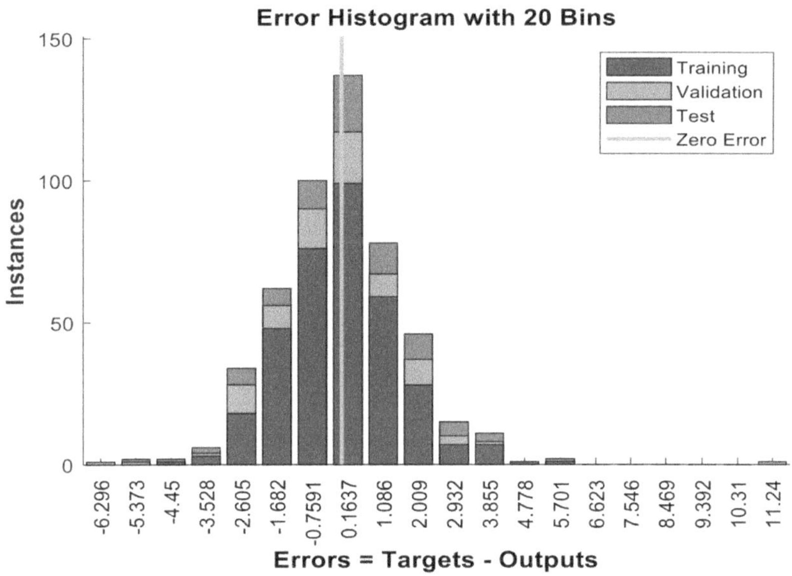

(d)

FIGURE 5.11 (*Continued*)

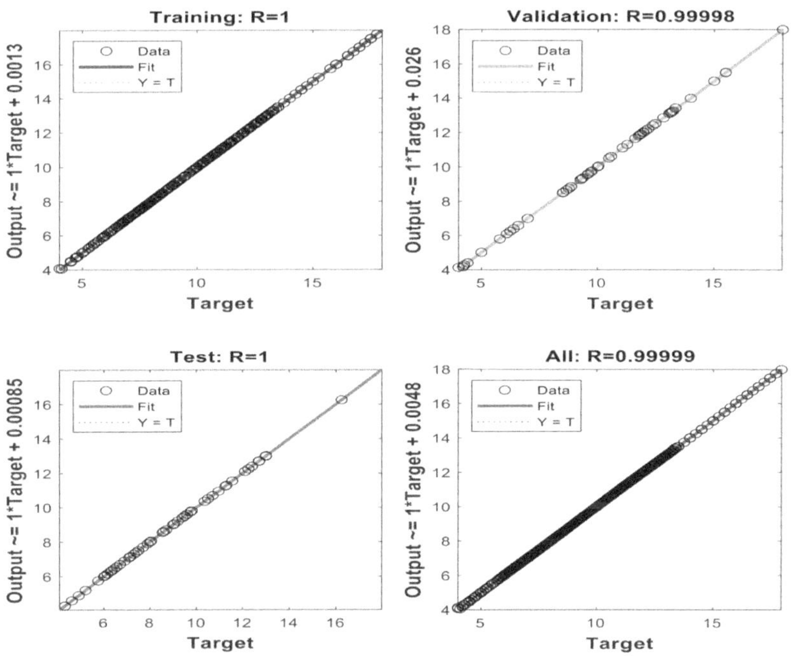

(a)

FIGURE 5.12 Mass transfer rate analysis (a) regression plot, (b) training state performance plot, (c) validation performance plot, and (d) error histogram.

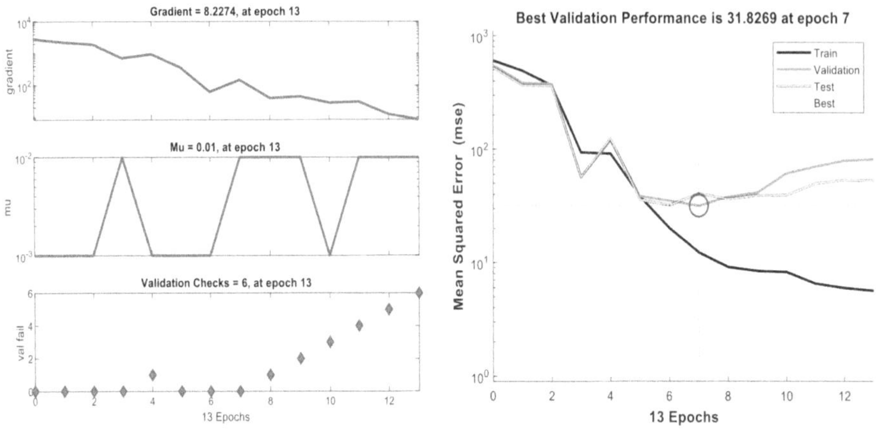

(b) (c)

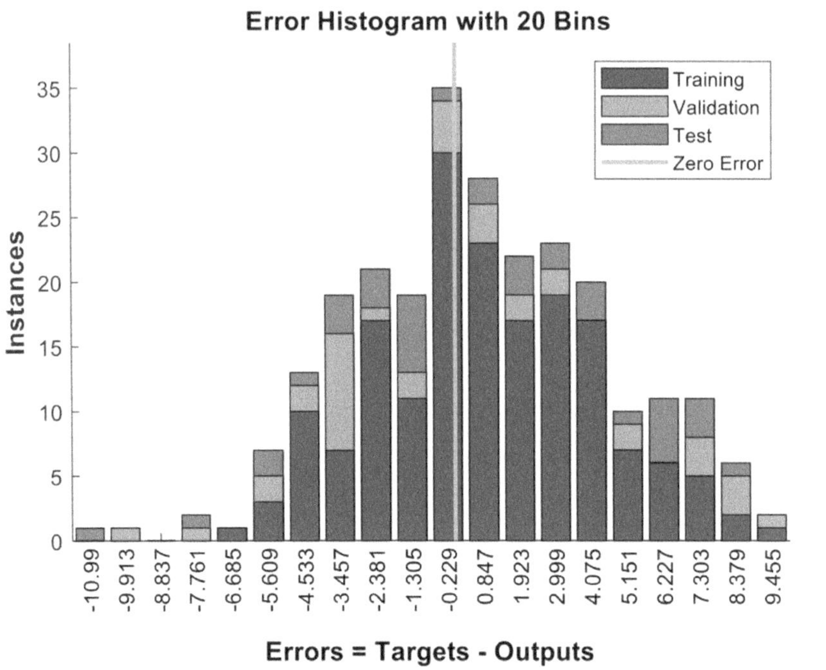

(d)

FIGURE 5.12 ***(Continued)***

these figures, while the outcomes of the ANN are shown on the other axis. This figure highlights three critical factors: the "bias *(B)*, slope value *(M)*, and correlation coefficient value *(R)*". The output of an ideal ANN model should coincide exactly with the desired values. In that case, the bias value will be zero, and the slope and correlation values will be one. Regression analysis of the (i) training, (ii) validation, (iii) testing, and (iv) total data concerning significant parameters is displayed in Figures 5.11(a) and 5.12(a).

Regression values are found to be either 1 or very near to 1. The data points nearly match the regression line when the regression (R) value is near 1. In addition, the graph's points show very little dispersal and are all exactly on the plane's bisector. This shows that the analyzed data and the intended outputs strongly correlate. This finding suggests that the machine learning algorithm (ANN) described in this study may predict the Nusselt number with remarkable accuracy and low error rates. The established ANN model's error distributions are presented through the error histograms in Figures 5.11(d) and 5.12(d). The graphic depiction shows how consistently low the error numbers are at every step of the ANN model. However, the zero-error line is where most of these faults are concentrated. The fact that this pattern appears in the error histogram graphs confirms that the ANN model was built to produce accurate predictions with little error.

5.7 CONCLUSIONS

The current chapter explores an in-depth examination of heat and mass transfer in ternary hybrid NFF past a stretching/shrinking channel. Several variables, including buoyancy effects, TPD, and heat radiation, are included in the mathematical model. Traditionally, parametric research and computationally demanding methodologies are used to tackle such a complicated problem. By using machine learning approaches based on the ANN algorithm, this work, in contrast, adopts a novel strategy to analyze the heat and mass transmission. Metrics, including mean-squared error (MSE), error histograms, and coefficient of regression, are used to assess the algorithm's performance.

According to the investigation, comparable physically connected problems can be resolved using the suggested machine learning technique that uses the ANN algorithm. The coefficient of regression (R) for every case studied using the ANN algorithm is greater than 0.97135, indicating a strong correlation between the numerical and predicted data. Throughout all examined situations, the ANN algorithm's MSE values range from 0 to 0.0048. These results show that the MSE values are very close to zero, suggesting that the machine learning (ANN) method developed can accurately predict the Sherwood and Nusselt numbers with little error. The study's concluding remarks also include the major results, such as a rise in the radiation parameter increases the warmth within the boundary layer of both stretching/shrinking convergent channels and also, as the thermophoretic parameter rises, the concentration of the fluid diminishes since when there is a significant thermal variance, the thermophoretic force draws more particles to the surface, which reduces the concentration.

The current chapter examines the flow behavior of a T-NF, focusing on the flow characteristics without considering the effect of the aggregation of nanoparticles. The effect of nanoparticle aggregation may also be included, broadening the research's potential use. This could be achieved by using accurate thermophysical correlations shown to model fluid flow accurately.

REFERENCES

1. A. Hamid, R. Naveen Kumar, R.J. Punith Gowda, R.S. Varun Kumar, S.U. Khan, M. Ijaz Khan, B.C. Prasannakumara, T. Muhammad, Impact of Hall current and homogenous–heterogenous reactions on MHD flow of GO-MoS2/water (H2O)-ethylene glycol (C2H6O2) hybrid nanofluid past a vertical stretching surface, Waves in Random and Complex Media 0 (2021) 1–18. https://doi.org/10.1080/17455030.2021.1985746.

2. T.A. Yusuf, R. Naveen Kumar, R.J. Punith Gowda, U.D. Akpan, Entropy generation on flow and heat transfer of a reactive MHD Sisko fluid through inclined walls with porous medium, International Journal of Ambient Energy 43 (2022) 6307–6316. https://doi.org/10.1080/01430750.2021.2013941.
3. T. Cyriac, B.N. Hanumagowda, M. Umeshaiah, V. Kumar, J.S. Chohan, R. Naveen Kumar, K. Karthik, Performance of rough secant slider bearing lubricated with couple stress fluid in the presence of magnetic field, Modern Physic Letters B (2023) 2450140. https://doi.org/10.1142/S0217984924501409.
4. B. Shilpa, V. Leela, B.C. Prasannakumara, P. Nagabhushana, Soret and Dufour effects on MHD double-diffusive mixed convective heat and mass transfer of couple stress fluid in a channel formed by electrically conducting and non-conducting walls, Waves in Random and Complex Media (2022) 1–22. https://doi.org/10.1080/17455030.2022.2119491.
5. M.A. Kumar, Y.D. Reddy, V.S. Rao, B.S. Goud, Thermal radiation impact on MHD heat transfer natural convective nanofluid flow over an impulsively started vertical plate, Case Studies in Thermal Engineering 24 (2021) 100826. https://doi.org/10.1016/j.csite.2020.100826.
6. R.J. Punith Gowda, H.M. Baskonus, R. Naveen Kumar, D.G. Prakasha, B.C. Prasannakumara, Evaluation of heat and mass transfer in ferromagnetic fluid flow over a stretching sheet with combined effects of thermophoretic particle deposition and magnetic dipole, Waves in Random and Complex Media 0 (2021) 1–19. https://doi.org/10.1080/17455030.2021.1969063.
7. Y. Wang, R.N. Kumar, S. Gouadria, M.M. Helmi, R.J.P. Gowda, E.R. El-Zahar, B.C. Prasannakumara, M.I. Khan, A three-dimensional flow of an Oldroyd-B liquid with magnetic field and radiation effects: An application of thermophoretic particle deposition, International Communications in Heat and Mass Transfer 134 (2022) 106007. https://doi.org/10.1016/j.icheatmasstransfer.2022.106007.
8. B.C. Prasannakumara, R.J.P. Gowda, Heat and mass transfer analysis of radiative fluid flow under the influence of uniform horizontal magnetic field and thermophoretic particle deposition, Waves in Random and Complex Media 0 (2022) 1–12. https://doi.org/10.1080/17455030.2022.2096943.
9. J.K. Madhukesh, R.S. Varun Kumar, R.J. Punith Gowda, B.C. Prasannakumara, S.A. Shehzad, Thermophoretic particle deposition and heat generation analysis of Newtonian nanofluid flow through magnetized Riga plate, Heat Transfer 51 (2022) 3082–3098. https://doi.org/10.1002/htj.22438.
10. M. Yasir, M. Khan, Z.U. Malik, Analysis of thermophoretic particle deposition with Soret-Dufour in a flow of fluid exhibit relaxation/retardation times effect, International Communications in Heat and Mass Transfer 141 (2023) 106577. https://doi.org/10.1016/j.icheatmasstransfer.2022.106577.
11. P. Nimmy, K.V. Nagaraja, P. Srilatha, K. Karthik, G. Sowmya, R.S.V. Kumar, U. Khan, S.M. Hussain, A.S. Hendy, M.R. Ali, Implication of radiation on the thermal behavior of a partially wetted dovetail fin using an artificial neural network, Case Studies in Thermal Engineering 51 (2023) 103552. https://doi.org/10.1016/j.csite.2023.103552.
12. S.B. Prakash, K. Chandan, K. Karthik, S. Devanathan, R.S.V. Kumar, K.V. Nagaraja, B.C. Prasannakumara, Investigation of the thermal analysis of a wavy fin with radiation impact: An application of extreme learning machine, Physica Scripta 99 (2023) 015225. https://doi.org/10.1088/1402-4896/ad131f.
13. F. Wang, S.P. Rani, K. Sarada, R.J. Punith Gowda, U. Khan, H.Y. Zahran, E.E. Mahmoud, The effects of nanoparticle aggregation and radiation on the flow of nanofluid between the gap of a disk and cone, Case Studies in Thermal Engineering 33 (2022) 101930. https://doi.org/10.1016/j.csite.2022.101930.

14. S. Nadeem, B. Ishtiaq, N. Abbas, Impact of thermal radiation on two-dimensional unsteady third-grade fluid flow over a permeable stretching Riga plate, International Journal of Modern Physics: Part B 37 (2023) 2350009. https://doi.org/10.1142/S0217979223500091.
15. D. Yu, R. Wang, An optimal investigation of convective fluid flow suspended by carbon nanotubes and thermal radiation impact, Mathematics 10 (2022) 1542. https://doi.org/10.3390/math10091542.
16. U. Khan, R. Naveen Kumar, A. Zaib, B.C. Prasannakumara, A. Ishak, A.M. Galal, R.J. Punith Gowda, Time-dependent flow of water-based ternary hybrid nanoparticles over a radially contracting/expanding and rotating permeable stretching sphere, Thermal Science and Engineering Progress 36 (2022) 101521. https://doi.org/10.1016/j.tsep.2022.101521.
17. F. Gamaoun, A. Abdulrahman, S.O. Paramesh, R. Kumar, P. G d, R.J. Punith Gowda, Impact of the exponential form of internal heat generation on water-based ternary hybrid nanofluid flow by capitalizing non-Fourier heat flux model, Case Studies in Thermal Engineering 38 (2022) 102332. https://doi.org/10.1016/j.csite.2022.102332.
18. M. Nagapavani, G.V. Ramana Reddy, A. Abdulrahman, R. Kumar, R.J. Punith Gowda, Three-dimensional swirling flow of a ternary composite nanofluid induced by the torsional motion of a cylinder considering non-Fourier law, Numerical Heat Transfer, Part A: Applications 0 (2023) 1–14. https://doi.org/10.1080/10407782.2023.2219834.
19. R.P. Sharma, J.K. Madhukesh, S. Shukla, F. Gamaoun, B.C. Prasannakumara, Numerical study of the thermophoretic velocity of ternary hybrid nanofluid in a microchannel bounded by the two parallel permeable flat plates, Journal of Thermal Analysis and Calorimetry 148 (2023) 14069–14080. https://doi.org/10.1007/s10973-023-12691-3.
20. K. Karthik, M. J K, S. Kiran, N. K V, B.C. Prasannakumara, G. Fehmi, Impacts of thermophoretic deposition and thermal radiation on heat and mass transfer analysis of ternary nanofluid flow across a wedge, International Journal of Modelling and Simulation 0 (2024) 1–13. https://doi.org/10.1080/02286203.2023.2298234.
21. M.A.Z. Raja, M. Shoaib, R. Tabassum, M.I. Khan, R.J.P. Gowda, B.C. Prasannakumara, M.Y. Malik, W.-F. Xia, Intelligent computing for the dynamics of entropy optimized nanofluidic system under impacts of MHD along thick surface, International Journal of Modern Physics: Part B 35 (2021) 2150269. https://doi.org/10.1142/S0217979221502696.
22. P. Srilatha, R.J.P. Gowda, J. Madhu, K.V. Nagaraja, F. Gamaoun, R.S.V. Kumar, K. Karthik, Designing a solid–fluid interface layer and artificial neural network in a nanofluid flow due to rotating rough and porous disk, Journal of Thermal Analysis and Calorimetry 149 (2024) 867–878. https://doi.org/10.1007/s10973-023-12706-z.
23. R.S.V. Kumar, M.D. Alsulami, I.E. Sarris, G. Sowmya, F. Gamaoun, Stochastic Levenberg–Marquardt neural network implementation for analyzing the convective heat transfer in a wavy fin, Mathematics 11 (2023) 2401. https://doi.org/10.3390/math11102401.
24. J. Suresh Goud, P. Srilatha, R.S. Varun Kumar, G. Sowmya, F. Gamaoun, K.V. Nagaraja, J. Singh Chohan, U. Khan, S.M. Eldin, Heat transfer analysis in a longitudinal porous trapezoidal fin by non-Fourier heat conduction model: An application of artificial neural network with Levenberg–Marquardt approach, Case Studies in Thermal Engineering 49 (2023) 103265. https://doi.org/10.1016/j.csite.2023.103265.
25. G. Sowmya, R.S.V. Kumar, Assessment of transient thermal distribution in a moving porous plate with temperature-dependent internal heat generation using Levenberg–Marquardt backpropagation neural network, Waves in Random and Complex Media (2023). https://www.tandfonline.com/doi/abs/10.1080/17455030.2023.2198040 (accessed January 31, 2024).

26. B. Shilpa, V. Leela, I.A. Badruddin, S. Kamangar, M.N. Bashir, M.M. Ali, Integrated neural network based simulation of thermo solutal radiative double-diffusive convection of ternary hybrid nanofluid flow in an inclined annulus with thermophoretic particle deposition. Case Studies in Thermal Engineering 62 (2024) 105158.
27. B. Shilpa, V. Leela, M.D. Alsulami, Artificial neural network approach for MHD mixed convection and entropy generation in a vertical annulus with time periodic thermal boundary conditions in the presence of radial and induced magnetic field. Numerical Heat Transfer, Part B: Fundamentals 85 (2024) 1072–1098.
28. B. Shilpa, V. Leela, An artificial intelligence model for heat and mass transfer in an inclined cylindrical annulus with heat generation/absorption and chemical reaction. International Communications in Heat and Mass Transfer 147 (2023) 106956.
29. H.P. Rani, B. Shilpa, V. Leela, R.G. Reddy, Numerical simulation and neural network model for hydromagnetic nanofluid convection in a porous wavy channel with thermal non-equilibrium model. Physica Scripta 99 (2024) 115219.
30. B. Shilpa, V. Leela, Galerkin finite element analysis of heat and mass transfer of Jeffrey, Maxwell and Oldroyd-B nanofluids in a vertical annulus with an induced magnetic field and a non–uniform heat source/sink, International Journal of Ambient Energy 44 (2023) 1887–1903. https://doi.org/10.1080/01430750.2023.2196988.
31. J. Rekha, S.P. Suma, B. Shilpa, U. Khan, S.M. Hussain, A. Zaib, A.M. Galal, Solute transport exponentially varies with time in an unsaturated zone using finite element and finite difference method, International Journal of Modern Physics: Part B 37 (2023) 2350089. https://doi.org/10.1142/S0217979223500893.
32. A.A. Pasha, N. Islam, W. Jamshed, M.I. Alam, A.G.A. Jameel, K.A. Juhany, R. Alsulami, Statistical analysis of viscous hybridized nanofluid flowing via Galerkin finite element technique, International Communications in Heat and Mass Transfer 137 (2022) 106244. https://doi.org/10.1016/j.icheatmasstransfer.2022.106244.
33. S.G. Bejawada, W. Jamshed, R. Safdar, Y.D. Reddy, M.M. Alanazi, H.Y. Zahran, M.R. Eid, Chemical reactive and viscous dissipative flow of magneto nanofluid via natural convection by employing Galerkin finite element technique, Coatings 12 (2022) 151. https://doi.org/10.3390/coatings12020151.
34. J. Bouslimi, A.A. Alkathiri, A.N. Alharbi, W. Jamshed, M.R. Eid, M.L. Bouazizi, Dynamics of convective slippery constraints on hybrid radiative Sutterby nanofluid flow by Galerkin finite element simulation, Nanotechnology Reviews 11 (2022) 1219–1236. https://doi.org/10.1515/ntrev-2022-0070.
35. B. Shilpa, I.A. Badruddin, R.G. Reddy, S. Kamangar, A.A. Khan, Exploration of linear and exponential heat source/sink with the significance of thermophoretic particle deposition on ZnO-SAE50 nano lubricant flow past a curved surface. Case Studies in Thermal Engineering 61 (2024) 104883.
36. B. Shilpa, P. Srilatha, U. Khan, N. Kumar, S.B. Ahmed, R. Kumar, Numerical study of thermal and solutal advancements in ZnO–SAE50 nanolubricant flow past a convergent/divergent channel with the effects of thermophoretic particle deposition. Nanoscale Advances 5 (2023) 6647–6658.
37. B. Shilpa, V. Leela, I.A. Badruddin, S. Kamangar, P. Ganesan, A.A. Khan, Exploration of Arrhenius activation energy and thermal radiation on MHD double-diffusive convection of ternary hybrid nanofluid flow over a vertical annulus with discrete heating. Case Studies in Thermal Engineering 65 (2025) 105593.
38. M. Turkyilmazoglu, Extending the traditional Jeffery-Hamel flow to stretchable convergent/divergent channels. Computers & Fluids 100 (2014) 196–203.

6 Physics-Informed Deep Learning Approaches for Industrial Heat Exchangers

Vishal Jadhav, Ritam Majumdar, Anirudh Deodhar, Shirish Karande, Lovekesh Vig, and Venkataramana Runkana

6.1 INTRODUCTION

Manufacturing and process industries transform raw materials into products through various interconnected operations employing a variety of equipment. Process monitoring, optimization and control are important to achieve the targeted performance goals such as maximizing profitability, minimizing emissions, ensuring safety of personnel, etc. Plant availability or equipment health is also intrinsically linked to these goals. Mathematical modeling has been employed not only for design of processes and equipment but also for optimization and control of operations. Rigorous physics-based models such as computational fluid dynamics (CFD) and discrete element models are used for design, simulation and analysis whereas simplified lumped parameter physics-based models, reduced order models and data-driven models are more commonly used for process optimization and control. As availability of data from manufacturing processes in digital form has improved considerably over the years, artificial intelligence (AI) techniques such as machine learning (ML) and deep learning (DL) have emerged as powerful tools for addressing the industrial needs, especially for process monitoring, simulation, analysis, optimization and control. Since rigorous physics-based models are computationally intensive and require considerable time, effort and domain expertise to develop, there is also a drive towards utilizing AI techniques for design as well. These techniques have evolved significantly since the initial use of artificial neural networks (ANNs) in the 1990s [1], with advances in computational hardware like graphic processing units (GPUs) propelling the development of more sophisticated models such as convolutional neural networks (CNNs) [2–5] and deep reinforcement learning (DRL) [6, 7]. The application of CNNs has proven particularly effective for high-dimensional problems, for example, analysis of image-like data to predict local heat flux and patterns related to turbulent heat transfer [5]. DRL, on the other hand, has shown promise in controlling thermal systems [6] by learning to stabilize thermal regimes under various conditions and in improving traditional control methods [7]. One of the key developments

DOI: 10.1201/9781032688121-6

is the application of AI techniques for development and deployment of soft sensors and digital twin systems that can enhance real-time monitoring, optimization and control of production operations [8–10].

Despite their successes, pure data-driven approaches have inherent limitations. These models rely heavily on quality and quantity of relevant data. Furthermore, predictions made by purely data-driven models can be in conflict with established physical principles, underscoring the need for hybrid models that integrate data-driven models with physics-based models to ensure physically plausible predictions and enhance reliability in practical applications [8, 11]. To overcome the limitations of data-driven models and leverage the availability of established physics-based models, hybrid digital twins that combine physics-based and data-driven models have been developed and deployed for industrial systems [8].

To address the need to combine physics-based and data-driven models, physics-informed deep learning (PIDL) has recently evolved into a new field of research. Physics-informed neural networks (PINNs) have been developed to address the limitations of pure data-driven models by incorporating physics-based models, typically expressed as partial differential equations (PDEs), directly into the neural network's architecture [12]. By including the residuals of governing PDEs and boundary conditions in the training loss function, PINNs ensure physically consistent solutions while learning from both data and underlying physical principles. This semi-supervised approach is particularly valuable for industrial problems, where complete datasets with good variability are not readily available. PINNs have been used for solving various scientific and engineering problems in several domains [13] including fluid flow [14] and heat transfer [15].

One of the main limitations of PINNs is the need for retraining of models as the system boundary conditions change either in space or in time. Recently, Cuomo et al. [13] reviewed the current work on PINNs and highlighted a few of their limitations. Even though the inference time for PINN models is low, high training time and significant convergence difficulties in complex scenarios limit their implementation for real-life applications [16]. Shukla et al. [17] suggested a distributed framework for training PINN models to reduce the training time. Moseley et al. [18] proposed the domain decomposition approach to solve large multiscale problems. Another limitation of the current PINN architectures is that they fail to generalize over dynamically changing boundary conditions [13] for the governing differential equations, a scenario often found in industrial applications. PINNs trained using a single set of boundary conditions cannot be used if the parameters change dynamically [19]. Wang et al. [19] proposed a 'train once use forever' algorithm comprising a combination of GFNet and mosaic flow predictor that enables one time training of a neural network that can generalize over arbitrary boundary conditions as well as arbitrary domain shapes. Chakraborty [20] suggested the use of transfer learning for training of multi-fidelity PINNs. Desai et al. [21] proposed application of transfer learning with pre-trained neural network for one-shot inference of a linear system of both ordinary and PDEs. Meta learning [22] and hypernetwork [23] approaches have also been suggested for adapting PINNs to dynamic boundary conditions.

Hypernetworks [24] integrated with PINNs (HxPINNs) [25] offer a solution to several limitations of basic PINNs. By separating the tasks of parameter space handling and solution function approximation, hypernetwork-based PINNs enable dynamic adaptation to new boundary conditions without retraining, significantly reducing computational costs and training time. Hypernetwork-based PINNs have been applied for prediction of internal temperatures in an air preheater (APH), quickly adjusting to changing operating conditions of the APH based on external sensor data while maintaining accuracy comparable to conventional physics-based simulations [14].

Neural operators belong to the general class of neural networks designed to learn mappings between function spaces, providing rapid, scalable, and efficient solutions for complex or high-dimensional PDEs. Current research in neural operators, particularly their physics-informed variant, physics-informed neural operators (PINOs) [26], focuses on addressing challenges such as training time, computational resources, and generalization to unseen data. Techniques like self-training and incorporation of physics-informed losses improve PINO's performance by leveraging both data-driven and physics-based approaches [27]. Recent studies have explored replacing predefined transform-based layers with learnable linear transformations, for simplifying the architecture and for reducing computational demands while maintaining performance in terms of accuracy [28].

Symbolic regression (SR) is another approach that is being attempted to solve PDEs. The aim is to uncover intrinsic mathematical relationships within data and generate interpretable models. Recent advancements such as physics-informed symbolic networks (PISNs) [29] utilize context-free grammar and differentiable programming to enforce physically valid solutions. The preliminary application of SR to model phenomena within an APH has shown promise in deriving practical mathematical models that elucidate heat transfer [30].

Heat exchangers play a vital role in enhancing the energy efficiency of unit operations in process and utility industries. Fouling of heat exchangers is a common problem across industries. It results in low thermal efficiency and, in some cases, expensive unplanned plant stoppages. According to Müller–Steinhagen and Malayeri [31], the total cost of fouling – including increased fuel consumption, additional maintenance and lost production due to equipment downtime – is estimated to be 0.25% of the GDP of industrialized countries. Fouling refers to the accumulation of unwanted materials on heat transfer surfaces, which diminishes heat transfer efficiency and increases pressure drop. Several factors like quality of fluids being processed, process operating conditions, and environmental conditions significantly influence fouling. Detecting fouling is essential for taking preventive measures in time. However, direct measurement is challenging due to the lack of appropriate inline fluid quality sensors and the difficulty of installing sensors to assess the equipment's internal condition. The present authors have been exploring the application of physics-based models and AI techniques for predicting fouling in industrial APHs. Different PIDL approaches that we have applied to study heat transfer in an APH are described in this chapter.

Since physics-based models are the heart of any PIDL approach, we start with the description of a detailed CFD model for heat transfer in the APH first in Section 6.2. The PINN approach is then demonstrated to predict the thermal profile for the APH in Section 6.3.1. Section 6.3.2 discusses the use of hypernetwork-based PINNs for quick inference for varying design and operation conditions of APHs. The symbolic regression approach for solving PDEs is explained in Section 6.3.3 and its application to predict APH thermal profiles is demonstrated. Finally, Section 6.4 summarizes the current work and discusses potential research directions for the future.

6.2 PHYSICS-BASED MODELING OF AN AIR PREHEATER

The APH is a critical heat exchanger in a thermal power plant. It is employed to recover waste heat from boiler flue gas and improve the overall plant efficiency. Generally, every 22°C rise in air temperature by APH is considered equivalent to an improvement of 1% in thermal efficiency of a boiler [32]. A typical regenerative APH consists of a central rotor housing a heat transfer matrix (metallic or ceramic), surrounded by a stationary housing divided into sectors for the counter-current flow of flue gas and air streams (see Figure 6.1). As the rotor slowly rotates, the matrix elements are alternately exposed to hot flue gas and cold air which facilitates efficient heat transfer. The flue gas enters the APH from the top, typically at high temperatures (300–400°C), while ambient air enters from the bottom. As the gas flows downward from the hot end of the matrix, it transfers heat to the matrix before exiting at a lower temperature, usually around 120–150°C. Simultaneously, the ambient air flows upward through the cold end of the matrix, absorbs heat from the matrix, and reaches temperatures in the range of 250–300°C before entering the boiler for combustion of coal.

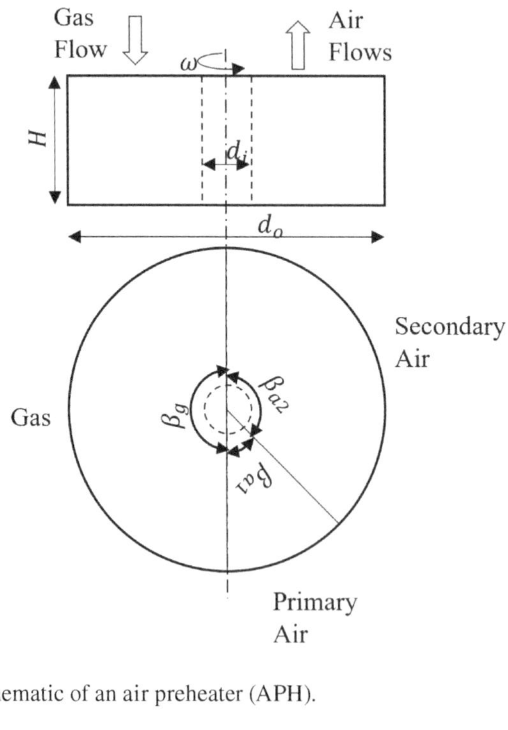

FIGURE 6.1 Schematic of an air preheater (APH).

The fouling of APH is primarily caused by the deposition of ammonium bi-sulphate (ABS) and fly ash on the internal heat transfer surfaces or the metal matrix. ABS forms through complex chemical reactions involving flue gas constituents like SO_3 and NH_3, in the temperature range of 190–240°C [33], predominantly through the mechanism shown below:

$$NH_3 + SO_3 + H_2O \rightarrow NH_4HSO_4 \tag{6.1}$$

$$SO_3 + H_2O \rightarrow H_2SO_4 \tag{6.2}$$

$$NH_3 + H_2SO_4 \rightarrow NH_4HSO_4 \tag{6.3}$$

As the flue gas cools further in the APH, the liquid ABS absorbs fly ash and condenses on the heat transfer surfaces, forming a solid fouling layer. This fouling impedes heat transfer, increases pressure drop, and can lead to unplanned plant shutdowns for cleaning, resulting in significant economic losses [34].

The formation of ABS and the resultant APH fouling are significantly influenced by the operating conditions within the APH such as internal temperatures and chemical composition of the gas that are largely governed by the stringent constraints of APH operation and the fuel used. The internal temperatures are also influenced by external conditions such as operating load and ambient air characteristics. Previous studies [33, 35] have shown that the internal thermal profile not only affects the magnitude of fouling but the location of fouling as well, which ultimately determines the real impact on the operation of APH [36]. Therefore, monitoring the internal temperature profile in real time is critical to decide appropriate preventive actions such as online cleaning (called soot-blowing) or water cleaning. More importantly, monitoring is essential to determine the appropriate time to initiate cleaning. It is difficult to install sensors to monitor the internal thermal or fouling conditions within the APH. The complex physico-chemical phenomena and the interplay of operating conditions makes fouling very difficult to predict intuitively even for an expert. In such a situation, a computational model-based monitoring tool would be of great utility in improving visibility in real time. However, modeling these complex interactions is challenging due to the lack of internal sensors, the slow progression of fouling, and the influence of numerous operating parameters and fluctuating ambient conditions that vary with time.

To study heat transfer and fouling phenomena in an APH, various modeling approaches were explored including pure physics-based modeling like CFD [37–39] and pure data-driven modeling such as ANNs [40, 41], recurrent neural networks (RNNs) like LSTM ([36, 42]) and ensemble DL [43]. While CFD models provide detailed insights into the complex thermal and fluid flow phenomena, they are computationally intensive and require accurate input parameters. Data-driven models, on the other hand, can learn from historical operational data without explicitly incorporating physical equations but they rely heavily on the quality and quantity of training data.

Hybrid approaches that combine the advantages of the first principles-based models with those of data-driven techniques [44] are gaining popularity. The utility of PINNs has been demonstrated recently, as they offer faster (mesh-free) inference and exhibit low reliance on sensor data for training [12]. Additionally, the ability to quickly learn from limited data while adhering to physical laws makes PINNs highly appealing. They have also been effective in solving inverse problems with limited data [12]. Their compatibility with generalization techniques such as transfer learning [20] makes them suitable alternatives to both first principles-based models and pure data-driven models for real-time applications. As mentioned earlier, we have explored PINNs coupled with transfer learning [11], hypernetworks [25] and symbolic regression [29, 30] to model heat transfer in an APH. A two-dimensional physics-based model for heat transfer in an APH was developed first [45] for benchmarking these approaches. Details of the physics-based model are provided next.

6.2.1 Governing Equations of an Air Preheater

A schematic of the APH with dimensions including height (H), outer diameter d_o, inner diameter d_i and sector angles β_g, β_{a1}, β_{a2} for the flow of flue gas, primary air, and secondary air, respectively, is depicted in Figure 6.1. The flue gas enters the APH from the top, while the primary air and secondary air enter from the bottom. The metallic packing rotates at a rate of ω revolutions per minute (rpm). The high-temperature flue gas heats up the rotating metallic packing, which transfers this heat to the cold ambient air entering from the bottom. While convection governs heat transfer between the fluids and the metal matrix, conduction significantly contributes to heat transfer within the matrix. We have adopted a two-dimensional formulation [46], considering the tangential and axial directions, to solve the governing equations within the APH, represented by Eqns. (6.4)–(6.9).

The following assumptions are made to simplify the model development:

- Heat transfer in the radial direction is constant and hence temperature does not vary along the radial direction.
- Convective heat transfer is more dominant than conductive heat transfer in the tangential direction within the metal matrix.
- Thermal properties of fluids and metal matrix vary with temperature.
- Convective heat transfer coefficients vary with temperature and the Reynolds number.
- Flow of air and gases through the metal matrix follow plug flow, and air leakages can be neglected.
- The effect of deposition on heat transfer efficiency is insignificant [47].

Heat transfer in an APH is represented by a set of coupled PDEs derived from energy conservation principles and the Fourier's law of heat conduction. [46] The following equations describe the spatial and temporal temperature variations in gas, air, and metal matrix domains for computational domain shown in Figure 6.2.

Gas sector:

$$\frac{\partial T_{m_g}}{\partial \phi} = NTU_{m_g}\left(T_g - T_{m_g}\right) + \frac{1}{Pe_{m_g}}\frac{\partial^2 T_{m_g}}{\partial z^2} \tag{6.4}$$

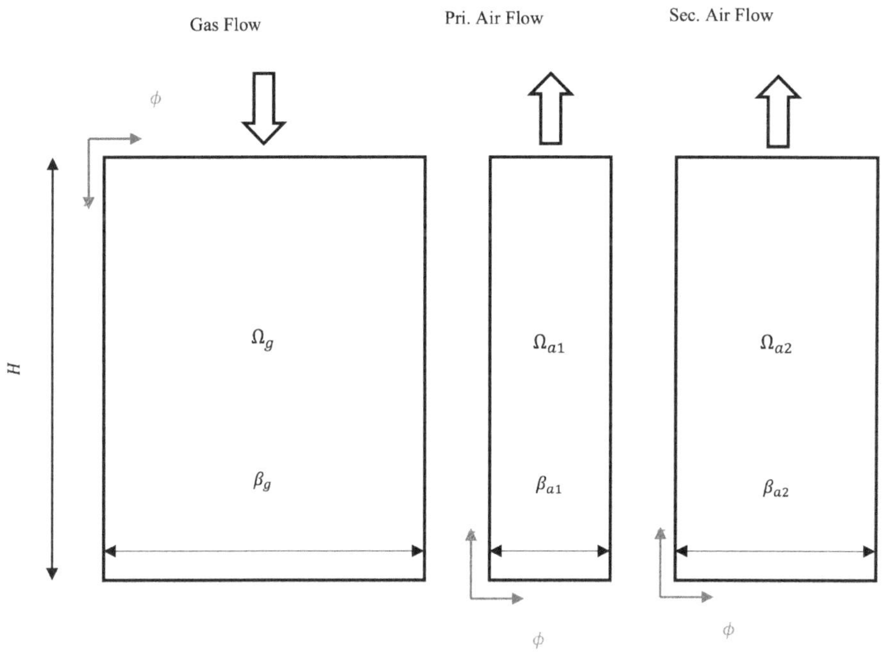

FIGURE 6.2 Computational domain used for the numerical simulation of the APH.

$$\frac{\partial T_g}{\partial z} = NTU_g\left(T_{m_g} - T_g\right) \tag{6.5}$$

Primary air sector:

$$\frac{\partial T_{m_{a1}}}{\partial \phi} = NTU_{m_{a1}}\left(T_{a1} - T_{m_{a1}}\right) + \frac{1}{Pe_{m_{a1}}}\frac{\partial^2 T_{m_{a1}}}{\partial z^2} \tag{6.6}$$

$$\frac{\partial T_{a1}}{\partial z} = NTU_{a1}\left(T_{m_{a1}} - T_{a1}\right) \tag{6.7}$$

Secondary air sector:

$$\frac{\partial T_{m_{a2}}}{\partial \phi} = NTU_{m_{a2}}\left(T_{a2} - T_{m_{a2}}\right) + \frac{1}{Pe_{m_{a2}}}\frac{\partial^2 T_{m_{a2}}}{\partial z^2} \tag{6.8}$$

$$\frac{\partial T_{a2}}{\partial z} = NTU_{a2}\left(T_{m_{a2}} - T_{a2}\right) \tag{6.9}$$

In the above equations, T represents temperature, θ is angular coordinate in the rotational direction, z is axial coordinate, and subscripts m, g, $a1$, and $a2$ denote metal matrix, gas, primary air, and secondary air, respectively. In addition, subscripts m_g, m_{a1}, and m_{a2} represent metallic packing within the gas, primary air, and secondary air sectors, respectively. The key parameters are Number of Transfer Units (NTU), which quantifies the heat transfer effectiveness, and the Peclet number (Pe), which characterizes the relative importance of convection and conduction. A set of boundary conditions, interface conditions, and gradient conditions must be specified to solve these governing PDEs. The boundary conditions define the inlet temperatures of gas and air streams (Eqn. 6.10–6.12) while the interface conditions ensure temperature continuity between adjacent sectors (Eqns. 6.13–6.15). The gradient conditions impose zero-gradient constraints on metal matrix temperatures at the inlet and outlet of each sector (Eqns. 6.16–6.18).

Boundary Conditions:

$$T_g\left(\phi, z=0\right) = T_{g_{in}} \tag{6.10}$$

$$T_{a1}\left(\phi, z=0\right) = T_{a1_{in}} \tag{6.11}$$

$$T_{a2}\left(\phi, z=0\right) = T_{a2_{in}} \tag{6.12}$$

Interface Conditions:

$$T_{m_g}\left(\phi=0,\ z\right) = T_{m_{a2}}\left(\phi=1,\ 1-z\right) \tag{6.13}$$

$$T_{m_g}\left(\phi=1,\ z\right) = T_{m_{a1}}\left(\phi=0,\ 1-z\right) \tag{6.14}$$

$$T_{m_{a1}}\left(\phi=1,\ z\right) = T_{m_{a2}}\left(\phi=0,\ z\right) \tag{6.15}$$

Gradient Conditions:

$$\frac{\partial T_{m_g}\left(\phi,\ z=0\right)}{\partial z} = \frac{\partial T_{m_g}\left(\phi, z=1\right)}{\partial z} = 0 \tag{6.16}$$

$$\frac{\partial T_{m_{a1}}\left(\phi,\ z=0\right)}{\partial z} = \frac{\partial T_{m_{a1}}\left(\phi, z=1\right)}{\partial z} = 0 \tag{6.17}$$

$$\frac{\partial T_{m_{a2}}\left(\phi,\ z=0\right)}{\partial z} = \frac{\partial T_{m_{a2}}\left(\phi, z=1\right)}{\partial z} = 0 \tag{6.18}$$

6.2.2 Model Validation with Experimental Data

The governing equations along with the boundary conditions are solved using the finite difference method ([45]) to predict the temperature profiles of gas, air, and metal matrix inside the APH. They are only solved in the axial and tangential directions using an equivalent Cartesian coordinate system. The design specifications and operating conditions of the APH and thermo-physical properties of the metallic packing used for simulation were taken from literature ([48]) and reproduced in Tables 6.1 and 6.2. The APH comprises of three layers of metallic packing in the axial direction and three sectors in the tangential direction. The metallic packing used in all layers and sectors was assumed to have a density of 753.35 kg/m^3, a thermal conductivity of 52.92 W/m-K, a heat capacity of 456 J/kg-K, and a specific surface area of 420.21 m^2/kg [48]. These values were consistently applied throughout the study to

TABLE 6.1
APH Design Specifications [48]

Parameter	Value
Matrix	
Inner radius (m)	1.63
Outer radius (m)	8.21
Height of matrix (m)	2.05
Sector Angles	
β_g	180°
β_{a1}	70°
β_{a2}	110°

maintain uniformity in the material properties. A computational grid comprising 57,600 cells was found to be sufficient based on the grid independence simulations.

The predictions from the thermal model were compared against the experimental results from literature [48] as shown in Table 6.2. The predictions from the thermal model are in good agreement with the results reported in the literature. The internal temperature profiles were also qualitatively similar to the profiles reported in the literature [49]. These results can be utilized for simulating the thermal performance and inferring the fouling behavior of an APH under various operating scenarios. This information is useful not only for designing the APH appropriately but also for optimizing its performance in actual operation, which ultimately leads to improved power plant efficiency and reduced emissions.

While such physics-based models are useful for simulation and analysis of the phenomena, their utility for real-time operations is limited because they are computationally intensive and require accurate values of input and output variables and model parameters. Moreover, convergence is also not guaranteed for all operating conditions of an industrial APH. Similarly, although rigorous CFD models have been used for equipment and process design, need for computationally efficient models is well-known because design involves a large number of simulation experiments with the models to explore the design space.

TABLE 6.2
APH Operating Data [48] and Predictions of the Present Model

	Flue Gas	Primary Air	Secondary Air
Flow rate, tons/h	2774.37	965.25	1629.12
Inlet temperature, °C	400	32	33
Outlet temperature, °C	118.3	360.2	340.6
Predicted outlet temperature, °C	**121.58**	**373.84**	**344.03**
Relative deviation, %	2.78	3.79	1.01

6.3 PHYSICS-INFORMED DEEP LEARNING APPROACHES

PIDL is evolving as a new field of study to leverage the well-established principles of physics and take advantage of the latest developments in AI in general and DL in particular for scientific and engineering applications. Recent work by the present authors on this topic is discussed in detail in this section with appropriate references to related literature.

6.3.1 Physics-Informed Neural Networks

PINNs [12] are gaining popularity as a viable alternative to address the limitations and to harness the power of both physics-based and data-driven models. PINNs belong to the class of universal function approximators that are trained by imposing governing PDEs as constraints. These constraints are applied by introducing governing equation residuals and boundary or initial conditions in the loss function. This approach enables incorporation of domain knowledge in the learning process and also in learning without any data in a semi-supervised manner. PINNs make the model flexible by eliminating a fixed mesh typically required for a physics-based numerical solver, while remaining physically consistent. Lagaris et al. [50] were the first to introduce neural networks to solve boundary value problems represented by PDEs. Raissi et al. [12] incorporated the governing PDEs into the loss function of a deep neural network (DNN) to solve both forward and inverse problems. Since then, PINNs have been used for solving various scientific and engineering problems in several domains [13, 14] and [15].

6.3.1.1 Mathematical Formulation and Training Strategies

PINNs leverage the universal function approximation capabilities of neural networks while incorporating physical laws, typically in the form of PDEs, directly into the network structure and in the training process. This allows PINNs to generate physically consistent solutions even with limited training data. PINNs leverage the power of automatic differentiation [51] that enables efficient computation of derivatives of the network outputs with respect to the inputs. The required derivatives of the PDEs can be calculated through automatic differentiation without the need for numerical discretization or manual derivation of the gradients. The resulting physics-informed loss function includes terms that penalize the residuals of the governing PDEs as well as deviations from the boundary and interface conditions. This allows PINNs to handle complex geometries and boundary conditions as well as to solve both forward and inverse problems. During training, PINNs learn to approximate the solution of the PDEs while respecting physical principles by minimizing the composite loss function.

The fundamental concept of PINNs is to represent the solution $u(x,t)$ of a PDE using a neural network and include the residual of the PDE in the loss function used for training the network. For example, consider a general PDE of the form:

$$\frac{\partial u}{\partial t} + f[u] = g(x,t) \tag{6.19}$$

where f is a linear or nonlinear differential operator and $g(x,t)$ is a known forcing function. In the PINNs framework, the solution $\boldsymbol{u}(\boldsymbol{x},t)$ is approximated by a neural network $\hat{u}(x,t;\theta)$ where θ denotes the set of network parameters (weights and biases). The residual of the PDE is then defined as:

$$\mathcal{F} := \frac{\partial \hat{u}}{\partial t} + f[\hat{u}] - g \tag{6.20}$$

This residual, evaluated at a set of collocation points in the domain, is included in the loss function along with the terms that capture the boundary and initial conditions:

$$MSE = MSE_{\mathcal{F}} + MSE_{bc} + MSE_{ini} \tag{6.21}$$

where $MSE_{\mathcal{F}}$ is mean squared residual, MSE_{bc} is mean squared error of the boundary conditions, and MSE_{ini} is mean squared error of the initial conditions. These MSE terms are evaluated using following equations (Eqns. 22–25):

$$MSE_{\mathcal{F}} = \frac{1}{N_{\mathcal{F}}} \sum_{i=1}^{N_{\mathcal{F}}} \left| \mathcal{F}(x_{\mathcal{F}}^{i}, t_{\mathcal{F}}^{i}) \right|^2 \tag{6.22}$$

$$MSE_{ini} = \frac{1}{N_{\text{ini}}} \sum_{i=1}^{N_{\text{ini}}} \left| \boldsymbol{u}_{ini} - \hat{\boldsymbol{u}}(\boldsymbol{x}_{ini}^{i}, \boldsymbol{t}_{ini}^{i}) \right|^2 \tag{6.23}$$

$$MSE_{bc} = \frac{1}{N_{bc}} \sum_{i=1}^{N_{bc}} \left| \boldsymbol{u}_{bc} - \hat{\boldsymbol{u}}(\boldsymbol{x}_{bc}^{i}, \boldsymbol{t}_{bc}^{i}) \right|^2 \tag{6.24}$$

The PINN simultaneously fits any available training data by minimizing this composite loss function while respecting the physical constraints encoded in the PDE (see Figure 6.3, for illustration of PINN training with governing equations of APH as loss function).

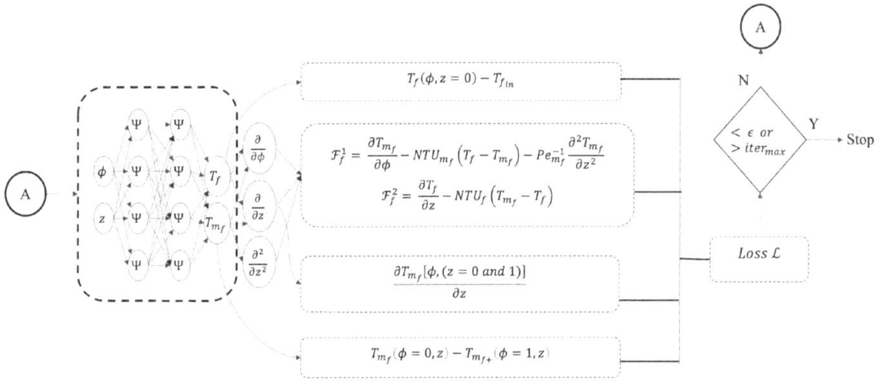

FIGURE 6.3 Illustrative architecture of the PINN for governing equations of APH.

One of the key advantages of PINNs is their mesh-free nature, that is, collocation points can be arbitrarily placed in the domain without the need for a structured grid. However, for complex problems with steep gradients or multiscale phenomena, many collocation points are needed which can make training computationally expensive. Domain decomposition strategies have been proposed to mitigate this issue. The idea is to split the computational domain into multiple subdomains, employ separate PINNs for each subdomain, and couple them at the interfaces. Methods like conservative-PINN (cPINN) [16] and extended PINN (xPINN) [52] follow this approach to enable the solution of complex nonlinear PDEs.

Another challenge with PINNs is incorporation of varying boundary conditions, material properties, and PDE parameters. The base PINN must be re-trained if any of these factors change. To overcome this limitation, approaches based on transfer learning [11] and hypernetworks [25] have recently been attempted. These approaches are discussed next.

6.3.1.2 Domain Decomposition of PINNs

Domain decomposition (DD) of PINNs [16, 52] leverages the inherent parallelism in large-scale scientific computing. By splitting the domain into smaller subdomains, DD promotes localized learning by lightening the load on individual networks and enhances learning dynamics. Each subdomain is modeled by a distinct neural network with appropriate interface conditions to ensure continuity and smoothness across adjacent subdomains. These conditions, enforced through penalties in the loss function, maintain solution coherence and continuity across domains. The DD approach not only facilitates parallelization but also enables independent training of networks on different subdomains, making it ideal for distributed computing. This is particularly advantageous for high-dimensional PDEs for which traditional PINNs face challenge of convergence. Additionally, DD allows networks to specialize for various subdomain which enhances accuracy and stability of the solution.

6.3.1.3 Transfer Learning of PINNs

Transfer learning enhances the efficiency of PINNs by repurposing a model developed for one task to another related task. This reduces computational costs and improves accuracy without the need for developing the PINN from the start all over again. It includes two key strategies: Initial Condition Transfer, where a pre-trained model is applied to new initial conditions while maintaining the same physical laws, useful for quick and accurate predictions under different scenarios; and Potential Transfer Learning, where the model adapts to both new initial conditions and changes in physical potential [20, 53, 54]. Common methods of implementing transfer learning in PINNs involve freezing the base layers of the network to speed up learning by reducing the number of parameters that need adjustment and applying Singular Value Decomposition (SVD) to simplify the network's weights. These methods help in generalization of the PINNs.

The basic PINNs, domain decomposition and transfer learning were applied for simulation of heat transfer in an APH. Details of implementation and the results obtained are presented next.

6.3.1.4 Application of PINNs for an Air Preheater

In the context of an APH, the governing PDEs, presented in Section 2, represent the spatio-temporal variations of temperature in the gas, air, and metal matrix domains. The PINN architecture employed for modeling heat transfer in an APH leverages a domain decomposition approach in which separate DNNs are used for each subdomain: gas (Ω_g), primary air (Ω_{a1}), and secondary air (Ω_{a2}). Coordinates of each subdomain are normalized from 0 to 1 for both axial and tangential directions. It should be noted that positive axial direction for the gas-side subdomain is from top to bottom which is same as the direction of gas flow. Similarly, for the primary and the secondary air-side subdomains, positive axial direction is from bottom to top which is same as the direction of flow for primary and secondary air. Positive tangential direction is considered from left to right for all subdomains, which is same as rotational direction of the metal matrix. For simplicity, all metal matrix layers are assumed to be made of a single homogeneous material. However, the current approach can be readily extended to an APH with a multi-material matrix as well as the matrix is discretized into separate layers and sectors.

6.3.1.4.1 Neural Network Architecture

The domain-decomposed PINN model consists of a DNN for each of the subdomains p in the APH (i.e., Ω_g, Ω_{a1}, Ω_{a2}). The spatial co-ordinates (ϕ, z) are the inputs to each network and the outputs are fluid temperature (T_f) (air or gas) and matrix temperature (T_m) for each subdomain, respectively. Let $\mathcal{N}^L : \mathbb{R}^{D_i} \to \mathbb{R}^{D_o}$ be a DNN of L layers and N_k neurons in k^{th} layer ($N_0 = D_i$ and $N_L = D_o$). The weight matrix and bias vector in the kth layer $(1 \le k \le L)$ are denoted by $\mathbf{W}^k \in \mathbb{R}^{N_k \times N_{k-1}}$ and $\mathbf{b}^k \in \mathbb{R}^{N_k}$, respectively. The input vector is denoted as $\mathbf{x} \in \mathbb{R}^{D_i}$, output vector at k^{th} layer is denoted as $\mathcal{N}^k(\mathbf{x})$ and $\mathcal{N}^0(\mathbf{x}) = \mathbf{x}$. The activation function is denoted as Ψ. DNN is defined by Eqn. (6.22):

$$\mathcal{N}^k(\mathbf{x}) = \Psi\left(\mathbf{W}^k \mathcal{N}^{k-1}(\mathbf{x}) + \mathbf{b}^k\right),\ 1 \le k \le L \tag{6.25}$$

Let $\Theta = \left\{\mathbf{W}^k, \mathbf{b}^k\right\}$ be a collection of all weights and biases. The output of neural network is then given by Eqn. (6.26).

$$u_\Theta(\mathbf{x}) = \mathcal{N}^L(\mathbf{x};\Theta) \tag{6.26}$$

A schematic of the domain-decomposed PINN architecture is shown in Figure 6.4, wherein three DNNs are used for each subdomain, respectively. For each subdomain, the output of individual DNN is given as:

$$u_{i\Theta}(\mathbf{x}) = \mathcal{N}_i^L(\mathbf{x};\Theta\ ),\ \forall\ i = 1,2,3 \tag{6.27}$$

The final solution is given by:

$$u_\Theta(\mathbf{x}) = \bigcup_{i=1}^{3} u_{i\Theta}(\mathbf{x}) \tag{6.28}$$

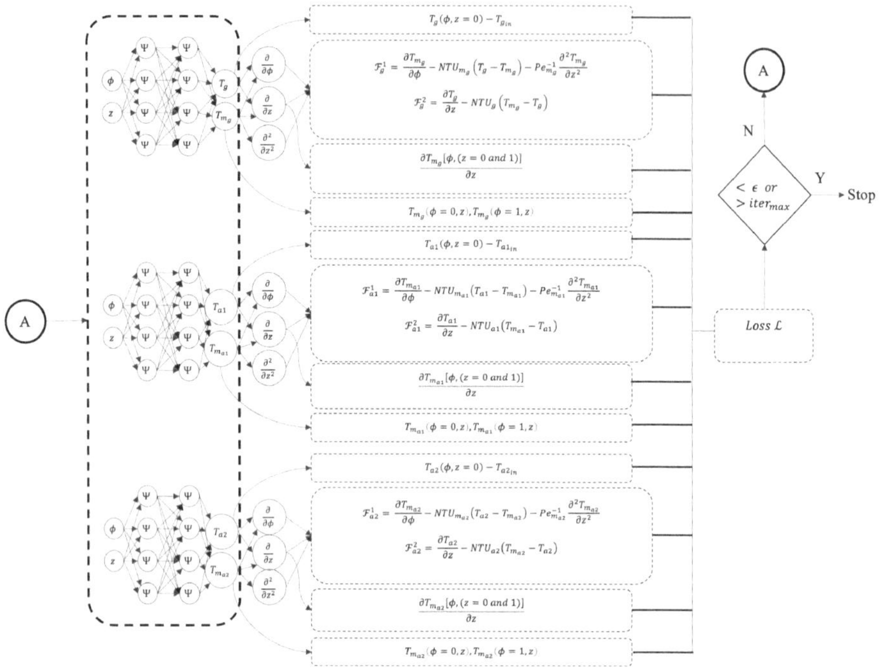

FIGURE 6.4 Architecture of domain decomposition-based PINN.

6.3.1.4.2 Subdomain Loss Function

The total loss for a PINN comprises of the mean-squared error (MSE) due to residuals of the governing equations calculated using the collocation points ($MSE_{\mathcal{F}_p}$), loss due to boundary conditions calculated using the boundary points (MSE_{bc_p}), MSE loss due to interface condition calculated at the interface points (MSE_{ic_p}) and MSE loss due to matrix temperature gradient calculated at the top and bottom points of each subdomain (MSE_{grad_p}). The mean squared error for different components of subdomain p is calculated using Eqns. (6.29–6.32):

$$MSE_{\mathcal{F}_p} = \frac{1}{N_{\mathcal{F}_p}} \sum_{i=1}^{N_{\mathcal{F}p}} \left| \mathcal{F}\left(\theta_{\mathcal{F}p}^{\ i}, z_{\mathcal{F}p}^{\ i}\right) \right|^2 \tag{6.29}$$

$$MSE_{ic_p} = \frac{1}{N_{ic_p}} \sum_{i=1}^{N_{icp}} \left| T_{mp}\left(\theta_{icp}^{\ i}, z_{icp}^{\ i}\right) - T_{mp^+}\left(\theta_{icp}^{\ i}, z_{icp}^{\ i}\right) \right|^2 \tag{6.30}$$

$$MSE_{bc_p} = \frac{1}{N_{bc_p}} \sum_{i=1}^{N_{bcp}} \left| T_{fp}^{i} - T_{fp}\left(\theta_{bcp}^{\ i}, z_{bcp}^{\ i}\right) \right|^2 \tag{6.31}$$

$$MSE_{grad_p} = \frac{1}{N_{grad_p}} \sum_{i=1}^{N_{grad_p}} \left| \frac{\partial T_{m_p}\left(\theta_{grad_p}^{i}, z_{grad_p}^{i}\right)}{\partial z} \right|^2 \tag{6.32}$$

where, $\mathcal{F}$ is residual of the governing PDEs. Subscript p^+ indicates the neighboring subdomain to subdomain p. $N_{\mathcal{F}_p}, N_{ic_p}$, N_{bc_p}, and N_{grad_p} represent number of collocation points, number of interface condition points, number of boundary condition points and number of matrix temperature gradient condition points in the pth subdomain, respectively. $\left(\theta_{\mathcal{F}}{}_p^i, z_{\mathcal{F}}{}_p^i\right)$, $\left(\phi_{bc}{}_p^i, z_{bc}{}_p^i\right)$ and $\left(\theta^i_{grad_p}, z^i_{grad_p}\right)$ represent coordinates of the residual points, boundary condition points and gradient condition points for the pth sub-domain, respectively. $\left(\theta_{ic}{}_p^i, z_{ic}{}_p^i\right)$ represents common interface points of two neighboring subdomains p and p^+. The loss for p^{th} subdomain is given by Eqn. (6.33).

$$\mathcal{L}(\Theta)_p = MSE_{\mathcal{F}_p} + MSE_{ic_p} + MSE_{bc_p} + MSE_{grad_p} \tag{6.33}$$

The total loss for the PINN is given by Eqn. (34) where subscripts $g, a1,$ *and* $a2$ represent subdomains for gas, primary air, and secondary air, respectively.

$$\mathcal{L}(\Theta) = \mathcal{L}(\Theta)_g + \mathcal{L}(\Theta)_{a1} + \mathcal{L}(\Theta)_{a2} \tag{6.34}$$

The DNN for each subdomain consists of one input layer (two neurons), two hidden layers (with 16 neurons in each layer) and one output layer (two neurons). The activation function used for the hidden layers and for the output layer is *tanh*. The gradients for evaluating the residual equation were calculated using the auto-differentiation feature [51]. The Adam optimizer was used to train the PINN model and mean-squared error (MSE) was used as the loss metric. In addition, reducing learning rate callback and early stopping callback features from *TensorFlow* were employed for better control while training. If the loss does not reduce compared to the best loss value for 50 epochs, learning rate was reduced by a factor of 0.1 with reducing learning rate callback. Also, early stopping callback was used to stop the training if the training loss did not improve for 100 epochs. All the computations were performed on a computer with the specifications: AMD Ryzen 5 2500U processor with Radeon Vega Mobile Gfx 2.00 GHz, RAM of 24 GB and 64-bit operating system.

The domain-decomposed PINN model was trained using the architecture described above with 15,000 collocation points, 15,000 boundary points and 15,000 subdomain interface points [8]. The training process was carried out for 788 epochs until an acceptable loss value was achieved, as determined by the early stopping criteria. The temperature profiles obtained using the physics-based model and the trained PINN model for the gas, primary air, secondary air and metal matrix domains are compared in Figure 6.5. The close agreement between the two solutions is evident, with the PINN model accurately capturing the thermal phenomena across all subdomains, including the interfaces. A quantitative assessment of the PINN model's accuracy revealed a mean absolute error of 8.1×10^{-3} and a maximum absolute error of 0.03 for the normalized temperature values. In terms of the non-normalized solution, these errors correspond to a mean absolute error of 2.7°C and a maximum absolute error of 10°C. While the finite difference method employed for solving the physics-based model required 1076 seconds to compute the temperature profiles, the trained PINN model generated predictions in just 1.8 seconds, a

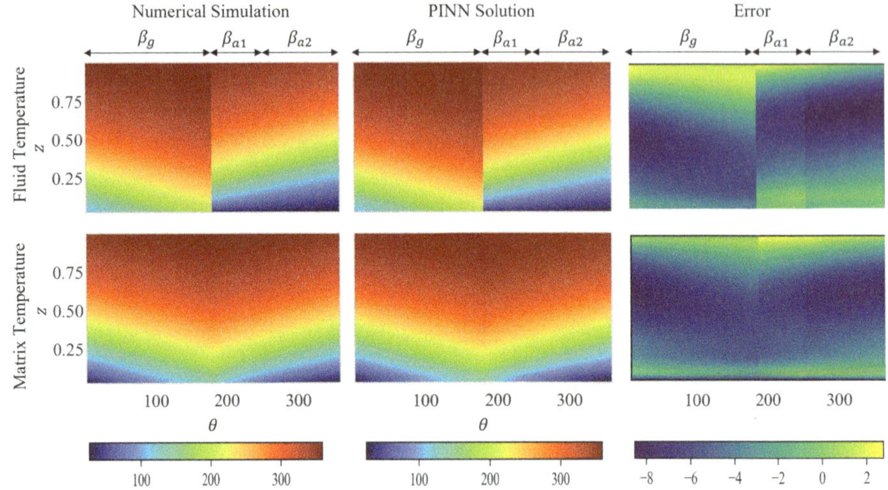

FIGURE 6.5 Comparison of internal temperature profiles obtained with the physics-based model through numerical simulation and with a domain decomposition-based PINN.

reduction of over 99%. This substantial improvement in computational efficiency highlights the potential of PINNs for real-time monitoring of industrial APHs.

To address the limitation of PINNs being trained only for specific boundary conditions, the transfer learning framework was employed. This framework leverages knowledge from a pre-trained base PINN to quickly adapt to new boundary conditions. In the specific case of APHs, the weights of a base PINN trained for a benchmark condition (e.g., inlet gas temperature of 359°C and inlet air temperature of 25°C) were used to initialize a new PINN. During retraining for the new boundary condition, the weights and biases of the input layer and the first hidden layer were frozen for each subdomain's DNN, based on the assumption that these layers capture the fundamental physical phenomena (see Figure 6.6). This approach significantly reduced the training time while maintaining accuracy comparable to the base PINN.

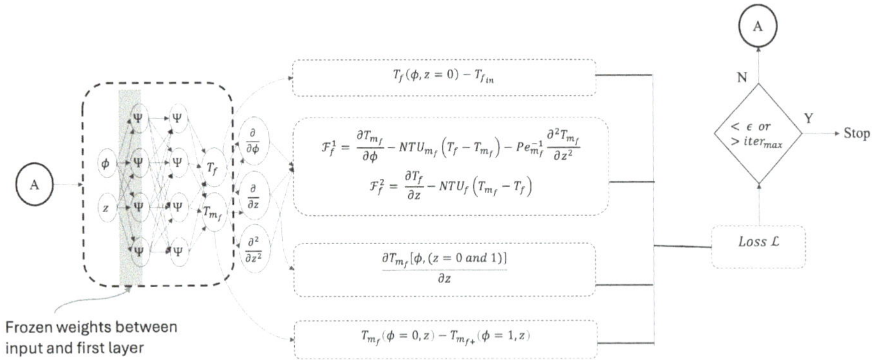

FIGURE 6.6 Illustrative architecture for transfer learning of PINNs.

TABLE 6.3
Inference Times for Numerical Simulation and for Transfer Learning of PINN

			Numerical Simulation	PINN Trained with Random Initialization (without using Base PINN)			Transfer Learning using Base PINN		
Case	Inlet Gas Temperature BC (°C)	Inlet Air Temperature BC (°C)	Inference Time (sec)	Training time (sec)	Inference time (sec)	MAE	Training time (sec)	Inference time (sec)	MAE
1	329	19	1028	3716	3.6	0.012	211	3.5	0.0105
2	354	48	922	2782	3.7	0.005	255	3.8	0.0052
3	331	38	1122	3314	3.9	0.030	187	3.6	0.0098
4	330	31	915	3174	4.0	0.008	190	3.4	0.0105
5	390	36	1066	2895	3.5	0.007	194	3.5	0.0033
6	320	37	1076	2847	3.3	0.016	268	3.5	0.0112
7	399	23	1179	3370	3.7	0.011	194	3.6	0.0056
8	388	13	1188	2878	3.7	0.020	195	3.9	0.0021
9	363	43	1038	2726	3.7	0.016	181	3.7	0.0043

As shown in Table 6.3, the transfer learning approach achieved an average combined training and inference time that is 78% lower than the corresponding numerical simulation, without compromising on accuracy. In contrast, PINNs trained from scratch often require more time than the numerical simulation. By enabling rapid adaptation to new boundary conditions, transfer learning enhances the practicality of PINNs for real-time monitoring and predictive maintenance of industrial heat exchangers, paving the way for more efficient and reliable operation.

6.3.2 Hypernetworks

Hypernetworks belong to the class of meta-learning methods that divide learning into two networks: a main network and a hypernetwork ([24, 55, 56]). The main network performs the task like a typical neural network. However, its parameters are not learned during training. Instead, the hypernetwork generates the main network's parameters at the time of evaluation. Formally, let $\mathcal{N}_{\Theta_h}$ be the hypernetwork with input x_h and $\mathcal{N}_{\Theta_m}$ be the main network with input x_m. The parameter Θ_m of the main network is given by:

$$\Theta_m = \mathcal{N}_{\Theta_h}(x_h),\ \hat{y} = \mathcal{N}_{\Theta_m}(x_m) \tag{6.35}$$

where Θ_h are learnable parameters of the hypernetwork. This structure enables the hypernetwork to meta-learn a space of tasks defined by x_h, generating an appropriate main network for each task.

The main idea behind hypernetwork-based PINNs (HxPINNs) is learning a mapping between the operating conditions and the parameters of the PINN model [25]. By training a hypernetwork on a diverse set of pre-trained PINNs corresponding to different boundary conditions, it is possible to capture the underlying relationships between the operating parameters and the PINN weights effectively.

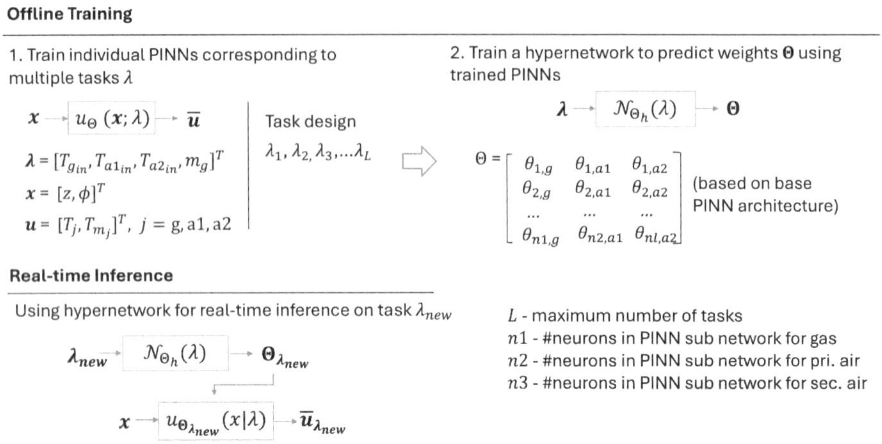

FIGURE 6.7 Training and inference methodology for hypernetwork-based PINNs.

During inference, the hypernetwork takes the new operating conditions as input and generates the appropriate PINN parameters on the fly, allowing the model to adapt to unseen scenarios with minimal computational overhead. Hypernetwork-based PINNs were explored recently for APH temperature profile predictions [25]. A brief summary of this work is presented here. The steps involved in training and inference scheme for the HxPINNs are shown in Figure 6.7. Initially, several base PINNs $u_\Theta(\boldsymbol{x};\lambda)$ are trained for specific tasks λ which represents typical operating conditions of an APH. These PINNs predict $\hat{\boldsymbol{u}}$, the thermal profile of APH. The parameters $\boldsymbol{\Theta}$ of these base PINNs are extracted. A hypernetwork $\mathcal{N}_{\Theta_h}$ meta-learns the parameters $\boldsymbol{\Theta}$ in a supervised manner as a function of the operating conditions λ. During the inference stage, a new set of tasks (operating conditions) λ_{new} is input to the hypernetwork, which predicts the parameters $\Theta_{\lambda_{new}}$ of the PINN corresponding to λ_{new}. These parameters when applied to the base PINN architecture predict the thermal profile $\boldsymbol{u}_{\Theta_{\lambda\text{new}}}$.

To generate training data for the hypernetwork, multiple domain decomposed base PINN models were created under varying operating conditions, as shown in Table 6.4. These conditions include inlet temperatures and flow rates of process

TABLE 6.4
Variables and Range of APH Operating Conditions Considered for Full Factorial Design

	Inlet Gas Temperature T_{gin}, (°C)	Inlet Air Temperature $T_{a1in} = T_{a2in}$, (°C)	Gas Flow Rate m_g (kg/s)
Minimum	200	10	600
Maximum	400	80	800
Step size	25	10	50

streams entering APH. Parameters from these PINNs were used to train the hypernetwork. The training tasks were based on a full factorial design of operating conditions for the APH. The considered operating conditions include inlet gas, primary air, and secondary air temperatures, which vary due to changing boiler and ambient conditions. Primary and secondary air temperatures were considered identical, as observed in the field. Gas flow rate, varying with plant load, was also an input. A total of 315 PINNs were trained for 315 different operating conditions. The hypernetwork architecture (see Figure 6.8) consisted of an input layer with 3 neurons (corresponding to the inlet temperatures and mass flow rates), 2 hidden layers with 256 neurons each, and 3 output layers with 354 neurons each (corresponding to the weights of the base PINNs). The hypernetwork was trained using the Adam optimizer with a learning rate of 1.0×10^{-4} and an early stopping criterion based on the validation loss. The weights of the pre-trained PINNs were extracted and paired with their corresponding operating conditions to form the input-output pairs for the hypernetwork.

The performance of the hypernetwork based PINN model was evaluated on a test set of 19 unseen operating conditions and compared against the physics-based model and the basic PINN. The hypernetwork based PINN demonstrated excellent accuracy, with a mean absolute error of less than 5°C and a mean absolute percentage error below 10% across all subdomains. Thermal profiles obtained with

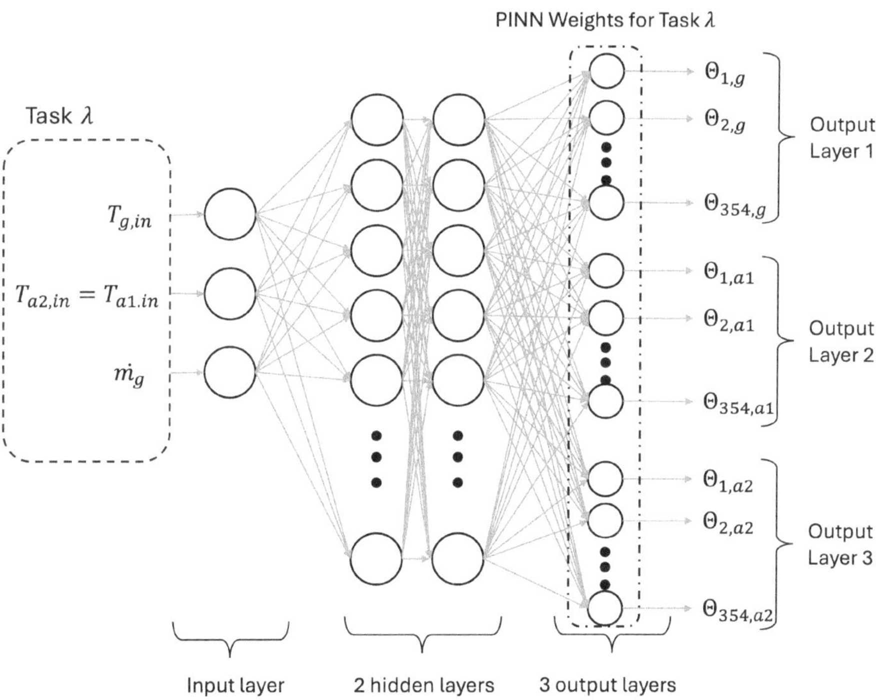

FIGURE 6.8 Architecture of hypernetwork for domain decomposition-based PINN of APH.

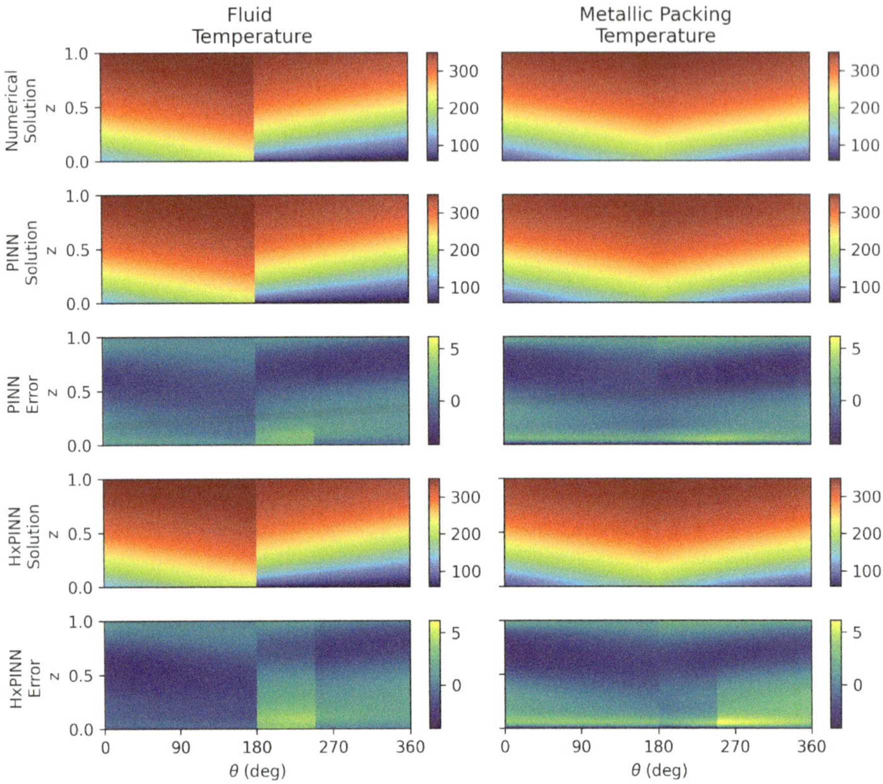

FIGURE 6.9 Comparison of APH internal temperatures predicted through numerical simulation, domain decomposed PINN and hypernetwork-based domain decomposed PINN (HxPINN).

the hypernetwork based PINN is shown in Figure 6.9. In terms of computational efficiency, the hypernetwork based PINN provided near-instantaneous inference, with a prediction time of approximately 5 seconds, compared to 200 seconds for the physics-based simulation and 180 seconds for the standalone PINN with transfer learning. Moreover, the hypernetwork based PINN exhibited remarkable adaptability to varying operating conditions, accurately predicting the thermal profiles without the need for retraining.

The successful application of hypernetwork based PINNs for real-time monitoring of the industrial APH has significant implications for fouling detection and predictive maintenance. By providing accurate and timely insights into the internal thermal conditions of the APH, the model enables early detection of fouling-prone zones and facilitates proactive cleaning interventions. This can help prevent unplanned shutdowns, reduce maintenance costs, and extend the lifespan of the APH. Furthermore, integrating the hypernetwork based PINN model with the plant's control system can enable dynamic optimization of operating parameters, such as air and gas flow rates, to mitigate fouling and improve overall plant efficiency. The case study highlights the potential of hypernetwork based PINNs as a powerful tool for real-time monitoring and optimization of industrial heat exchangers, paving the way for more efficient, reliable, and sustainable power generation.

The integration of hypernetworks with PINNs offers several significant advantages for industrial heat exchanger applications. First, it considerably reduces the computational costs associated with retraining PINNs for every new set of boundary conditions, making the approach more efficient and scalable. Second, hypernetworks enable real-time inference capabilities, as they can instantly generate PINN parameters for any given operating condition, enabling swift decision-making and control. This is especially useful for monitoring and optimizing the performance of heat exchangers in dynamic industrial environments. Furthermore, the meta-learning nature of hypernetworks allows them to leverage knowledge from a wide range of operating conditions, potentially improving the generalization and robustness of the PINN models.

While hypernetwork based PINNs have demonstrated significant potential for real-time monitoring and optimization of industrial heat exchangers, several challenges and opportunities remain for further development and deployment. One key limitation of current hypernetwork based PINN approaches is the need for a comprehensive bank of pre-trained PINNs spanning a wide range of operating conditions. Generating this training data can be computationally expensive and time-consuming, requiring numerous physics-based simulations or experimental measurements. Moreover, the accuracy of the hypernetwork predictions relies significantly on the quality and diversity of the base PINNs. Any errors or biases present in the base PINNs can propagate through the hypernetwork, leading to suboptimal performance. These limitations will require further research with focus on developing efficient sampling strategies for selecting the most informative and representative operating conditions, minimizing the required number of base PINNs while maintaining adequate coverage of the parameter space.

Hypernetworks often struggle when dealing with base networks that possess an excess of weights, given that predicting the weights of a neural network involves addressing a high-dimensional regression problem. This issue becomes particularly evident when hypernetworks are tasked with predicting parameters for large base networks. However, there exists an alternative approach wherein one can represent the adaptation of a large network through parameter-efficient tuning [57, 58]. This method involves training a hypernetwork using adaptation parameters rather than the entire network, as demonstrated in prior works on image generation [59] and instruction tuning of large language models (LLMs) [60]. A case study in the field of material science conducted by the current authors highlights the efficacy of such techniques in rapidly adapting solutions of parameterized PDEs from one instance to another [61]. This case study showcased that optimal low rank adapted PINNs outperform others.

6.3.3 Symbolic Regression

Symbolic regression (SR) is the process of identifying a mathematical expression that most accurately represents data. This technique is pivotal in deciphering complex relationships within data across various domains such as scientific research [62, 63], engineering [64, 65], and financial forecasting [66]. SR not only simplifies model structures but also enhances the interpretability of solutions, thus fostering greater transparency in AI-driven systems. Recent studies, including those by Virgolin et al. [67], have classified SR as an NP-hard problem, underscoring its computational intensity. Traditionally, approaches to SR have included genetic programming [68], DL techniques like sequence generation [69, 70], tree search methods [71–73], and hybrids of DL with genetic programming [74, 75]. Applications of SR are diverse, ranging from discovering equations for PDEs using genetic programming [76] to the innovation of replacing neural network activation functions with primitive functions [77] and employing transformers for sequence-to-sequence equation generation [78].

A significant advancement in the field was marked by the introduction of PISNs by Majumdar et al. [29]. Inspired by the work on differentiable programs [79], PISN utilizes a context-free grammar to approximate production rules linearly, enhancing the ability to generate analytical expressions for PDEs. This approach not only streamlines the derivation of mathematical models for complex systems but also aids in the practical application of these models, such as modeling the thermal profiles of APHs. This section further explores the intricacies of PISNs, elaborating on the differentiable program architecture (DPA) that underpins symbolic regression and its practical applications to temperature profile expression estimation for APH.

6.3.3.1 Physics-Informed Symbolic Networks

Physics-informed symbolic networks (PISNs) integrate physics-informed methodologies with the capabilities of symbolic regression. This integration aims to solve complex systems of differential equations, particularly PDEs by embedding inherent physical laws into the learning mechanisms of neural networks. The objective is not only to predict but also to ensure that predictions are consistent with the underlying physical principles, thus enhancing the models' reliability and interpretability. PISNs comprise symbolic networks designed to discover and represent mathematical expressions that adhere to a specified grammar of algebraic operations, such as sine, exponential, addition, and multiplication. These expressions are dynamically structured and refined during the training process to best encapsulate the target physical phenomena. The approach leverages a context-free grammar that outlines the potential forms expressions might take, facilitating a systematic exploration of the symbolic space.

Mathematical formulation of PISNs is described below:

Context-free grammar, $\mathbb{G}$ consists of symbolic expressions defined by Eqn. 36, where x,y,t,c are terminal symbols. A continuous relaxation of $\mathbb{G}$ is achieved by taking a weighted sum over all the production rules as illustrated in Eqn. 37.

Grammar Definition

$$\alpha ::= \sin(\alpha_1) \mid \exp(\alpha_1) \mid \mathit{Add}(\alpha_1, \alpha_2) \mid \mathit{Multiply}(\alpha_1, \alpha_2) \mid x \mid y \mid t \mid c \tag{6.36}$$

$$\begin{aligned} \alpha ::= {} & w_1 \sin(\alpha_1) + w_2 \exp(\alpha_2) + w_3 (\alpha_3 + \alpha_4) + w_4 (\alpha_5 \times \alpha_6) \\ & + w_5 x + w_6 y + w_7 t + w_8 c \end{aligned} \tag{6.37}$$

Network Architecture

Eqn. 37 facilitates the induction of a symbolic network with a depth $d+2$, where d represents the total number of hidden layers. The network configuration is further detailed as below:

Input and weight configuration

- Input $\boldsymbol{x} = [x, y, t, 1]$
- First hidden layer weights $\boldsymbol{W_k^0} = \left[w_{k1}^0, w_{k2}^0, w_{k3}^0, w_{k4}^0\right]$
- Subsequent hidden layer weights
  $\boldsymbol{W_k^j} = \left[w_{k1}^j, w_{k2}^j, w_{k3}^j, w_{k4}^j, w_{k5}^j, w_{k6}^j, w_{k7}^j, w_{k8}^j\right]$
- Output layer weights $\boldsymbol{W}^d$

Layer Configuration

- Output of the network $u_{pisn} = \boldsymbol{W}^d \cdot \boldsymbol{h}^d$ (6.38a)
- First hidden layer $\boldsymbol{h}^0 = \Big[\sin\left(\boldsymbol{W}_1^0 . \boldsymbol{x}\right), \exp\left(\boldsymbol{W}_2^0 \cdot \boldsymbol{x}\right), \boldsymbol{W}_3^0 \cdot \boldsymbol{x}$
  $+ \boldsymbol{W}_4^0 \cdot \boldsymbol{x}, \boldsymbol{W}_5^0 . \boldsymbol{x} \times \boldsymbol{W}_6^0 . \boldsymbol{x}, \boldsymbol{x}, y, t\Big]$ (6.38b)
- Subsequent Hidden layers $\boldsymbol{h}^{j+1} = \Big[\sin\left(\boldsymbol{W}_1^j . h^j\right), \exp\left(\boldsymbol{W}_2^j \cdot h^j\right), \boldsymbol{W}_3^j \cdot \boldsymbol{h}^j$
  $+ \boldsymbol{W}_4^j \cdot \boldsymbol{h}^j, \boldsymbol{W}_5^j . \boldsymbol{h}^j \times \boldsymbol{W}_6^j . \boldsymbol{h}^j, x, y, t\Big]$
  (6.38c)

Majumdar et al. [18] designed a network, as described by Eqn. 38, that ensures a linear increase in parameter count with network depth. This approach sacrifices the capacity to perfectly represent wider expressions, potentially limiting the network's ability to handle complex symbolic forms. This design was chosen over the DPA based on prior studies [79], which, although capable of exact expression representation, result in an exponential increase in parameters with increasing network depth. It was reported that, while PISNs performed on par with PINNs, the analytical expressions generated were large and weren't interpretable.

6.3.3.2 Pruned Differentiable Programs

In a later work by Majumdar et al. [34], it was demonstrated that symbolic regression, when coupled with DPAs and executed with pruning in a depth-first manner, results in sparse and easily interpretable expressions. This approach enhances the interpretability of mathematical models derived from complex systems, particularly those described by PDEs. The methodology employed by Majumdar et al. [34] involves a sequential two-step process (as illustrated in Algorithm # 1 (Figure 6.10) and #2,

Algorithm 1 Symbolic Regression for Partial Differential Equation

1: Train a PDE solver to solve for the PDE
2: Generate input-output data points using the learned PDE solver
3: Perform Regression on generated data points using Differentiable Program Architecture
4: Prune the Differentiable Program Architecture

FIGURE 6.10 Symbolic regression for partial differential equations with pruned differentiable program architecture.

Algorithm 2 Depth First Search Pruning Strategy of Differentiable Program Architecture (DPA)

```
Global initialize W ← Unpruned DPA weights, loss ← SCORE (W)
Function DFS (node)
Initialize visited ← node, children ← children of node sorted by absolute value
if children is None then
    W′ ← PRUNE (node)
    W′ ← FINETUNE (W′)
    finetuned-loss ← SCORE (W′)
    if finetuned-loss ≤ loss then
        loss ← finetuned-loss
        W ← W′
    end if
    Return W,loss
end if
for all child in children do
    if child not in visited then
        DFS(child)
    end if
end for
```

FIGURE 6.11 Depth first search pruning strategy for differentiable program architecture.

(Figure 6.11)), beginning with the solution of a given PDE setup using a PINN. This choice of solver is not restrictive, any suitable numerical PDE solver can be utilized depending on the specific requirements of the problem.

Following the data generation, symbolic regression is implemented using a DPA. This architecture is constructed based on a context-free grammar, which includes unary operators such as sine, exponential, logarithmic functions and power functions, as well as binary operators such as addition and multiplication (as shown in Eqn. 39). An illustrative example of this expansion process to a depth of two is provided here, which showcases the potential complexity and reach of the architecture through basic operations and terminal symbols (*x*, *y*, *t*, *c*) (refer to Figure 6.12).

$$\alpha ::= \sin(\alpha_1) \; |exp(\alpha_1)| \log\alpha_1 | \; pow2\ \alpha_1 | pow3\ \alpha + 1 \; | \; Add(\alpha_1, \alpha_2)| \\ Multiply(\alpha_1, \alpha_2)| \; x \,|\, y \,|\, t \,|\, c \tag{6.39}$$

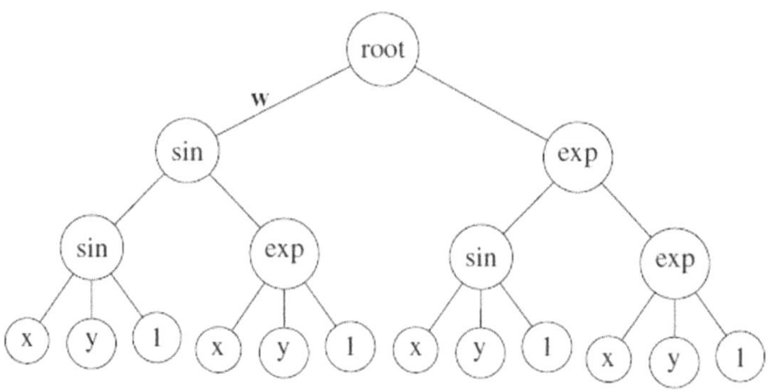

FIGURE 6.12 Symbolic regression with pruned differentiable program architecture.

TABLE 6.5
Comparison of Fluid and Metal Matrix Temperature Predictions Obtained Using Domain Decomposed PINN and Unpruned DPA and Pruned DPA Architecture

	PINN	DPA-Unpruned	DPA-Pruned
T_g	2.03	2.14	2.06
T_{mg}	2.53	2.55	2.54
T_{a1}	3.08	3.27	3.09
T_{ma1}	2.81	3.02	2.85
T_{a2}	2.97	3.1	3
T_{ma2}	3.02	3.08	3.04

Table 6.5 showcases the comparative performance of the DPA in the context of APH temperature profile predictions, assessing the efficacy of the pruned versus unpruned models against a trained PINN. The evaluation is anchored on the L2 relative error metrics derived from numerical simulations using the finite difference method ([45]), serving as the ground truth for temperature distribution. The results, presented in Table 6.5 indicate nuanced differences in performance across the various models:

- *PINN*: Shows consistent baseline performance across all temperature variables.
- *Unpruned DPA*: Generally exhibits slightly higher error rates, suggesting some overfitting or model complexity.
- *Pruned DPA*: Demonstrates error rates that are comparable and occasionally superior to the PINN, reflecting the effectiveness of the pruning process in enhancing model accuracy and simplicity.

The pruned expressions derived for the APH temperature profile (shown in Figure 6.13) are illustrative of the simplification achieved through the pruning process.

$$T_g = 0.99sin(0.22z + 0.83e^{-0.21z+0.61\phi+0.66})$$

$$T_{mg} = (0.52sin(1.56\phi) + 1.18)(0.83e^{-0.44\phi} + 0.18z\phi - 0.23\phi^2 - 0.27\phi)$$

$$T_{a1} = (-0.11z + 0.77\phi - 1.59 + 0.27(z\phi))(0.03z + 0.22\theta - 0.60 + 0.37\phi^2)$$

$$T_{ma1} = (-1.14sin(0.07z + 0.88\phi - 1.06))(0.06z + 0.69\phi + 0.97)$$

$$T_{a2} = 0.86sin(0.34z + 1.26\phi + 0.55)(-0.10z - 1.02\phi)(0.11z + 1.10\phi) + 0.96$$

$$T_{ma2} = (0.13z + 1.30sin(1.08\phi) + 1.25)(0.08z - 0.68\phi + 0.77)$$

FIGURE 6.13 Symbolic expressions obtained with pruned differentiable program architecture for governing equations of APH.

6.4 SUMMARY AND OUTLOOK

Mathematical models are useful for the design and operation of manufacturing processes. Physics-based models conform to the physical laws governing the process, but they are computationally intensive, require accurate input data and face convergence issues. They are more suitable for design, simulation and off-line optimization. Data-driven models, on the other hand, are more suitable for manufacturing operations and take advantage of the availability of real-time sensor data. However, their performance depends heavily on the quality and availability of relevant data. PIDL represents a hybrid approach that combines physics-based models with DNNs.

Various PIDL approaches such as PINNs, domain decomposition based PINNs, transfer learning, hypernetworks, symbolic regression with DPAs, PISNs were described in this chapter. Their application for an APH, an important heat exchanger in thermal power plants, was demonstrated with data from an industrial unit. In order to evaluate their performance, a rigorous CFD model for heat transfer in an APH was also presented and results of the PIDL models were compared with the CFD model. The predictions of various PIDL models were fairly close to the numerical simulation results obtained with the CFD model. Expectedly, the inference times of the PIDL models were much lower than the simulation time required for the CFD model.

A domain decomposition-based PINN was developed to capture thermal phenomena within an APH, using a segmented approach where different subdomains were modelled using individual DNNs that are unified through a common loss function which incorporates both governing differential equations and continuity constraints at the subdomain interfaces. Transfer learning significantly enhances the model's responsiveness to dynamic boundary conditions. It not only provides rapid predictions compared to traditional numerical solvers but also can be integrated seamlessly into digital twin technologies for continuous system performance optimization.

Hypernetworks facilitate the learning of physical behavior without the need for retraining when boundary conditions change, offering substantial improvements in inference times while maintaining accuracy. The application of domain-decomposed, hypernetwork-based PINNs to complex industrial heat exchangers like the APH demonstrates the potential of these models to enhance real-time monitoring within industrial settings. Such advancements underscore the adaptability of PINNs to the rigorous demands of industrial applications, paving the way for broader adoption in digital twin environments.

Furthermore, the application of PISNs to discover analytical expressions for solutions to PDEs was attempted and preliminary results were presented. Exploration of DPA highlights the innovative use of pruning strategies to simplify the expressions generated by DPAs. Although the current pruning approach, which is based on a depth-first search, may be sub-optimal due to its greedy nature, it provides a foundational technique for reducing model complexity. This strategy underscores the ongoing need to refine computational techniques to achieve more concise and computationally efficient models, essential for real-time applications.

Despite the power of advanced models such as PINNs and hypernetwork-based PINNs, applying them for industrial processes is still expertise-intensive (both domain and DL knowledge). Future research should focus on simplifying and automating the application of these techniques for complex engineering systems. Special attention should be given to reducing the computational footprint of these models by intelligent architecture search. With rapid advances in the field of DL, it is imperative that the power of advanced techniques, such as transformers and neural operators, is explored to improve the modeling of industrial processes.

It is expected that AI techniques that combine the advantages of physics and data science will continue to advance. Neural network models for entire industrial systems, e.g., a complete power plant or a pharmaceutical manufacturing plant, will be desirable. However, several challenges will have to be solved to achieve such a goal. The scale of the models will be large and hence modularity as well as composition of trained neural network modules will be important. Hypernetworks are required to make a high-dimensional prediction and to overcome this drawback, parameter efficient representation of model adaptation (e.g., using low-rank adaptation [61]) will have to be explored. Foundational models, and possibly instruction tuned hypernetworks, which require only description of the problem as input, are also likely to evolve. The search for explainable models will advance beyond just regression and classification as LLMs are likely to play an important role in future.

6.5 LIST OF SYMBOLS

d_i	inner diameter of the APH
d_o	outer diameter of the APH
H	height of the APH
MSE_{bcp}	mean squared error associated with boundary conditions
MSE_{icp}	mean squared error associated with interface equality constraints
$MSE_{\mathcal{F}_p}$	mean squared error of the residuals of the governing equations
MSE_{grad_p}	mean squared error associated with gradient constraints
N_{bc_p}	number of boundary condition points in subdomain p
N_{ic_p}	number of interface condition points in subdomain p
$N_{\mathcal{F}_p}$	number of collocation points in subdomain p for the residual $\mathcal{F}$
N_{grad_p}	number of metallic packing temperature gradient condition points in subdomain p
$\mathcal{N}_{\Theta_h}(\lambda)$	hypernetwork mapping λ to PINN weights
T_f	fluid temperature (air or gas)
T_m	metallic packing temperature
$\left(\phi_{bcp}^{i}, z_{bcp}^{i}\right)$	coordinates of the boundary condition points in subdomain p
$\left(\phi_{icp}^{i}, z_{icp}^{i}\right)$	coordinates of the common interface points of subdomains p and p^+
$\left(\phi_{\mathcal{F}p}^{i}, z_{\mathcal{F}p}^{i}\right)$	coordinates of the residual points in subdomain p
$\left(\phi_{grad_p}^{i}, z_{grad_p}^{i}\right)$	coordinates of the gradient condition points in subdomain p
$\mathbf{u}$	thermal profile of the heat exchanger
$u_{\Theta\lambda_{\text{new}}}$	thermal profile predicted by the base PINN for λ_{new}
z	axial coordinate
$\mathbb{G}$	context free grammar
x,y,t,c	terminal symbols

Greek Symbols

β_{a1}	sector angle for the flow of primary air
β_{a2}	sector angle for the flow of secondary air
β_g	sector angle for the flow of flue gas
λ	set of operating conditions
λ_{new}	new set of operating conditions
$\mathcal{F}$	residual of the governing partial differential equations
Ω	computational domain
Ω_{a1}	primary air subdomain of the computational domain
Ω_{a2}	secondary air subdomain of the computational domain
Ω_g	gas subdomain of the computational domain
Θ	parameters of domain decomposed PINNs $u_\Theta(x;\lambda)$
$\Theta_{\lambda new}$	parameters of the PINN for λ_{new}
Θ_h	parameters of the hypernetwork $\mathcal{N}_{\Theta h}(\lambda)$
ϕ	tangential coordinate
ω	rotational speed of the metallic packing (in revolutions per minute)

Subscripts

$a1$	primary air
$a2$	secondary air
g	gas
m	metallic matrix packing
m_{a1}	metallic matrix packing within the primary air sector
m_{a2}	metallic packing within the secondary air sector
m_g	metallic packing within the gas sector
p	subdomain in domain Ω
p^+	neighboring subdomain to subdomain p
bc_p	boundary condition in subdomain p
$grad_p$	gradient condition in subdomain p
ic_p	interface condition in subdomain p
$\mathcal{F}_p$	residual of the governing PDEs in subdomain p

ACKNOWLEDGMENTS

The authors thank the management of Tata Consultancy Services Limited to publish this article and Dr. Harrick Vin, Mr. K. Ananth Krishnan and Dr. Gautam Shroff for their encouragement and support.

REFERENCES

1. K. Jambunathan, S. Hartle, S. Ashforth-Frost and V. Fontama, "Evaluating convective heat transfer coefficients using neural networks," *International Journal of Heat and Mass Transfer*, vol. 39, no. 11, pp. 2329–2332, 1996.
2. E. Fonda, A. Pandey, J. Schumacher and K. R. Sreenivasan, "Deep learning in turbulent convection networks," *Proceedings of the National Academy of Sciences*, vol. 116, no. 18, pp. 8667–8672, 2019.

3. Y. Li, H. Wang and X. Deng, "Image-based reconstruction for a 3D-PFHS heat transfer problem by ReConNN," *International Journal of Heat and Mass Transfer*, vol. 134, pp. 656–667, 2019.
4. M. Edalatifar, M. B. Tavakoli, M. Ghalambaz and F. Setoudeh, "Using deep learning to learn physics of conduction heat transfer," *Journal of Thermal Analysis and Calorimetry*, vol. 146, pp. 1435–1452, 2021.
5. J. Kim and C. Lee, "Prediction of turbulent heat transfer using convolutional neural networks," *Journal of Fluid Mechanics,* vol. 882, p. A18, 2020.
6. G. Beintema, A. Corbetta, L. Biferale and F. Toschi, "Controlling Rayleigh–Bénard convection via reinforcement learning," *Journal of Turbulence*, vol. 21, pp. 585–605, 2020.
7. E. Hachem, H. Ghraieb, J. Viquerat, A. Larcher and P. Meliga, "Deep reinforcement learning for the control of conjugate heat transfer," *Journal of Computational Physics*, vol. 436, p. 110317, 2021.
8. S. H. Nistala, K. Singh, P. Vakkantham, A. Pareek, V. Masampally, R. Kumar and V. Runkana, "Virtual Sinter®: Digital Twin for Integrated Sinter Plants," *Transactions of the Indian National Academy of Engineering*, vol. 7, no. 3, pp. 943–955, 2022.
9. V. S. Masampally, A. Pareek, N. R. V. Nadimpalli and V. Runkana, "Artificial Intelligence for Monitoring and Optimization of an Integrated Mineral Processing Plant," *Transactions of the Indian Institute of Metals,* Vol. 77, 4231–4240, 2024.
10. B. Deivendran, V. S. Masampally, N. R. V. Nadimpalli and V. Runkana, "Virtual metrology for chemical mechanical planarization of semiconductor wafers," *Journal of Intelligent Manufacturing*, Vol. 36, pp. 1923–1942, 2025.
11. V. Jadhav, A. Deodhar, A. Gupta and V. Runkana, "Physics informed neural network for health monitoring of an air preheater," in *PHM Society European Conference*, Italy, 2022.
12. M. Raissi, P. Perdikaris and G. E. Karniadakis, "Physics-informed neural networks: A deep learning framework for solving forward and inverse problems involving nonlinear partial differential equations," *Journal of Computational physics,* vol. 378, pp. 686–707, 2019.
13. S. Cuomo, V. S. Di Cola, F. Giampaolo, G. Rozza, M. Raissi and F. Piccialli, "Scientific machine learning through physics–informed neural networks: Where we are and what's next," *Journal of Scientific Computing*, vol. 92, p. 88, 2022.
14. S. Cai, Z. Mao, Z. Wang, M. Yin and G. E. Karniadakis, "Physics-informed neural networks (PINNs) for fluid mechanics: A review," *Acta Mechanica Sinica*, vol. 12, pp. 1727–1738, 2021.
15. S. Cai, Z. Wang, S. Wang, P. Perdikaris and G. E. Karniadakis, "Physics-informed neural networks for heat transfer problems," *American Society of Mechanical Engineers*, vol. 143, p. 060801, 2021.
16. A. D. Jagtap, E. Kharazmi and G. E. Karniadakis, "Conservative physics-informed neural networks on discrete domains for conservation laws: Applications to forward and inverse problems," *Computer Methods in Applied Mechanics and Engineering*, vol. 365, p. 113028, 2020.
17. K. Shukla, A. D. Jagtap and G. E. Karniadakis, "Parallel physics-informed neural networks via domain decomposition," *Journal of Computational Physics*, vol. 447, p. 110683, 2021.
18. B. Moseley, A. Markham and T. Nissen-Meyer, "Finite basis physics-informed neural networks (FBPINNs): a scalable domain decomposition approach for solving differential equations," *Advances in Computational Mathematics*, Vol. 49, article number 62, 2023.
19. H. Wang, R. Planas, A. Chandramowlishwaran and R. Bostanabad, "Mosaic flows: A transferable deep learning framework for solving PDEs on unseen domains," *Computer Methods in Applied Mechanics and Engineering*, vol. 389, p. 114424, 2022.
20. S. Chakraborty, "Transfer learning based multi-fidelity physics informed deep neural network," *Journal of Computational Physics*, vol. 426, p. 109942, 2021.

21. S. Desai, M. Mattheakis, H. Joy, P. Protopapas and S. Roberts, "One-Shot Transfer Learning of Physics-Informed Neural Networks," in *arXiv*, 2022.
22. M. Penwarden, S. Zhe, A. Narayan and R. M. Kirby, "A metalearning approach for Physics-Informed Neural Networks (PINNs): Application to parameterized PDEs," *Journal of Computational Physics*, vol. 477, p. 111912, 2023.
23. F. d. A. Belbute-Peres, Y.-F. Chen and F. Sha, "HyperPINN: Learning parameterized differential equations with physics-informed hypernetworks," https://arxiv.org/abs/2111.01008, 2021.
24. D. Ha, A. Dai and Q. V. Le, "Hypernetworks," *arXiv preprint arXiv:1609.09106,* 2016.
25. R. Majumdar, V. Jadhav, A. Deodhar, S. Karande, L. Vig and V. Runkana, "HxPINN: A hypernetwork-based physics-informed neural network for real-time monitoring of an industrial heat exchanger," *Numerical Heat Transfer, Part B: Fundamentals*, pp. 1–22, 2024.
26. Z. Li, H. Zheng, N. Kovachki, D. Jin, H. Chen, B. Liu, K. Azizzadenesheli and A. Anandkumar, "Physics-informed neural operator for learning partial differential equations," *ACM/JMS Journal of Data Science*, 2021.
27. R. Majumdar, A. Varhade, S. Karande and L. Vig, "Can Physics informed Neural Operators Self Improve?," in *The Symbiosis of Deep Learning and Differential Equations III*, arXiv preprint arXiv:2311.13885, 2023.
28. R. Majumdar, S. Karande and L. Vig, "How important are specialized transforms in Neural Operators?," in *1st Workshop on the Synergy of Scientific and Machine Learning Modeling@ ICML2023*, 2023.
29. R. Majumdar, V. Jadhav, A. Deodhar, S. Karande, L. Vig and V. Runkana, "Physics Informed Symbolic Networks," *arXiv preprint arXiv:2207.06240,* 2022.
30. R. Majumdar, V. S. Jadhav, A. Deodhar, S. Karande, L. Vig and V. Runkana, "Symbolic Regression for PDEs using Pruned Differentiable Programs," in *ICLR 2023 Workshop on Physics for Machine Learning*, 2023.
31. H. Müller-Steinhagen, M. R. Malayeri and A. P. Watkinson, "Heat Exchanger Fouling: Environmental Impacts," *Heat Transfer Engineering*, vol. 30, no. 10–11, pp. 773–776, 2009.
32. S. Yuanhao, W. Jingcheng and W. Jie, "On-line monitoring of ash fouling for air pre-heaters in coal-fired power plant boiler," in *29th Chinese Control and Decision Conference (CCDC)*, 2017.
33. C. Zhou, L. Zhang, Y. Deng and S.-C. Ma, "Research progress on ammonium bisulfate formation and control in the process of selective catalytic reduction," *Environmental Progress and Sustainable Energy*, vol. 35, no. 6, pp. 1664–1672, 2016.
34. J. Guffre, "Eliminating air heater plugging and corrosion caused by SCR/SNCR systems for NOx control on coal-fired boilers.," *Power Engineering*, vol. 111, no. 10, pp. 84–87, 2007.
35. L. Muzio, S. Bogseth, R. Himes, Y.-C. Chien and D. Dunn-Rankin, "Ammonium bisulfate formation and reduced load SCR operation," *Fuel*, vol. 206, pp. 180–189, 2017.
36. A. Gupta, V. Jadhav, M. Patil, A. Deodhar and V. Runkana, "Forecasting of Fouling in Air Pre-Heaters Through Deep Learning," in *ASME Power Conference*, 2021.
37. A. Heidari-Kaydan, E. Hajidavalloo and S. Mehrzad, "Three-Dimensional Simulation of Leakages in Rotary Air Preheater of Steam Power Plant," *Heat Transfer Engineering*, vol. 43, no. 17, pp. 1512–1528, 2022.
38. B. Drobnič, J. Oman and M. Tuma, "A numerical model for the analyses of heat transfer and leakages in a rotary air preheater," *International journal of heat and mass transfer*, vol. 49, pp. 5001–5009, 2006.
39. Y. Bu, L. Wang, X. Chen, X. Wei, L. Deng and D. Che, "Numerical analysis of ABS deposition and corrosion on a rotary air preheater," *Applied Thermal Engineering*, vol. 131, pp. 669–677, 2018.
40. M. Z. A. Khan, M. Aziz and A. T. Wijayanta, "Prediction of heat transfer enhancement of delta-wing tape inserts using artificial neural network," *Case Studies in Thermal Engineering*, vol. 27, p. 101322, 2021.

41. M. Z. A. Khan, H. A. Khan and M. Aziz, "Performance optimization of heat-exchanger with delta-wing tape inserts using machine learning," *Applied Thermal Engineering*, vol. 216, p. 119135, 2022.
42. A. Gupta, V. Jadhav, A. Deodhar and V. Runkana, "Physics-Assisted Long-Short-Term-Memory Network for Forecasting of Fouling in a Regenerative Heat Exchanger," in *Heat Transfer Summer Conference*, 2022.
43. S. Sundar, M. C. Rajagopal, H. Zhao, G. Kuntumalla, Y. Meng, H. C. Chang, C. Shao, P. Ferreira, N. Miljkovic, S. Sinha and others, "Fouling modeling and prediction approach for heat exchangers using deep learning," *International Journal of Heat and Mass Transfer*, vol. 159, p. 120112, 2020.
44. A. Karpatne, G. Atluri, J. H. Faghmous, M. Steinbach, A. Banerjee, A. Ganguly, S. Shekhar, N. Samatova and V. Kumar, "Theory-guided data science: A new paradigm for scientific discovery from data," *IEEE Transactions on knowledge and data engineering*, vol. 29, no. 10, pp. 2318–2331, 2017.
45. C.-H. Li, "A Numerical Finite Difference Method for Performance Evaluation of a Periodic-Flow Heat Exchanger," *Journal of Heat Transfer*, vol. 105, no. 3, pp. 611–617, 1983.
46. T. Skiepko, "The effect of matrix longitudinal heat conduction on the temperature fields in the rotary heat exchanger," *International journal of heat and mass transfer*, vol. 31, no. 11, pp. 2227–2238, 1988.
47. J. Burke and K. L. Johnson, "Ammonium sulfate and bisulfate formation in air preheaters. Final report Oct 80-Oct 81," Radian Corp., Austin, TX (USA), 1982.
48. L. Wang, Y. Bu, D. Li, C. Tang and D. Che, "Single and multi-objective optimizations of rotary regenerative air preheater for coal-fired power plant considering the ammonium bisulfate deposition," *International Journal of Thermal Sciences*, vol. 136, pp. 52–59, 2019.
49. A. Heidari-Kaydan and E. Hajidavalloo, "Three-dimensional simulation of rotary air preheater in steam power plant," *Applied Thermal Engineering*, vol. 73, no. 1, pp. 399–407, 2014.
50. I. Lagaris, A. Likas and D. Fotiadis, "Artificial neural networks for solving ordinary and partial differential equations," *IEEE Transactions on Neural Networks,* vol. 9, no. 5, pp. 987–1000, 1998.
51. A. G. Baydin, B. A. Pearlmutter, A. A. Radul and J. M. Siskind, "Automatic differentiation in machine learning: a survey," in *arXiv*, 2018.
52. A. D. Jagtap and G. E. Karniadakis, "Extended physics-informed neural networks (xpinns): A generalized space-time domain decomposition based deep learning framework for nonlinear partial differential equations," *Communications in Computational Physics*, vol. 28, no. 5, pp. 2002–2041, 2020.
53. R. Pellegrin, B. Bullwinkel, M. Mattheakis and P. Protopapas, "Transfer Learning with Physics-Informed Neural Networks for Efficient Simulation of Branched Flows," in *arXiv*, 2022.
54. Y. Gao, K. C. Cheung and M. K. Ng, "SVD-PINNs: Transfer Learning of Physics-Informed Neural Networks via Singular Value Decomposition," in *2022 IEEE Symposium Series on Computational Intelligence (SSCI)*, 2022, pp. 1443–1450.
55. T. Galanti and L. Wolf, "On the Modularity of Hypernetworks," 2020.
56. V. K. Chauhan, J. Zhou, P. Lu, S. Molaei and D. A. Clifton, "A Brief Review of Hypernetworks in Deep Learning," 2023.
57. E. J. Hu, Y. Shen, P. Wallis, Z. Allen-Zhu, Y. Li, S. Wang, L. Wang and W. Chen, "LoRA: Low-Rank Adaptation of Large Language Models," in *arXiv*, 2021.
58. X. L. Li and P. Liang, "Prefix-Tuning: Optimizing Continuous Prompts for Generation," in *arXiv*, 2021.
59. N. Ruiz, Y. Li, V. Jampani, W. Wei, T. Hou, Y. Pritch, N. Wadhwa, M. Rubinstein and K. Aberman, "HyperDreamBooth: HyperNetworks for Fast Personalization of Text-to-Image Models," in *arXiv*, 2023.

60. H. Ivison, A. Bhagia, Y. Wang, H. Hajishirzi and M. Peters, "HINT: Hypernetwork Instruction Tuning for Efficient Zero- & Few-Shot Generalisation," in *arXiv*, 2023.
61. R. Majumdar, V. Jadhav, A. Deodhar, S. Karande, L. Vig and V. Runkana, "HyperLoRA for PDEs," in *arXiv preprint arXiv:2308.09290,* 2023.
62. Y. M. El Hasadi and J. T. Padding, "Solving fluid flow problems using semi-supervised symbolic regression on sparse data," *AIP Advances*, vol. 11, no. 9, 2019.
63. M. Ansari, H. A. Gandhi, D. G. Foster and A. D. White, "Iterative symbolic regression for learning transport equations," *AIChE Journal*, vol. 68, no. 6, 2022.
64. K. Danai and W. G. La Cava, "Controller design by symbolic regression," *Mechanical Systems and Signal Processing*, vol. 151, p. 107348, 2021.
65. H. Vaddireddy, A. Rasheed, A. E. Staples and O. San, "Feature engineering and symbolic regression methods for detecting hidden physics from sparse sensor observation data," *Physics of Fluid*, vol. 32, no. 1, 2020.
66. M. Mahsa, "A Study on Financial Time Series Forecasting and Symbolic Regression by means of a Hybrid Probabilistic Model-Building Cartesian Genetic Programming Methodology," Toronto Metropolitan University, 2013.
67. M. Virgolin and S. P. Pissis, "Symbolic Regression is NP-hard," in *arXiv*, 2022.
68. B. He, Q. Lu, Q. Yang, J. Luo and Z. Wang, "Taylor genetic programming for symbolic regression," in *Proceedings of the Genetic and Evolutionary Computation Conference*, New York, NY, USA, 2022.
69. M. Valipour, B. You, M. Panju and A. Ghodsi, "SymbolicGPT: A Generative Transformer Model for Symbolic Regression," *CoRR,* vol. abs/2106.14131, 2021.
70. J. Devlin, J. Uesato, S. Bhupatiraju, R. Singh, A.-r. Mohamed and P. Kohli, "RobustFill: Neural Program Learning under Noisy I/O," *CoRR,* vol. abs/1703.07469, 2017.
71. L. Li, M. Fan, R. Singh and P. Riley, "Neural-Guided Symbolic Regression with Semantic Prior," *CoRR,* vol. abs/1901.07714, 2019.
72. S.-M. Udrescu and M. Tegmark, "AI Feynman: a Physics-Inspired Method for Symbolic Regression," 2019.
73. S.-M. Udrescu, A. K. Tan, J. Feng, O. Neto, T. Wu and M. Tegmark, "AI Feynman 2.0: Pareto-optimal symbolic regression exploiting graph," 2020.
74. B. K. Petersen, M. L. Larma, T. N. Mundhenk, C. P. Santiago, S. K. Kim and J. T. Kim, "Deep symbolic regression: Recovering mathematical expressions from data via risk-seeking policy gradients," in *International Conference on Learning Representations*, 2021.
75. D. A. Abolafia, M. Norouzi and Q. V. Le, "Neural Program Synthesis with Priority Queue Training," 2018.
76. M. Maslyaev, A. Hvatov and A. Kalyuzhnaya, "Data-Driven Partial Derivative Equations Discovery with Evolutionary Approach," in *Computational Science – ICCS 2019*, Springer International Publishing, 2019, p. 635–641.
77. M. Zhang, S. Kim, P. Y. Lu and M. Soljačić, "Deep Learning and Symbolic Regression for Discovering Parametric Equations," *IEEE Transactions on Neural Networks and Learning Systems,* 2023.
78. G. Lample and F. Charton, "Deep Learning for Symbolic Mathematics," in *arXiv*, 2019.
79. C. Guofeng and H. Zhu, "Differentiable Synthesis of Program Architectures," in *Advances in Neural Information Processing Systems*, 2021.

7 AI-Based Analysis for Optimizing Radiative Jeffery–Hamel Flow for Cross-Diffusion Effects

A Physics-Informed Machine Learning

Muhammad Naeem Aslam, Nadeem Shaukat, and Arshad Riaz

7.1 INTRODUCTION

The behavior of fluid flow in Jeffery–Hamel (JH) problem specifically in converging and diverging (CD) configurations pertains to the passage of fluids within channels that either converge or diverge. This fluid flow pattern is used in numerous scientific and engineering systems, and comprehending its characteristics is essential for numerous applications in different scientific era. The investigation of CD fluid flow involves studying the fluid dynamics for velocity, temperature, and concentration properties in these channels. CD JH fluid flows play an important role in the best designing of nozzles. The JH fluid flow phenomena in both channels were firstly introduced by Jeffery [1] and Hamel [2]. This flow laminar or turbulence pattern holds significance in various scientific fields including fluid mechanics, mechanical engineering, chemical engineering, and bio-mechanical engineering. JH flow finds applications in various areas such as chemical vapor deposition reactors [3, 4], high-current arcs in plasma generators, industrial machines [5, 6], gas compressors and pipe sections [7]. The JH flow has garnered considerable attention nowadays by using machine learning algorithms because the governing Navier–Stokes equations have no exact solution.

Sobamowo et al. [8] examined the magnetohydrodynamic (MHD) flow characteristics of an upper convective Maxwell viscoelastic nanofluid via a permeable medium. The broader significance of MHD is evident in numerous scientific applications. Sucharitha et al. [9] explored the utilization of MHD in nanomaterial flow within an aligned channel, with a particular emphasis on Joule heating scenario. Extending this exploration, Sucharitha et al. [10] analyzed on the impact of MHD on Jeffery nanofluid peristaltic flow via a flexible channel. In addition, Reddy et al.

DOI: 10.1201/9781032688121-7

[11] carried out an analysis of double diffusion features in nanofluid flow by including a magnetized field. Ullah et al. [12] studied the effects of Lorentz forces on magnetized nanofluids and Coriolis forces through a porous space. Reddy et al. [13] further focused to the discourse by examining the flow characteristics of radiation Williamson nanomaterial utilizing MHD.

Furthermore, both CD channels hold significance across various thermal and flow era. Barzegar Gerdroodbary et al. [14] investigated on the influence of thermal radiation on these CD channels to comprehend its impact on thermal and configurations under numerous flow conditions. Yarmand et al. [15] analyzed a hybrid nanofluid comprising activated carbon and grapheme though in CD channels. In addition, Makinde examined fluid flow irreversibility in CD channels with viscosity and non-uniform temperatures [16], while Beǵ and Makinde [17] delved into the inherent irreversibility in non-uniform channels.

Rosenhead [18] initiated early investigations into the traditional JH flow, formulating the solution using Jacobian elliptic functions. Millsaps and Pohlhausen [19], along with Riley [20], obtained exact solutions for the energy equation when incorporating heat transfer within the JH flow. In a broader asymmetric context, Fraenkel [21] delved into the existence of multiple solutions. Terrill [22] demonstrated the behavior of slow laminar flow in a CD channel, considering suction at one wall and blowing at the other. Schlichting [23] briefly explored these issues. Roy and Nayak [24] investigated continuous two-dimensional incompressible laminar viscoelastic flow in CD channels, accounting for suction and injection. The three-dimensional extension of the JH flow was thoroughly explored and realized in reference [25]. Numerous experiments have been conducted to gain insights into the physical mechanisms underlying the JH flow and its hydrodynamic stability properties.

At present, various methods such as the least square method (LSM), differential transform method (DTM), weighted-residual method (WRM), and homotopy analysis method (HAM) have been employed for investigating MHD JH flow [26, 27]. Researchers have explored the significance of fluid in CD channels in separate studies [28–31]. Ullah et al. [32] conducted a study examining the influence of chemical reactive and Soret/Dufour characteristics on MHD JH flow through CD channels. The investigation includes the incorporation of radiation, Lorentz forces, and heat source effects to analyze thermal aspects in the CD channel. In addition, cross-diffusion aspects arising from concentration and temperature profiles are considered. The system of transformed differential equations (DEs) solves using NDSolve.

In recent years, researcher tackle to find the solution of JH fluid flow in both channels that converge and diverge using machine learning techniques. If the system of DEs becomes highly nonlinear, then we can't find the solution using traditional analytical or numerical methods to address this complexity. In such cases, researchers applied artificial neural networks (ANNs) for solution of these complexities [33]. Raja and Samar [34] studied the JH problem using unsupervised neural networks integrated with interior point optimization algorithm, and compared the obtained results with other well-established techniques for validation of the ANNs prediction. Chandra and Das [35] investigated the JH problem using machine learning technique merged with ANNs with two meta-heuristic optimization hybrid techniques in CD channels with copper nanoparticles. Aslam et al. [36] studied the MHD JH flow problem with Soret and Dufour effect using ANNs with artificial bee colony hybrid with recently introduced neural networks algorithm (NNA) and compare their results with Keller box method for validation of the neural networks prediction.

Recent advancements in global optimization techniques especially stochastic optimization algorithms, particularly within the domain of artificial intelligence, have unlocked new possibilities for tackling intricate DEs, especially fractional derivatives [37, 38]. Stochastic optimization techniques have found utility across various scientific disciplines, such as astrophysics [39], plasma physics [40], bioinformatics modeling [41], fluid dynamics [42–45], and so on. Nature-inspired algorithms, notably particle swarm optimization (PSO), genetic algorithms (GA), have emerged as a potent approach for stochastic optimization. GA represents global optimization approach and has been extensively applied to solve a variety of nonlinear problems in engineering field [46–49]. These approaches come with several advantages in different fields including ease of implementation, broad stability, applicability, prevention of divergence, and high reliability. The integration of stochastic optimization within artificial intelligence frameworks has not only broadened the scope of problem-solving but has also significantly enhanced the efficiency and reliability of solutions across diverse scientific domains.

While the existing literature extensively covers CD channels, no attempt has been made to investigate the interplay of Joule heating, chemical reactions, and Soret/Dufour effects in the flow of JH viscous fluid through these channels using machine learning algorithms for accuracy. In this chapter, we are applying the ANNs along with nature-inspired algorithms especially PSO hybrid with NNA for better accuracy. The prediction of ANNs results is compared with NDSolve from Mathematica 11 for accuracy and validation of the hybrid approach. Also, we are conducting 200 independent runs for efficiency of the proposed approach.

7.2 PROBLEM FORMULATION

Study the two-dimensional (2D) MHD JH flow from a source via channels that are expanding as well as contracting at an angle of $2a$ (Figure 7.1). The following presumptions are included in the current problem:

- It is anticipated that the walls of the channels shall expand or contract radially by the radial velocity $u = Uw = sr - 1$, where s denotes the rate of expansion and contraction.
- The channels are thought to be expanding or contracting at an angle of 2α.
- Since the velocity field only runs parallel to the radial direction and depends on r and θ, we may proceed with the liquid velocity as $V = \left(u(r, \theta), 0, 0\right)$;
- Furthermore, it is assumed that the channels are either divergent or convergent, meaning that $\alpha > 0$ *and* $\alpha < 0$ correspondingly.
- A stable, 2D, incompressible viscous fluid is taken into consideration.
- A magnetic field perpendicular to the divergent and convergent walls is applied.

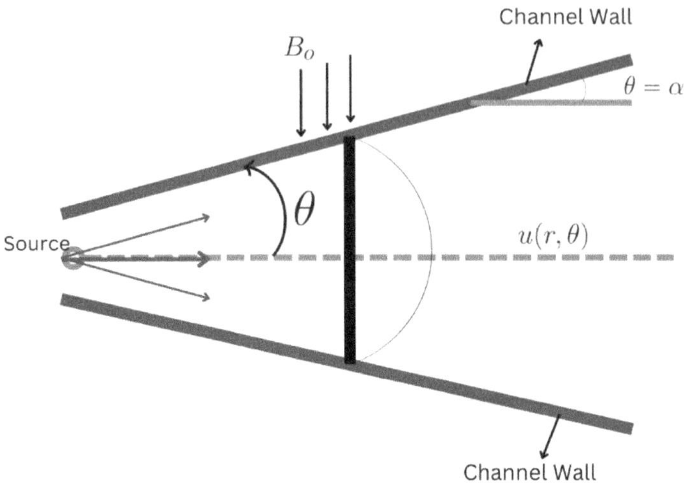

FIGURE 7.1 Geometry of the problem.

- We address the heat source, Joule heating, Soret and Dufour traits, and chemical reactions.
- *Tw and* C_w^* stand for the channel wall temperature and concentration, respectively.

$\Theta = \alpha$

Given the assumptions indicated before, the determining expressions are:

$$\frac{\rho}{r}\frac{\partial}{\partial r}(ru) = 0, \tag{7.1}$$

$$u\frac{\partial u}{\partial r} = -\frac{1}{\rho}\frac{\partial P}{\partial r} + v\left[\frac{\partial^2 u}{\partial r^2} + \frac{1}{r}\frac{\partial u}{\partial r} + \frac{1}{r^2}\frac{\partial^2 u}{\partial \theta^2} - \frac{u}{r^2} - \frac{\sigma B_0^2}{\rho}u\right], \tag{7.2}$$

$$0 = -\frac{1}{\rho r}\frac{\partial P}{\partial \theta} + \frac{2v}{r^2}\frac{\partial u}{\partial \theta}, \tag{7.3}$$

$$\begin{aligned} u\frac{\partial T}{\partial r} = {} & \frac{k}{(\rho C_p)}\left(\frac{\partial^2 T}{\partial r^2} + \frac{1}{r}\frac{\partial T}{\partial r} + \frac{1}{r^2}\frac{\partial^2 T}{\partial \theta^2}\right) - \frac{1}{r(\rho C_p)}\frac{\partial}{\partial r}(rq_{r,rad}) \\ & - \frac{1}{r^2(\rho C_p)}\frac{\partial}{\partial \theta}(q_{\theta,rad}) + \frac{\mu}{(\rho C_p)}\left[2\left\{\left(\frac{\partial u}{\partial r}\right)^2 + \left(\frac{u}{r}\right)^2\right\} + \left(\frac{1}{r}\frac{\partial u}{\partial \theta}\right)^2\right] \\ & + \frac{Q_0}{r^2(\rho C_p)}T + \frac{\sigma B_0^2}{r^2(\rho C_p)}u^2 + \frac{DK_T\rho}{C_s(\rho C_p)}\left(\frac{\partial^2 C^*}{\partial r^2} + \frac{1}{r}\frac{\partial C^*}{\partial r} + \frac{1}{r^2}\frac{\partial^2 C^*}{\partial \theta^2}\right), \end{aligned} \tag{7.4}$$

$$u\frac{\partial C^*}{\partial r} = D\left(\frac{\partial^2 C^*}{\partial r^2} + \frac{1}{r}\frac{\partial C^*}{\partial r} + \frac{1}{r^2}\frac{\partial^2 C^*}{\partial \theta^2}\right) - KC^* + \frac{DK_T}{T_m}\left(\frac{\partial^2 T}{\partial r^2} + \frac{1}{r}\frac{\partial T}{\partial r} + \frac{1}{r^2}\frac{\partial^2 T}{\partial \theta^2}\right). \tag{7.5}$$

In this case, the variables P, T, *and* ρ stand for pressure, temperature, and fluid density, respectively. The variables μ and v represent fluid dynamic viscosity and kinematic viscosity measurement; and heat generation/absorption Q, magnetic field B_0, mass diffusivity coefficient D, fluid mean temperature Tm, chemical reaction coefficient K, and thermal diffusion ratio KT are used. Similarly, cp and q stand for specific heat and heat flux, respectively. The two concepts that determine heat flux are conduction and radiation.

$$q = q_{r,rad} + q_{\theta,rad}, \tag{7.6}$$

$$q_{r,rad} = -\frac{4\sigma^*}{3K^*}\frac{\partial T^4}{\partial r} \text{ and } q_{\theta,rad} = -\frac{4\sigma^*}{3K^*}\frac{\partial T^4}{\partial \theta} \tag{7.7}$$

The radiative heat flux along (r, θ) coordinates is represented as $q_{r,rad}$ $q_{\theta,rad}$, and σ^* indicates the Stephan–Boltzmann constant in this case. K^* stands for the mean absorption coefficient. After applying the Taylor series and decomposing, we obtain:

$$T^4 = T_\infty^4 + 4T_\infty^3(T - T_\infty) + 6T_\infty^2(T - T_\infty)^2 + \ldots, \tag{7.8}$$

dismissing the more complex terms, one has

$$T^4 \cong T_\infty^4 + 4T_\infty^3 T - 3T_\infty^4, \tag{7.9}$$

In Eqn. (7.9) and Eqn. (7.7) becomes:

$$q_r = -\frac{16T_\infty^3\sigma^*}{3K^*}\frac{\partial T}{\partial r}. \tag{7.10}$$

Assuming at centerline $(\theta = 0)$ and channels are symmetrical then the obtained boundary conditions are:

$$\frac{\partial u}{\partial \theta} = \frac{\partial T}{\partial \theta} = 0,\ u = \frac{u_c}{r},\ \frac{\partial C}{\partial \theta} = 0 \tag{7.11}$$

Wall conditions at the problem are $(\theta = \pm\alpha)$:

$$T = \frac{T_w}{r^2} = 0,\ u = u_w = \frac{s}{r},\ C^* = \frac{C_w}{r^2}. \tag{7.12}$$

The dimensionless variables are defined as:

$$F(\omega)=\frac{f(\theta)}{u_c},\ \omega=\frac{\theta}{a},\ \Theta(\omega)=r^2\frac{T}{T_w},\ \phi(\omega)=\frac{C^*r^2}{C_w}. \tag{7.13}$$

By substituting these variables in the governing equations, we obtained set of ordinary differential equations (ODEs):

$$F'''(\omega)+2\alpha ReF(\omega)F'(\omega)+(4-M)\alpha^2F'(\omega)=0, \tag{7.14}$$

$$\begin{aligned}(1+N_r)\Theta''(\omega)+2\alpha^2(2+N_r+PrF(\omega))\Theta(\omega)+\frac{PrEc}{Re}\left(4\alpha^2F^2(\omega)+F'^2(\omega)\right)\\ -\alpha^2H_s\Theta(\omega)+\alpha^2\frac{PrMEc}{Re}F^2(\omega)+PrD_u\left(\phi''(\omega)+4\alpha M\right)\phi(\omega)=0,\end{aligned} \tag{7.15}$$

$$\phi''(\omega)+2\alpha^2\left(2+S_rF(\omega)-\frac{\lambda S_c}{2}\right)\phi(\omega)=0. \tag{7.16}$$

$$F(0)=1,\ F'(0)=0,\ \Theta'(0)=0,\ \phi'(0)=0,\ F(\pm1)=C,\ \Theta(\pm1)=1,\ \phi(\pm1)=0. \tag{7.17}$$

In this case, $C=\frac{S}{U_c}$ acts to stretch when $C>0$ and to shrink when $C<0$. $Ec=\frac{\alpha U_c^2}{C_pT_w}$ stands for Eckert number, while $Pr=\frac{U_c\rho C_p}{K}$ stands for Prandtl number. The Reynold number is $Re=\frac{\alpha U_c}{\nu}$, the Dufour number is $Du=\frac{DK_TC_w\alpha}{\rho U_cC_pC_sT_w}$, the chemical reaction parameter is $\lambda=\frac{K}{U_c}$, the radiation variable is $Nr=\frac{kK^*}{4\sigma^*T_\infty^3}$, $Sr=\frac{DK_TT_w}{U_cT_mC_w}$ is the soret number the heat source/sink variable is $Hs=\frac{Q_0}{k_1}$, the magnetic strength is $M=\frac{\sigma B_0^2}{\mu}$, and the Schmidt number is $Sc=\frac{U_c}{D}$.

7.3 SOLUTION OF THE PROBLEM

We are using a combination of machine learning algorithm especially physics-informed machine learning and evolutionary meta-heuristic algorithms to solve a set of DEs governing hydro-magnetic thermal transport with Joule heating, Soret and Dufour effects in CD channels. The solution methodology consists of the following steps:

- Formulating the set of nonlinear partial differential equations (PDEs) by physical phenomena of the problems and transforming the PDEs into ODEs using similarity transformation.
- Utilizing unsupervised physics-informed machine leaning methodology to transform the ODEs into fitness function.
- Optimizing the fitness function through PSO hybrid with NNA as optimizer for determining the optimal weights and biases within the range $[-10, 10]$.
- Further, ANN-BLMT will be utilized for deep analysis of the problem.

7.3.1 Unsupervised Physics-Informed Neural Networks

In the advancement of feed-forward physics-informed neural networks (PINNs), a widely adopted approach is to approximate the solution of DE. The solution of the DEs and its nth time derivatives are represented as $F'(\omega)$, $F''(\omega), F'''(\omega), \ldots, F^{n}(\omega)$, $\Theta'(\omega)$, $\Theta''(\omega), \Theta'''(\omega), \ldots, \Theta^{n}(\omega)$ and $\phi'(\omega)$, $\phi''(\omega), \phi'''(\omega), \ldots, \phi^{n}(\omega)$ for a given set of DEs. (7.14–7.16) that transform into physics-informed machine leaning model. These continuous mappings (transform) serve as fundamental architectural components of the neural network, enabling it to effectively emulate the desired approximated solution of $F(\omega)$, $\Theta(\omega)$ and $\phi(\omega)$. The unsupervised physics-informed machine learning based on sigmoid as an activation functions are following equations for velocity:

$$
\left.
\begin{aligned}
\hat{F}(\omega) &= \sum_{i=1}^{t} \left(a_{F_i} \frac{1}{1+e^{-b_{F_i} - w_{F_i}\omega}} \right), \\
\hat{F}'(\omega) &= \sum_{i=1}^{t} \left(a_{F_i} w_{F_i} \frac{e^{-b_{F_i} - w_{F_i}\omega}}{\left(1+e^{-b_{F_i} - w_{F_i}\omega}\right)^2} \right), \\
\hat{F}''(\omega) &= \sum_{i=1}^{t} \left(a_{F_i} w_{F_i}^{\,2} \left(\frac{2e^{-2b_{F_i} - 2w_{F_i}\omega}}{\left(1+e^{-b_{F_i} - w_{F_i}\omega}\right)^3} - \frac{e^{-b_{F_i} - w_{F_i}\omega}}{\left(1+e^{-b_{F_i} - w_{F_i}\omega}\right)^2} \right) \right), \\
\hat{F}'''(\omega) &= \sum_{i=1}^{t} a_{F_i} \, w_{F_i}^{\,3} \left(\frac{6e^{-3b_{F_i} - 3w_{F_i}\omega}}{\left(1+e^{-b_{F_i} - w_{F_i}\omega}\right)^4} - \frac{6e^{-2b_{F_i} - 2w_{F_i}\omega}}{\left(1+e^{-b_{F_i} - w_{F_i}\omega}\right)^3} + \frac{e^{-b_{F_i} - w_{F_i}\omega}}{\left(1+e^{-b_{F_i} - w_{F_i}\omega}\right)^2} \right).
\end{aligned}
\right\}
\tag{7.18}
$$

The physics-informed machine learning transformation for temperature and concentration equations are following:

$$
\left.
\begin{aligned}
\hat{\Theta}(\omega) &= \sum_{i=1}^{t} a_{\Theta_i} \frac{1}{1+e^{-b_{\Theta_i} - w_{\Theta_i}\omega}}, \\
\hat{\Theta}'(\omega) &= \sum_{i=1}^{t} \left(w_{\Theta_i} \frac{e^{-b_{\Theta_i} - w_{\Theta_i}\omega}}{\left(1+e^{-b_{\Theta_i} - w_{\Theta_i}\omega}\right)^2} \right) a_{\Theta_i}, \\
\hat{\Theta}''(\omega) &= \sum_{i=1}^{t} \left(w_{\Theta_i}^{\,2} \left(\frac{2e^{-2b_{\Theta_i} - 2w_{\Theta_i}\omega}}{\left(1+e^{-b_{\Theta_i} - w_{\Theta_i}\omega}\right)^3} - \frac{e^{-b_{\Theta_i} - w_{\Theta_i}\omega}}{\left(1+e^{-b_{\Theta_i} - w_{\Theta_i}\omega}\right)^2} \right) \right) a_{\Theta_i}, \\
\hat{\Theta}'''(\omega) &= \sum_{i=1}^{t} \left(a_{\Theta_i} \, w_{\Theta_i}^{\,3} \left(\frac{6e^{-3b_{\Theta_i} - 3w_{\Theta_i}\omega}}{\left(1+e^{-b_{\Theta_i} - w_{\Theta_i}\zeta}\right)^4} - \frac{6e^{-2b_{\Theta_i} - 2w_{\Theta_i}\omega}}{\left(1+e^{-b_{\Theta_i} - w_{\Theta_i}\omega}\right)^3} + \frac{e^{-b_{\Theta_i} - w_{\Theta_i}\omega}}{\left(1+e^{-b_{\Theta_i} - w_{\Theta_i}\omega}\right)^2} \right) \right).
\end{aligned}
\right\}
\tag{7.19}
$$

$$\left.\begin{aligned}
\hat{\phi}(\omega) &= \sum_{i=1}^{t}\left(a_{\phi_i}\frac{1}{1+e^{-b_{\phi i}-w_{\phi_i}\omega}}\right),\\
\hat{\phi}'(\omega) &= \sum_{i=1}^{t}\left(a_{\phi_i}w_{\phi_i}\frac{e^{-b_{\phi i}-w_{\phi_i}\omega}}{\left(1+e^{-b_{\phi i}-w_{\phi_i}\omega}\right)^2}\right),\\
\hat{\phi}''(\omega) &= \sum_{i=1}^{t}\left(a_{\phi_i}w_{\phi_i}{}^2\left(\frac{2e^{-2b_{\phi i}-2w_{\phi_i}\omega}}{(1+e^{-b_{\phi i}-w_{\phi_i}\omega})^3}-\frac{e^{-b_{\phi i}-w_{\phi_i}\omega}}{(1+e^{-b_{\phi i}-w_{\phi_i}\omega})^2}\right)\right),\\
\hat{\phi}'''(\omega) &= \sum_{i=1}^{t}\left(a_{\phi_i}\;w_{\phi_i}{}^3\left(\frac{6e^{-3b_{\phi i}-3w_{\phi_i}\omega}}{\left(1+e^{-b_{\phi i}-w_{\phi_i}\omega}\right)^4}-\frac{6e^{-2b_{\phi i}-2w_{\phi_i}\omega}}{\left(1+e^{-b_{\phi i}-w_{\phi_i}\omega}\right)^3}+\frac{e^{-b_{\phi i}-w_{\phi_i}\omega}}{\left(1+e^{-b_{\phi i}-w_{\phi_i}\omega}\right)^2}\right)\right),
\end{aligned}\right\} \tag{7.20}$$

The above Eqns. (7.18–7.20) provide the representation of an activation function and its derivatives for velocity, temperature, and concentration. $W=\left[a_{F_i},\, w_{F_i},b_{F_i},\, a_{\Theta_i},\, w_{\Theta_i},b_{\Theta_i},\, a_{\phi_i},\, w_{\phi_i},b_{\phi_i}\right]$ are a set of neurons weights and biases of corresponding ANN architecture. The ANN weights and biases, denoted as set W, play an important role in physics-informed ANNs architecture that define the behavior and accuracy of the fitness function based approximating the solution to the JH nonlinear DEs. Utilizing of these weights and biases into the fitness function, expressions converge to zero, the fitness function for velocity, temperature and concentration with associated boundary condition are follows:

$$\varepsilon_1 = \frac{1}{N}\sum_{i=1}^{N}\left(\hat{F}'''(\omega)+2\alpha Re\hat{F}(\omega)\hat{F}'(\omega)+(4-M)\alpha^2\hat{F}'(\omega)\right)^2, \tag{7.21}$$

$$\begin{aligned}
\varepsilon_2 = \frac{1}{N}\sum_{i=1}^{N}\Bigg((1+N_r)\hat{\Theta}''(\omega)+2\alpha^2\left(2+N_r+\Pr\hat{F}(\omega)\right)\hat{\Theta}(\omega)+\frac{\Pr Ec}{\mathrm{Re}}\Big(4\alpha^2\hat{F}^2(\omega)\\
+\hat{F}'^2(\omega)\Big)-\alpha^2H_s\hat{\Theta}(\omega)+\alpha^2\frac{\Pr MEc}{\mathrm{Re}}\hat{F}^2(\omega)+\Pr D_u\left(\phi'''(\omega)+4\alpha M\right)\phi(\omega)\Bigg)^2,
\end{aligned} \tag{7.22}$$

$$\varepsilon_3 = \frac{1}{N}\sum_{i=1}^{N}\left(\hat{\phi}''(\omega)+2\alpha^2\left(2+S_r\hat{F}(\omega)-\frac{\lambda S_c}{2}\right)\hat{\phi}(\omega)\right)^2, \tag{7.23}$$

$$\begin{aligned}
\varepsilon_4 = \frac{1}{N}\sum_{i=1}^{N}\Bigg(\left(\hat{F}(0)-1\right)^2+\left(\hat{F}'(0)\right)^2+\left(\hat{F}'^{(\pm1)}-C\right)^2+\left(\hat{\Theta}(\pm1)-1\right)^2\\
+\left(\widehat{\Theta}'(0)\right)^2+\left(\hat{\phi}(\pm1)\right)^2+\left(\hat{\phi}'(0)\right)^2\Bigg),
\end{aligned} \tag{7.24}$$

$$E = \varepsilon_1+\varepsilon_2+\varepsilon_3+\varepsilon_4. \tag{7.25}$$

7.3.1.1 Evolutionary Algorithm (Meta-heuristic)

Evolutionary algorithms tackle the complex optimization problems. These algorithms are nature-inspired and find the approximate solution of the problems. Among the various available evolutionary techniques, swarm intelligence (SI) has gained popularity as a preferred method for addressing optimization problems. Several algorithms have made significant advancements in this field, including the genetic algorithm, firefly algorithm, water cycle algorithm, PSO, ant colony optimization, Levy flight, artificial bee colony, hunting search, simulated annealing, NNA, and others. In this study, our aim is to enhance the optimization of Eqn. (7.25) by synergistically combining the PSO and NNA (PSO-NNA) through hybridization. Our proposed PSO-NNA hybridization technique holds promise for achieving improved performance in solving these nonlinear DEs.

7.3.1.2 Particle Swarm Optimization

PSO is a nature-inspired meta-heuristic (evolutionary) algorithm especially inspired by the collective dynamical behavior of flocks of birds serving as an effective approach for finding optimal solutions of problems. It utilizes a population of swarms (particles), each representing a potential solution, which traverse the search domain, in this study from −10 to 10 and continuously update their positions based on the best solutions found individually and globally within the swarm (particles) population. The algorithm uses two main parameters, the cognitive component (c_1) reflecting the particle's memory of its personal best solution and the social component (c_2) that finds the best solution found by other particles. PSO has become widely used in different scientific and engineering domains, including machine learning, data analytics, and technical design. The PSO algorithm versatility is evident in its useful applications, such as high-dimensional data clustering, parameter estimation of chaotic systems, and optimization tasks related to different engineering fields including nuclear reactors, reactive power distribution, and optical properties of thin films [50–60].

The swarm's initial particles are chosen randomly from −10 to 10, and each iteration updates the positions and velocities of each particle if stopping criteria are not met, based on its most recent best local P_{LB}^{x-1} and global P_{GB}^{x-1} positions. The two general updating formulas for particle position and velocity for PSO as follows:

$$v_i^x = wv_i^{t-1} + c_1 r_1 \left(P_{LB}^{x-1} - X_i^{x-1}\right) + c_2 r_2 \left(P_{GB}^{x-1} - X_i^{x-1}\right), \tag{7.29}$$

$$X_i^x = X_i^{x-1} + v_i^{x-1}. \tag{7.30}$$

In the swarm, $i = 1, \ldots, p$ where p is the total number of particles. The vectors X_i and V_i represent the ith particle's position and velocity, respectively. V denotes inertia weight, c_1 and c_2 are local and global acceleration constants, and the weight linearly decreases between 0 and 1. Random vectors r_1 and c_2 have values ranging from $[0, 1]$.

7.3.1.3 Neural Network Algorithm

The NNA leverages innovatively ANNs and biological nervous phenomena. Unlike traditional evolutionary methods, NNA relies solely on population size and termination criteria, eliminating the need for additional parameters [61]. Leveraging the inherent architecture (structure) of neural networks, NNA exhibits robust global optimization, making it suitable for addressing various complex optimization problems in scientific domains. The NNA is characterized by four crucial important components that are discussed in given subsections [61].

7.3.1.3.1 Update Population

Through NNA scenarios, the population $Y_t = \left\{ y_1^t,\ y_2^t,\ y_3^t, \ldots, y_M^t \right\}$ undergoes updates via the weight matrix $w_t^i = \left\{ w_{i,1}^t,\ w_{i,2}^t,\ w_{i,3}^t, \ldots,\ w_{i,M}^t \right\}$ that represents the weight matrix of the ith individual and $y_t^i = \left\{ y_{i,1}^t,\ y_{i,2}^t,\ y_{i,3}^t, \ldots,\ y_{i,E}^t \right\}$ that represents the position of the jth individual. Where, E represents the number of variables. The generation of a new population of NNA as follows:

$$y_{new,j}^t = \sum_{j=1}^{M} w_{j,k}^t \times y_j^t,\ j = 1,2,3,\ldots,M,\ k = 1,2,3,\ldots,M, \tag{7.31}$$

$$y_j^t = y_j^t + y_{new,j}^t,\ j = 1,\ 2,\ 3,\ldots,\ M. \tag{7.32}$$

M denotes the total population size and t represents the current iteration number. The solution for j^{th} individual at time t is denoted as y_j^t, and $y_{new,j}^t$ signifies the solution for the jth individual at the equal time point, computed with best weights.

$$\sum_{j=1}^{M} w_{j,k}^t = 1,\ 0 < w_{j,k}^t < 1,\ j = 1,\ 2,\ 3,\ldots,M,\ k = 1,\ 2,\ 3,\ldots,\ M \tag{7.33}$$

7.3.1.3.2 Update Weight Matrix

The weights matrix W^t plays a crucial role in the NNA processed of generating a new population. The dynamics of the weight matrix W^tcan be enhanced follows:

$$w_j^t = \left| w_i^t + 2 \times \lambda_2 (w_{obj}^t - w_j^t) \right|,\ j = 1,2,3,\ldots,M \tag{7.34}$$

Where λ_2 is a randomly generated value from a uniform distribution [0, 1], and w_{obj}^t represents the objective weight vector. It is crucial to note that both w_{obj}^t and the target solution x_{obj}^t have corresponding indices. Specifically, if x_{obj}^t matches x_v^t, $(v \in [1,\ \mathrm{M}])$ at time t, then w_{obj}^t is equivalently aligned with w_v^t.

7.3.1.3.3 *Bias Operator*

The bias operator in NNA enhances its global best exploration capabilities. The modification factor, represented as β_1, is crucial for measuring the extent of bias introduced, and it undergoes updates through:

$$\beta_1^{t+1} = 0.99\beta_1^t. \tag{7.35}$$

In the bias population operator, two variables are involved – a randomly create number M_P and denoted as P. Suppose $L = (l_1, l_2, l_3, \ldots, l_D)$ and $U = (u_1, u_2, u_3, \ldots, u_D)$ represent the lower bound and upper bound of the JH ANN based fitness function variables, respectively. M_p is determined as $\beta_1^t \times E$, representing the ceiling value of the product of β_1^t and E. The set P consists of M_p randomly selected integers from the range $[0, E]$. Therefore, the bias population can be precisely mathematically defined as follows.

$$y_{j,P(S)}^t = l_{P(S)}^t + \left(u_{P(S)} - l_{P(S)}\right) \times \lambda_3,\ S = 1,2,3,\ldots,\ M_P \tag{7.36}$$

In this context, λ_3 signifies a uniformly distributed random value within the range [0, 1]. The bias matrix introduces two variables – M_w, a randomly generated number and R, a set. The calculation of M_w involves taking the ceiling of $\left[\beta_1^t \times M\right]$. Simultaneously, the set R comprises M_w randomly selected integers from the interval [0, M]. This precise definition characterizes the bias weight matrix.

$$w_{j,R(r)}^t = \lambda_4,\ r = 1,2,3,\ldots,\ M_w \tag{7.37}$$

Where λ_4,is a random number between $[0, 1]$ with respect to uniform distribution.

7.3.1.3.4 *Transfer Operator*

The transfer operator aims to refine the current optimal solution by emphasizing the local search capability of NNA, as illustrated in the following equation. This operator plays a crucial role in generating a superior solution within the local context.

$$y_j^{t+1} = x_j^t + 2\lambda_5\left(y_{obj}^t - y_j^t\right),\ j = 1,\ 2,\ 3,\ldots,\ M \tag{7.38}$$

where λ_5 is a random number from $[0, 1]$ uniform distribution, like the other metaheuristic optimization algorithms NNA is initialized by

$$y_{j,k}^t = l_k + (u_k - l_k) \times \lambda_6.\ j = 1,\ 2,\ 3,\ldots,\ M \text{ and } k = 1,\ 2,\ 3,\ldots,\ E. \tag{7.39}$$

where $\lambda_6$is a randomly from $[0, 1]$. The hybridization flow chart for PSO and NNAs is structured in Figure 7.2.

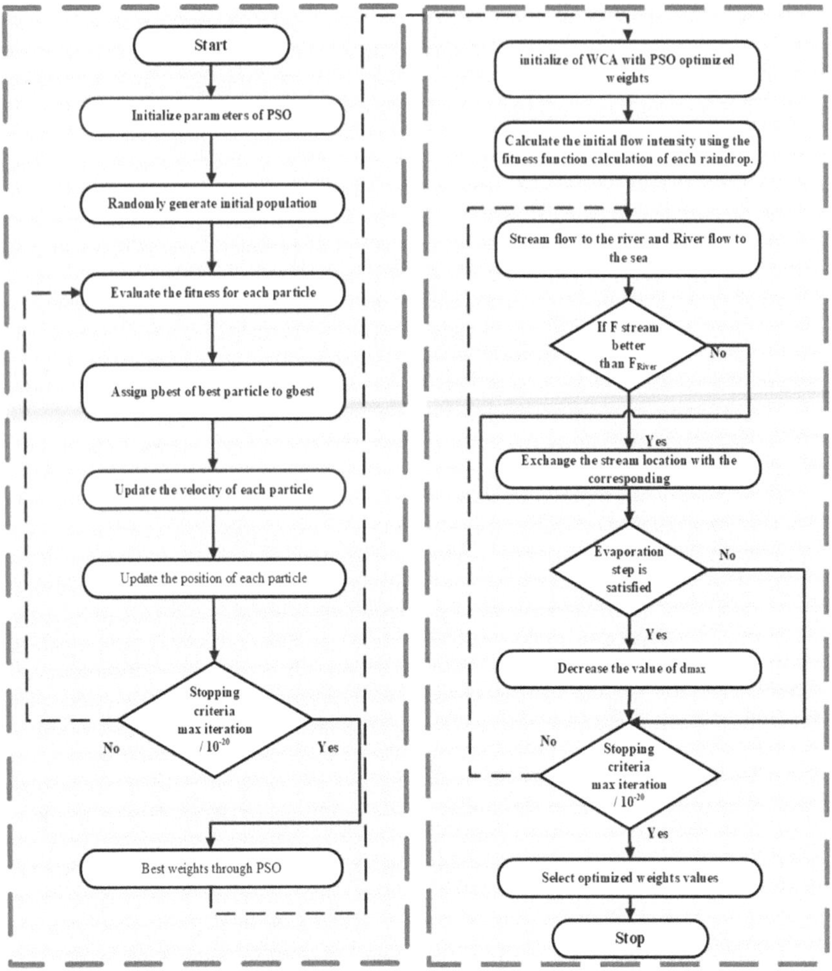

FIGURE 7.2 Flow chart of hybrid PSO-NNA.

7.4 RESULTS AND DISCUSSION

In addressing the challenges posed by complex systems where exact solutions are elusive, an intelligent approach of PINNs is employed. The combination of PSO and NNA is applied for finding the best optimal weights values and biases values of the PINNs. The flowchart of PSO-NNA is presented in Figure 7.2. The system considers various parameters, including Reynolds number (Re), Magnetic number (M), and Channel angle (α) in the Eqns. (7.24–7.26). The PINNs-based objective function was formulated in Eqn. (7.28) and is optimized using hybrid PSO and NNAs. Reference solution is obtained from NDSolve and compared with proposed PINNs method for validation. Solutions are computed for inputs within the range from 0 to 1 with a discretization size of h = 0.05 and n = 21. The optimized weights and biases, presented in Figure 7.3b, d, and f and also tabulated in numerical form in Tables 7.1, 7.3, and 7.5, are used for computation of numerical solutions $\hat{f}$, $\hat{\theta}$ and $\hat{\phi}$.

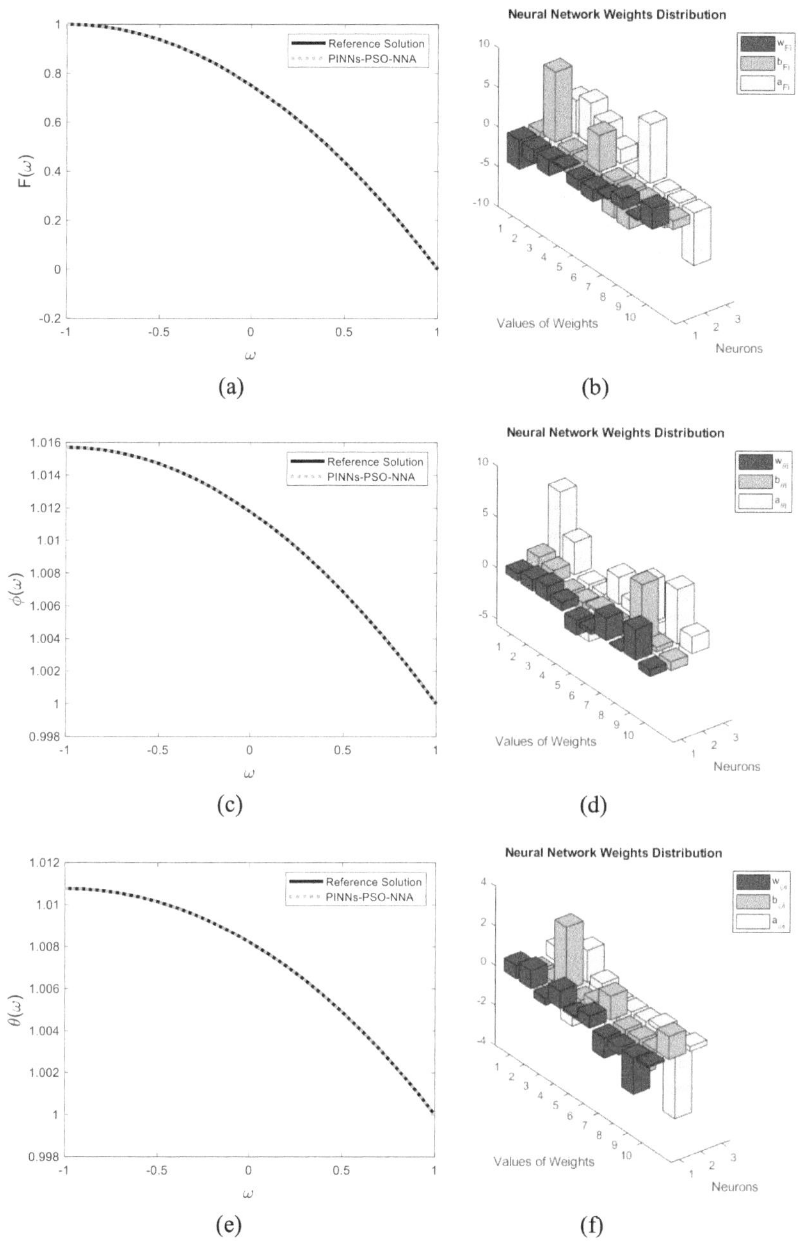

FIGURE 7.3 (a) Comparison solution of velocity obtained by NDSolve and ANN-PSO-NNA for $\boldsymbol{Re = 0.10, \alpha = -2^{o}, C = 0, M = 1.}$ (b). Optimized weights and biases for velocity through PSO-NNA for $\boldsymbol{Re = 0.10, \alpha = -2^{\circ}, C = 0, M = 1}$. (c, e) Comparison solution of temperature and concentration obtained by NDSolve and ANN-PSO-NNA for $\boldsymbol{Re = 5, \alpha = -5^{o}, C = 0.1, M = 3, N_r = 0.3, Pr = 0.1, Ec = 0.2, H_s = 0.0, D_u = 0.1, S_r = 0.1, S_c = 1, \lambda = 0.1}$., (d, f). Optimized weights and biases for temperature and concentration through PSO-NNA for $\boldsymbol{Re = 5, \alpha = -5^{o}, C = 0.1, M = 3, N_r = 0.3, Pr = 0.1, Ec = 0.2, H_s = 0.0, D_u = 0.1, S_r = 0.1, S_c = 1, \lambda = 0.1}$.

TABLE 7.1
Optimized Weights and Biases for Velocity Obtained by PSO-NNA for Parameters $Re = 0.10$, $\alpha = -2^{o}$, $C = 0$, $M = 1$, $N_r = 0.3$, $Pr = 0.1$, $Ec = 0.2$, $H_s = 0.0$, $D_u = 0.1$, $S_r = 0.1$, $S_c = 1$, $\lambda = 0.1$

w_{F_i}	b_{F_i}	a_{F_i}
−3.647630552	−4.679318131	−0.772735378
−1.690672496	4.265722189	8.833444819
−1.975406623	5.158730993	−2.851381666
0.197781304	3.734482538	−3.269232025
−1.286886332	1.761267312	4.565379397
−1.501443485	−1.706829574	−4.609807877
0.163283806	7.410238917	−4.745892814
1.322990657	−4.371927229	−1.013395992
−0.169980086	−2.695325741	−1.690275110
2.603127037	−6.661544752	−1.092343909

Figure 7.3a, c, and d depicts a comparison between the reference solution and the PINNs approaches, revealing higher accuracy in velocity, temperature, and concentration prediction. The numerical outcomes for the CD channel problem for velocity, temperature and concentration obtained through the proposed machine learning are detailed in Tables 7.2, 7.4, and 7.6 including their absolute errors (AE). The results from ANN-PSO-NNA closely align with the reference solution that is presented in

TABLE 7.2
Comparative Analysis for Velocity between Reference Solution and PINNs-PSO-NNA for Parameters $Re = 0.10$, $\alpha = -2^{o}$, $C = 0$, $M = 1$, $N_r = 0.3$, $Pr = 0.1$, $Ec = 0.2$, $H_s = 0.0$, $D_u = 0.1$, $S_r = 0.1$, $S_c = 1$, $\lambda = 0.1$

ω	Reference	PINNs-PSO-NNA	$ANN-BLMT$	|Reference − $(ANN-BLMT)$|	|Reference − (PINNs − PSO −NNA)|
0.00	1.00000000000000	0.999999844913744	1.000258672	2.59E-04	1.55086E-07
0.05	0.997500390476175	0.997500343400385	0.997536889	3.65E-05	4.70758E-08
0.10	0.990001568611728	0.990001621484040	0.989982107	1.95E-05	5.28723E-08
0.15	0.977503465428300	0.977503590929566	0.977455451	4.80E-05	1.25501E-07
0.20	0.960005980703891	0.960006135325184	0.959997606	8.38E-06	1.54621E-07
0.25	0.937508979003804	0.937509123463903	0.937549006	4.00E-05	1.44460E-07
0.30	0.910012295666169	0.910012408524854	0.909976839	3.55E-05	1.12859E-07

(Continued)

TABLE 7.2 (*Continued*)
Comparative Analysis for Velocity between Reference Solution and PINNs-PSO-NNA for Parameters $Re = 0.10$, $\alpha = -2^{o}$, $C = 0$, $M = 1$, $N_r = 0.3$, $Pr = 0.1$, $Ec = 0.2$, $H_s = 0.0$, $D_u = 0.1$, $S_r = 0.1$, $S_c = 1$, $\lambda = 0.1$

ω	Reference	PINNs-PSO-NNA	$ANN-BLMT$	|Reference − $(ANN-BLMT)$|	|Reference − (PINNs − PSO −NNA)|
0.35	0.877515737100772	0.877515822460589	0.87755073	3.50E-05	8.53598E-08
0.40	0.840019095066305	0.840019172272715	0.840013395	5.70E-06	7.72064E-08
0.45	0.797522150848288	0.797522242101227	0.797497408	2.47E-05	9.12529E-08
0.50	0.750024684635332	0.750024802554009	0.750059348	3.47E-05	1.17919E-07
0.55	0.697526491755715	0.697526626660294	0.697528622	2.13E-06	1.34905E-07
0.60	0.640027390341845	0.640027510373760	0.639991355	3.60E-05	1.20032E-07
0.65	0.577527233811016	0.577527294755557	0.577541255	1.40E-05	6.09445E-08
0.70	0.510025934356012	0.510025886873493	0.510047754	2.18E-05	4.74825E-08
0.75	0.437523463038780	0.437523277069077	0.437491122	3.23E-05	1.85970E-07
0.80	0.360019886357699	0.360019551549468	0.360038031	1.81E-05	3.34808E-07
0.85	0.277515379539382	0.277514901212397	0.277507131	8.25E-06	4.78327E-07
0.90	0.190010237403873	0.190009630140203	0.190023394	1.32E-05	6.07264E-07
0.95	0.0975049035367433	0.0975041702103549	0.097468204	3.67E-05	7.33326E-07
1.00	0.0000000000000000	-8.88352212763E-07	6.81E-05	6.81E-05	8.88351E-07

TABLE 7.3
Optimized Weights and Biases for Concentration Obtained by PSO-NNA for Parameters $Re = 5$, $\alpha = -5^{o}$, $C = 0.1$, $M = 3$, $N_r = 0.3$, $Pr = 0.1$, $Ec = 0.2$, $H_s = 0.0$, $D_u = 0.1$, $S_r = 0.1$, $S_c = 1$, $\lambda = 0.1$

w_{F_i}	b_{F_i}	a_{F_i}
0.755930056	7.165823747	1.540743058
1.115608939	3.169072176	1.193270545
1.410135592	-5.581073811	-1.201736701
0.783947086	-0.678821547	-0.826068083
-1.429326042	2.647470054	0.222525369
-0.388255944	1.105045381	-1.367613035
2.189775908	4.127485694	-0.695686163
-0.114890556	-2.518871061	5.609507842
2.998365134	5.327353536	0.448680948
-0.676959995	1.661599030	-0.888750171

TABLE 7.4
Comparative Analysis for Concentration between Reference and PINNs-PSO-NNA for Parameters $Re = 5$, $\alpha = -5^{\circ}$, $C = 0.1$, $M = 3$, $N_r = 0.3$, $Pr = 0.1$, $Ec = 0.2$, $H_s = 0.0$, $D_u = 0.1$, $S_r = 0.1$, $S_c = 1$, $\lambda = 0.1$

ω	Reference	$PINNs-PSO-NNA$	$ANN-BLMT$	|Reference − (ANN − BLMT)|	|Reference − (PINNs − PSO − NNA)|
0.00	1.01570413769372	1.01570381297004	1.015724486	2.03E-05	3.24724E-07
0.05	1.01566449638475	1.01566416874698	1.015664834	3.37E-07	3.27638E-07
0.10	1.01554558443284	1.01554526958642	1.01554355	2.03E-06	3.14846E-07
0.15	1.0153474355164	1.01534715314078	1.015327461	1.99E-05	2.82376E-07
0.20	1.01507010621707	1.01506986375918	1.015115512	4.54E-05	2.42458E-07
0.25	1.01471367580658	1.01471346608963	1.014623526	9.01E-05	2.09717E-07
0.30	1.01427824636534	1.01427805236529	1.014369791	9.15E-05	1.94000E-07
0.35	1.01376394289018	1.01376374513829	1.013669171	9.47E-05	1.97752E-07
0.40	1.01317091342068	1.01317069695694	1.013228482	5.75E-05	2.16464E-07
0.45	1.01249932918985	1.01249908823458	1.012468644	3.06E-05	2.40955E-07
0.50	1.01174938478843	1.01174912432932	1.011740213	9.17E-06	2.60459E-07
0.55	1.01092129834097	1.01092103264115	1.010985875	6.45E-05	2.65700E-07
0.60	1.01001531168779	1.01001506033328	1.009910827	1.04E-04	2.51355E-07
0.65	1.0090316905772	1.00903147309549	1.009197715	1.66E-04	2.17482E-07
0.70	1.00797072484424	1.00797055518535	1.007773111	1.98E-04	1.69659E-07
0.75	1.0068327285902	1.00683261080760	1.007049195	2.16E-04	1.17783E-07
0.80	1.00561804034617	1.00561796672089	1.005411482	2.07E-04	7.36253E-08
0.85	1.00432702321158	1.00432697579510	1.004485625	1.59E-04	4.74165E-08
0.90	1.00296006496651	1.00296002108251	1.002811966	1.48E-04	4.38840E-08
0.95	1.00151757814372	1.00151751981471	1.001625351	1.08E-04	5.83290E-08
1.00	1.00000000005289	0.99999992659812	0.999934994	6.50E-05	7.34548E-08

TABLE 7.5
Optimized Weights and Biases for Temperature Obtained by PSO-NNA for Parameters $Re = 5$, $\alpha = -5^{\circ}$, $C = 0.1$, $M = 3$, $N_r = 0.3$, $Pr = 0.1$, $Ec = 0.2$, $H_s = 0.0$, $D_u = 0.1$, $S_r = 0.1$, $S_c = 1$, $\lambda = 0.1$

w_{F_i}	b_{F_i}	a_{F_i}
0.669997033	0.713350949	-0.983617391
0.901939055	-2.711528116	-0.541352473
-0.343240500	1.598406179	3.182125093
0.860085002	0.427740786	-1.108645511
0.107581279	-2.100302502	0.400505869
0.607421917	-0.909304547	1.193471632
-1.010385212	-0.682520842	-0.946979245
-0.350878792	0.225634814	-1.777794943
-1.863854098	-3.899741850	-0.170253424
0.099723793	0.249896918	1.160769605

TABLE 7.6
Comparative Analysis for Temperature between Reference and PINNs-PSO-NNA for Parameters $Re = 5$, $\alpha = -5^{\circ}$, $C = 0.1$, $M = 3$, $N_r = 0.3$, $Pr = 0.1$, $Ec = 0.2$, $H_s = 0.0$, $D_u = 0.1$, $S_r = 0.1$, $S_c = 1$, $\lambda = 0.1$

ω	Reference	*PINNs – PSO – NNA*	*ANN – BLMT*	|Reference – ANN – BLMT|	|Reference – PSO – NNA|
0.00	1.01077593588135	1.01077617513591	1.010887433	1.11E-04	2.39255E-07
0.05	1.01075073692105	1.01075098516125	1.010753098	2.36E-06	2.48240E-07
0.10	1.01067509406122	1.01067535676267	1.010662757	1.23E-05	2.62701E-07
0.15	1.01054886693966	1.01054913464235	1.010529502	1.94E-05	2.67703E-07
0.20	1.01037182021684	1.01037207440852	1.010410556	3.87E-05	2.54192E-07
0.25	1.01014362118802	1.01014384061448	1.010092279	5.13E-05	2.19426E-07
0.30	1.00986383646353	1.00986400282616	1.009917434	5.36E-05	1.66363E-07
0.35	1.00953192761233	1.00953203002668	1.009486908	4.50E-05	1.02414E-07
0.40	1.00914724633291	1.00914728370108	1.009197457	5.02E-05	3.73682E-08
0.45	1.00870902863094	1.00870900996591	1.008614038	9.50E-05	1.86650E-08
0.50	1.00821638838489	1.00821633111961	1.00820331	1.31E-05	5.72653E-08
0.55	1.00766831036179	1.00766823698966	1.007738892	7.06E-05	7.33721E-08
0.60	1.00706364291460	1.00706357644431	1.007037668	2.60E-05	6.64703E-08
0.65	1.00640109009357	1.00640104941975	1.006478658	7.76E-05	4.06738E-08
0.70	1.00567920430387	1.00567919979052	1.005504324	1.75E-04	4.51335E-09
0.75	1.00489637875975	1.00489640938191	1.00502118	1.25E-04	3.06222E-08
0.80	1.0040508406148	1.00405089339012	1.003904621	1.46E-04	5.27753E-08
0.85	1.00314064501213	1.00314069743930	1.003311883	1.71E-04	5.24272E-08
0.90	1.00216367044693	1.00216369646616	1.002041597	1.22E-04	2.60192E-08
0.95	1.00111761589229	1.00111759558296	1.001187116	6.95E-05	2.03093E-08
1.00	1.00000000025082	0.99999993302936	0.999846685	1.53E-04	6.72215E-08

Tables 7.2, 7.4, and 7.6. To comprehensively analyze the algorithm's performance, 200 independent trials were run that offering insights into its convergence behavior. Figure 7.4a, b, c, and d presents the fitness plots and mean-squared error (MSE) for velocity, temperature, and concentration, illustrating the evolution of fitness scores throughout these 200 independent trials.

The statistical analysis of the numerical values for velocity, temperature, and concentration are shown in Tables 7.7, 7.8, and 7.9 that offer a comprehensive investigation of the PINNs algorithm's performance. For validation of ANN-PSO-NNA, important metrics include mean value, minimum value, maximum value, and standard deviation value calculated from 200 independent runs. These metrics provide a quantitative analyzed overview of the proposed ANN-PSO-NNA algorithm consistency and effectiveness over multiple (200) runs, aiding in a thorough understanding of its performance characteristics.

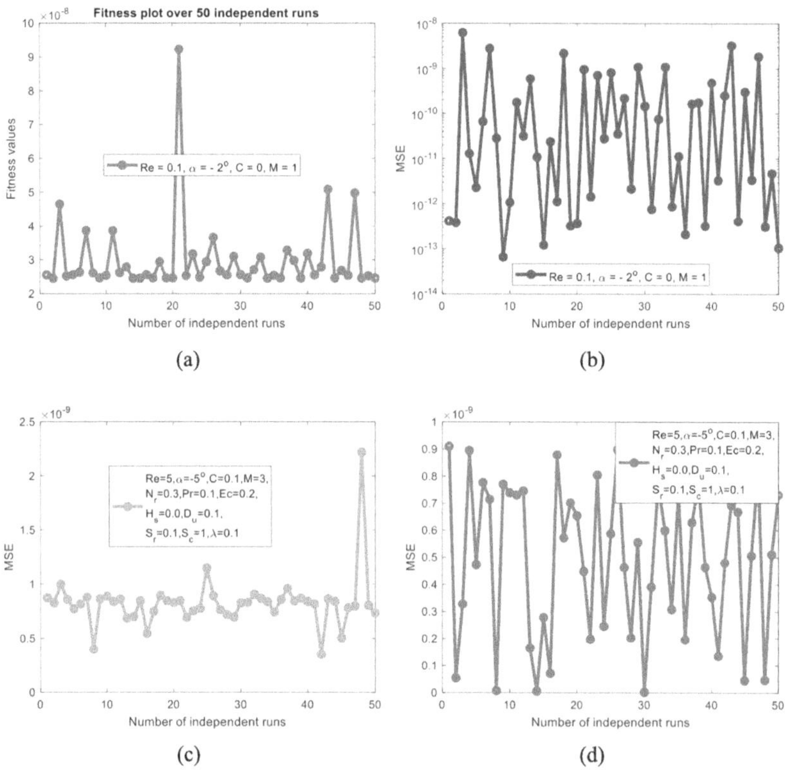

FIGURE 7.4 (a) Fitness function evaluation with (200) independence runs. (b, c, and d) Mean squared error comparison between NDSolve and ANN-PSO-NNA over 200 independence runs for velocity, temperature, and concentration, respectively.

TABLE 7.7
Statistical Analysis for Velocity of Absolute Errors (AE) between NDSolve and PINNs-PSO-NNA Algorithm for 200 Independence Runs for $Re = 110$, $\beta = 3°$, $We = 0$, $n = 1$, $M = 0$.

ω	Minimum	Maximum	Average	Standard Deviation
0.00	1.00705E-08	9.18193E-05	5.89845E-06	1.22751E-05
0.05	1.46466E-09	9.39584E-05	6.10711E-06	1.27273E-05
0.10	7.21790E-09	9.49881E-05	6.54979E-06	1.31574E-05
0.15	9.32920E-09	9.49211E-05	6.93685E-06	1.35480E-05
0.20	9.69796E-09	9.37827E-05	7.24577E-06	1.38115E-05
0.25	3.66235E-10	9.15778E-05	7.41926E-06	1.39080E-05
0.30	7.26703E-09	9.02178E-05	7.44100E-06	1.37961E-05
0.35	1.01189E-09	8.86845E-05	7.29931E-06	1.34420E-05
0.40	3.06530E-09	8.55659E-05	7.00843E-06	1.28130E-05
0.45	6.48176E-10	8.08456E-05	6.52661E-06	1.19222E-05
0.50	3.12579E-09	7.45206E-05	5.87016E-06	1.07697E-05

(Continued)

TABLE 7.7 (*Continued*)
Statistical Analysis for Velocity of Absolute Errors (AE) between NDSolve and PINNs-PSO-NNA Algorithm for 200 Independence Runs for $Re = 110$, $\beta = 3^{o}$, $We = 0$, $n = 1$, $M = 0$.

ω	Minimum	Maximum	Average	Standard Deviation
0.55	2.23323E-10	6.66025E-05	5.03206E-06	9.39733E-06
0.60	1.28755E-08	5.71064E-05	4.05358E-06	7.87167E-06
0.65	8.03540E-09	4.60433E-05	3.21064E-06	6.27092E-06
0.70	8.67161E-09	3.37287E-05	2.81453E-06	5.01496E-06
0.75	8.91602E-09	2.92400E-05	3.06375E-06	5.09378E-06
0.80	2.31442E-08	4.10687E-05	4.22352E-06	6.79906E-06
0.85	5.80133E-08	5.41565E-05	6.00867E-06	9.58229E-06
0.90	1.87290E-08	6.95921E-05	8.21932E-06	1.29988E-05
0.95	1.67532E-08	8.77497E-05	1.07818E-05	1.68474E-05
1.00	5.57029E-09	0.000107359	1.36142E-05	2.10707E-05

TABLE 7.8
Statistical Analysis for Temperature of Absolute Errors (AE) between NDSolve and PINNs-PSO-NNA Algorithm for 200 Independence Runs for $Re = 5$, $\alpha = -5^{o}$, $C = 0.1$, $M = 3$, $N_r = 0.3$, $Pr = 0.1$, $Ec = 0.2$, $H_s = 0.0$, $D_u = 0.1$, $S_r = 0.1$, $S_c = 1$, $\lambda = 0.1$.

ω	Minimum	Maximum	Average	Standard Deviation
0.00	6.17508E-06	9.78261E-05	2.35007E-05	6.87088E-06
0.05	6.33604E-06	9.63904E-05	2.34756E-05	6.80744E-06
0.10	6.65956E-06	9.52031E-05	2.35376E-05	6.69504E-06
0.15	7.15758E-06	9.35414E-05	2.37205E-05	6.54361E-06
0.20	7.67049E-06	9.09335E-05	2.38976E-05	6.30906E-06
0.25	7.97612E-06	8.71353E-05	2.3882E-05	5.99251E-06
0.30	7.86065E-06	8.21063E-05	2.34871E-05	5.62339E-06
0.35	7.15953E-06	7.59676E-05	2.25614E-05	5.24224E-06
0.40	5.77609E-06	6.89556E-05	2.10059E-05	4.88804E-06
0.45	3.69147E-06	6.13767E-05	1.87804E-05	4.59069E-06
0.50	9.54101E-07	5.35672E-05	1.59043E-05	4.36701E-06
0.55	1.80919E-06	4.58612E-05	1.24778E-05	4.17858E-06
0.60	9.69749E-07	3.85685E-05	8.66427E-06	3.96615E-06
0.65	8.31965E-08	3.19620E-05	4.86564E-06	3.59003E-06
0.70	1.17608E-08	2.62751E-05	1.92258E-06	3.12870E-06
0.75	2.05711E-07	2.17081E-05	3.91035E-06	2.94047E-06
0.80	1.33992E-06	1.86026E-05	6.86772E-06	3.04205E-06
0.85	1.31476E-06	1.94240E-05	8.65406E-06	2.61296E-06
0.90	1.43884E-06	1.83096E-05	8.64242E-06	2.32324E-06
0.95	6.49614E-08	1.84399E-05	6.10241E-06	2.58306E-06
1.00	4.02843E-09	2.26098E-05	1.49087E-06	3.29312E-06

TABLE 7.9
Statistical Analysis for Concentration of Absolute Errors (AE) between NDSolve and PINNs-PSO-NNA for 200 Independence Runs for $Re = 5$, $\alpha = -5^{o}$, $C = 0.1$, $M = 3$, $N_r = 0.3$, $Pr = 0.1$, $Ec = 0.2$, $H_s = 0.0$, $D_u = 0.1$, $S_r = 0.1$, $S_c = 1$, $\lambda = 0.1$.

ω	Minimum	Maximum	Average	Standard Deviation
0.00	8.70000E-08	5.14E-05	6.01072E-06	1.60092E-05
0.05	1.06124E-08	5.82E-05	7.10495E-06	1.81025E-05
0.10	4.58255E-08	6.36E-05	8.38901E-06	1.97367E-05
0.15	9.13199E-08	6.76E-05	9.41160E-06	2.09804E-05
0.20	1.71250E-07	7.01E-05	1.01285E-05	2.18061E-05
0.25	1.74160E-07	7.11E-05	1.05310E-05	2.21848E-05
0.30	1.24616E-07	7.08E-05	1.06160E-05	2.20916E-05
0.35	7.11732E-08	6.89E-05	1.03769E-05	2.15045E-05
0.40	2.88587E-08	6.56E-05	9.80836E-06	2.04129E-05
0.45	3.37137E-09	6.07E-05	8.90639E-06	1.88195E-05
0.50	1.12183E-08	5.43E-05	7.67150E-06	1.67476E-05
0.55	2.23323E-10	4.63E-05	6.10465E-06	1.42621E-05
0.60	1.68607E-08	3.68E-05	4.47478E-06	1.14051E-05
0.65	2.06202E-08	2.57E-05	3.91061E-06	8.19050E-06
0.70	1.10352E-08	1.87E-05	3.51977E-06	6.67083E-06
0.75	2.54155E-08	3.76E-05	3.50415E-06	9.06297E-06
0.80	3.08376E-08	5.67E-05	6.73731E-06	1.30395E-05
0.85	5.80133E-08	9.88E-05	1.03047E-05	1.84654E-05
0.90	1.17662E-07	6.85E-05	1.42050E-05	2.48266E-05
0.95	1.93392E-07	8.41E-05	1.84324E-05	3.19195E-05
1.00	2.62101E-07	2.67E-05	2.29814E-05	3.96518E-05

Furthermore, the discussion focuses on the cross-diffusion effects for convergent and divergent channels presented in Eqn. (7.14) to Eqn. (7.16), employing the supervised framework of backpropagated Levenberg–Marquardt technique (ANN-BLMT). The ANN-BLMT utilizes for velocity, temperature, and concentration prediction. Datasets for velocity, temperature, and concentration profiles are generated using the NDSolve scheme with a step size of 0.001. This scheme is applied to all three dynamics of the convergent and divergent channels with inputs ranging from 0 to 1. The resulting datasets for velocity, temperature, and concentration profiles serve as reference or targets solutions for ANN-BLMT. ANN-BLMT is trained using the "nftool" command of the neural network toolbox in MATLAB software to approximate solutions for the convergent and divergent channels. A total of 1001 input and target data points are employed for training, validation, and testing the learning process. These data points are randomly divided into 70% (701 points) for training, 15%

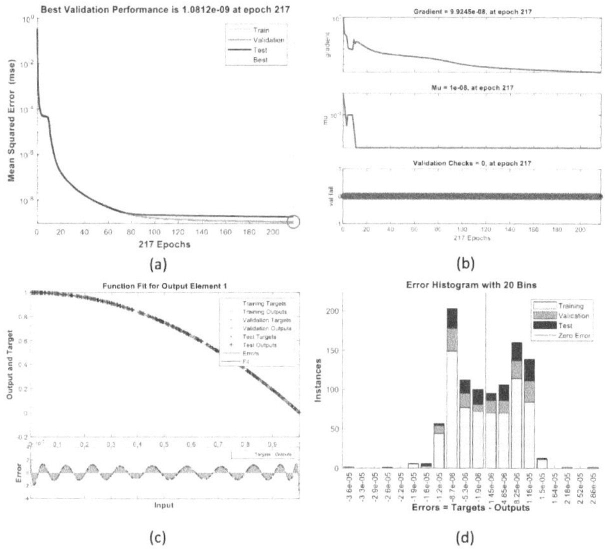

FIGURE 7.5 (a) Mean squared error for velocity from ANN-BLMT for parameters $\boldsymbol{Re = 0.10, \alpha = -2°, C = 0, M = 1, N_r = 0.3, Pr = 0.1, Ec = 0.2, H_s = 0.0, D_u = 0.1, S_r = 0.1, S_c = 1, \lambda = 0.1}$. (b) Transition states for velocity from ANN-BLMT for parameters $\boldsymbol{Re = 0.10, \alpha = -2°, C = 0, M = 1, N_r = 0.3, Pr = 0.1, Ec = 0.2, H_s = 0.0, D_u = 0.1, S_r = 0.1, S_c = 1, \lambda = 0.1}$ (c) Performance of prediction and target with error for velocity from ANN-BLMT for parameters $\boldsymbol{Re = 0.10, \alpha = -2°, C = 0, M = 1, N_r = 0.3, Pr = 0.1, Ec = 0.2, H_s = 0.0, D_u = 0.1, S_r = 0.1, S_c = 1, \lambda = 0.1}$. (d) Error histogram for velocity from ANN-BLMT for parameters $\boldsymbol{Re = 0.10, \alpha = -2°, C = 0, M = 1, N_r = 0.3, Pr = 0.1, Ec = 0.2, H_s = 0.0, D_u = 0.1, S_r = 0.1, S_c = 1, \lambda = 0.1}$.

(150 points) for testing, and 15% (150 points) for validation of the learning performance.

The convergence analysis and learning performance of ANN-BLMT optimization algorithms are examined the mean squared error (MSE) and transition states of Jeffery Hamel fluid flow. Figures 7.5a, b, 7.7a, b, and 7.9a, b illustrate the dynamical analysis for velocity, concentration and temperature. In addition, graphical representations comparing approximate results with referenced solutions from the NDSolve approach, along with error histogram plots, are depicted in Figures 7.5c, d, 7.7c, d, and 7.9c, d, respectively, of JH flow. Regression curves are presented in Figures 7.6, 7.8 and 7.10. Moreover, numerical outcomes computed from training 1.04687E-09, 2.39889E-09 and 8.72481E-09, validation 1.30182E-09, 2.75862E-09 and 7.74916E-09 and testing 1.58153E-09, 2.38561E-09 and 8.38135E-09 of data sets, as well as total epochs and backpropagation parameters like gradient 9.88E-08, 4.12E-08 and 6.90E-08 and step size 1.0010E-09, 1.00E-11 and 1.00E-08 for velocity, concentration and temperature respectfully. The convergence analysis reveals that for three scenarios like velocity, concentration and temperature, exhibits MSE values around

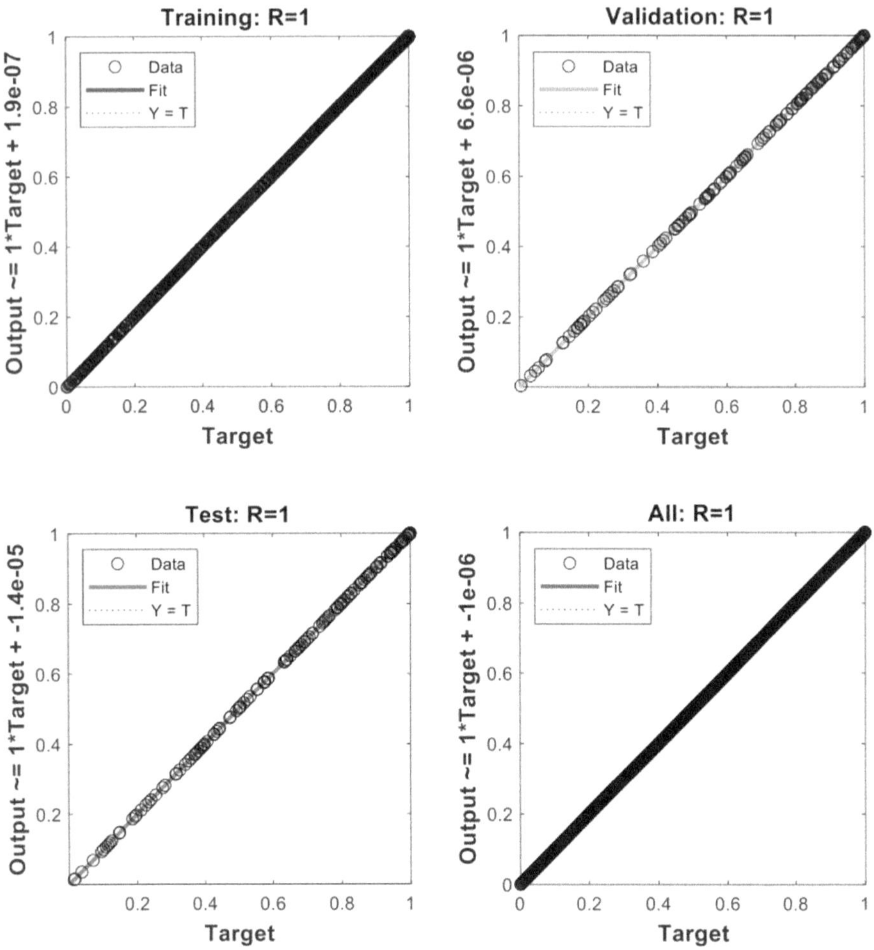

FIGURE 7.6 Regression analysis for velocity from ANN-BLMT for parameters $\boldsymbol{Re = 0.10, \alpha = -2°, C = 0, M = 1, N_r = 0.3, Pr = 0.1, Ec = 0.2, H_s = 0.0, D_u = 0.1, S_r = 0.1, S_c = 1, \lambda = 0.1}$.

1.0812E−09, 1.1071E−08, 7.460E−08. Error histograms from ANN-BLMT plots are provided for one of the three profiles in as depicted in Figures 7.5d, 7.7d, and 7.9d and Tables 7.2, 7.4 and 7.6. The efficiency of ANN-BLMT is further assessed through comparison with NDSolve and ANN-PSO-NNA approaches for velocity, concentration and temperature. These results demonstrate the convergence and efficacy of the proposed ANN-PSO-NNA approach is more accurate as compare to ANN-BLMT for DE based on JH fluid flow. The proposed ANN-PSO-NNA algorithm does not require the target dataset. If DEs for engineering problems are highly

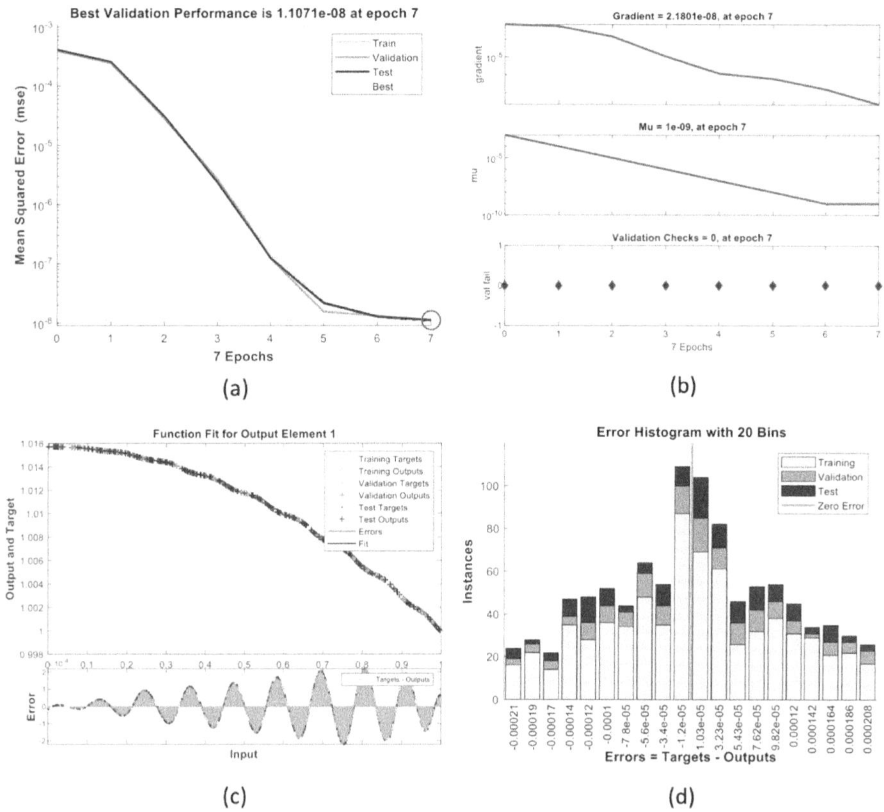

FIGURE 7.7 (a) Mean squared error for temperature from ANN-BLMT for parameters $\boldsymbol{Re = 5, \alpha = -5°, C = 0.1, M = 3, N_r = 0.3, Pr = 0.1, Ec = 0.2, H_s = 0.0, D_u = 0.1, S_r = 0.1, S_c = 1, \lambda = 0.1}$. (b) Transition states for temperature from ANN-BLMT for parameters $\boldsymbol{Re = 5, \alpha = -5°, C = 0.1, M = 3, N_r = 0.3, Pr = 0.1, Ec = 0.2, H_s = 0.0, D_u = 0.1, S_r = 0.1, S_c = 1, \lambda = 0.1}$. (c) Performance of prediction and target with error for temperature from ANN-BLMT for parameters $\boldsymbol{Re = 5, \alpha = -5°, C = 0.1, M = 3, N_r = 0.3, Pr = 0.1, Ec = 0.2, H_s = 0.0, D_u = 0.1, S_r = 0.1, S_c = 1, \lambda = 0.1}$. (d) Error histogram for temperature from ANN-BLMT for parameters $\boldsymbol{Re = 5, \alpha = -5^o, C = 0.1, M = 3, N_r = 0.3, Pr = 0.1, Ec = 0.2, H_s = 0.0, D_u = 0.1, S_r = 0.1, S_c = 1, \lambda = 0.1}$.

nonlinear, then analytical and numerical methods are not working to find the dataset for ANN-BLMT scheme. Thus proposed ANN-PSO-NNA is best for solving DE-based JH engineering problem.

The velocity characteristics for various scenarios are depicted in Figures 7.11 and 7.12 during the stretching processes through PINNs-PSO-NNA. Furthermore, the absolute error between the reference solutions for different scenarios are presented in Figures 7.11b, 7.12b, 7.13b, 7.14b, 7.15b, 7.16b, 7.17b, 7.18b, 7.19b, 7.20b, 7.21b, and 7.22b. This illustration reveals that velocity intensifies with

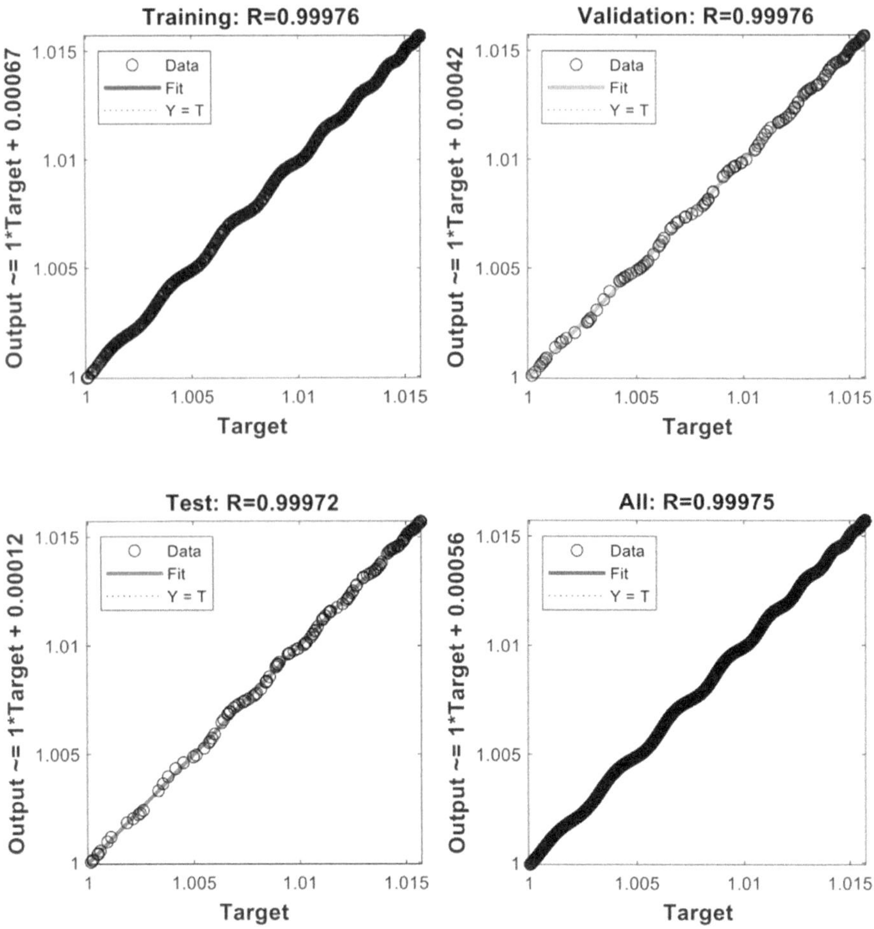

FIGURE 7.8 Regression analysis for temperature from ANN-BLMT for parameters $Re = 5$, $\alpha = -5°$, $C = 0.1$, $M = 3$, $N_r = 0.3$, $Pr = 0.1$, $Ec = 0.2$, $H_s = 0.0$, $D_u = 0.1$, $S_r = 0.1$, $S_c = 1$, $\lambda = 0.1$.

convergence (Figure 7.11a), while an inverse trend is observed for velocity in the case of divergence, as depicted in Figure 7.12a. Moreover, the highest velocity is observed in the starting region of the channel, gradually diminishing towards the upper or lower walls. The gradual decrease in velocity near the channel walls is attributed to the wall friction force. Furthermore, in a divergent channel, the velocity decreases as the opening angle increases, owing to the amplified back-flow of nanofluids. Conversely, in the case of convergence, the velocity exhibits an opposite trend due to the narrowing of the channel walls. Figures 7.13a and 7.14a present the shrinking process in both CD channels, the velocity profiles increase

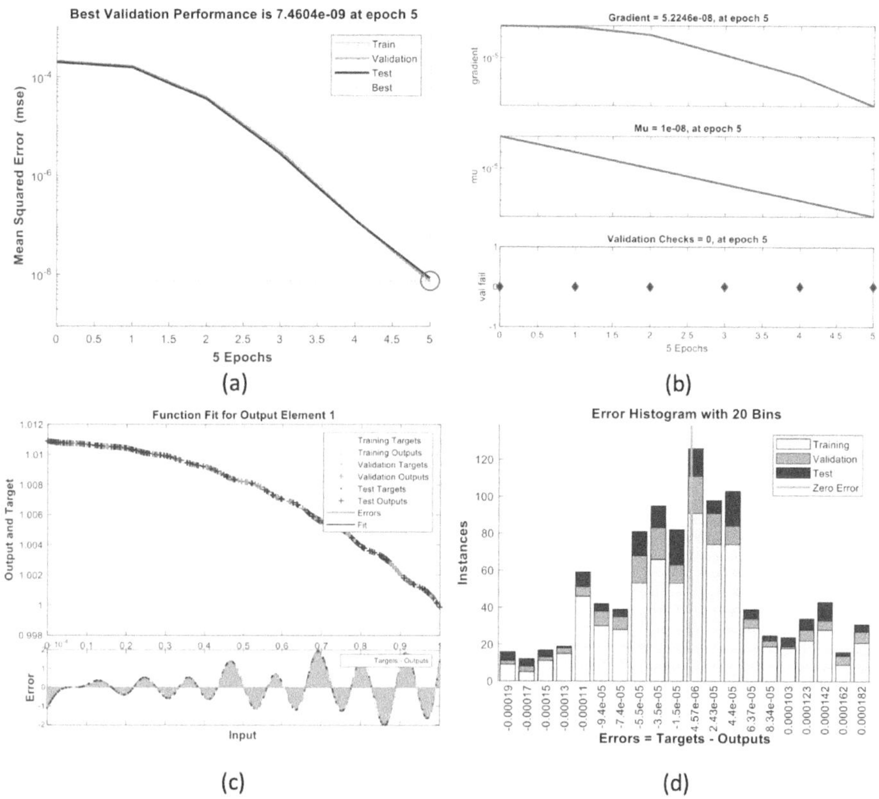

FIGURE 7.9 (a) Mean squared error for concentration from ANN-BLMT for parameters $\boldsymbol{Re = 5, \alpha = -5°, C = 0.1, M = 3, N_r = 0.3, Pr = 0.1, Ec = 0.2, H_s = 0.0, D_u = 0.1, S_r = 0.1, S_c = 1, \lambda = 0.1}$. (b) Transition states for concentration from ANN-BLMT for parameters $\boldsymbol{Re = 5, \alpha = -5°, C = 0.1, M = 3, N_r = 0.3, Pr = 0.1, Ec = 0.2, H_s = 0.0, D_u = 0.1, S_r = 0.1, S_c = 1, \lambda = 0.1}$. (c) Performance of prediction and target with error for concentration from ANN-BLMT for parameters $\boldsymbol{Re = 5, \alpha = -5°, C = 0.1, M = 3, N_r = 0.3, Pr = 0.1, Ec = 0.2, H_s = 0.0, D_u = 0.1, S_r = 0.1, S_c = 1, \lambda = 0.1}$. (d) Error histogram for concentration from ANN-BLMT for parameters $\boldsymbol{Re = 5, \alpha = -5°, C = 0.1, M = 3, N_r = 0.3, Pr = 0.1, Ec = 0.2, H_s = 0.0, D_u = 0.1, S_r = 0.1, S_c = 1, \lambda = 0.1}$

when we increase the angle in convergent channel, but the inverse behavior is observed for divergent channels.

The velocity characteristics for stretching and shrinking cases are examined in Figures 7.13a and 7.14a for the effect of Reynolds number (Re) in CD channels. Remarkably, the velocity patterns appear nearly identical for both scenarios. Physically, elevating the Re induces an increase in drag force, leading to enhanced resistance to fluid movement away from the channel walls during backward motion. Figures 7.15a–7.18a represent the temperature profiles against parameters D_u and H_s

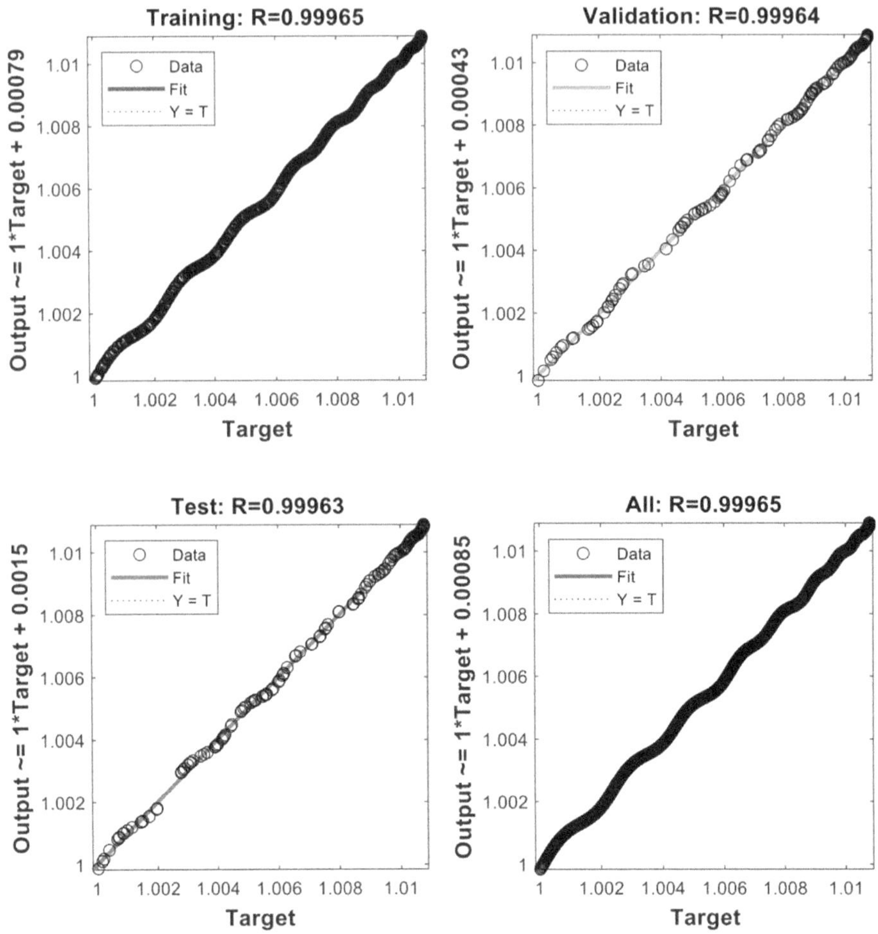

FIGURE 7.10 Regression analysis for concentration from ANN-BLMT for parameters $Re = 5$, $\alpha = -5°$, $C = 0.1$, $M = 3$, $N_r = 0.3$, $Pr = 0.1$, $Ec = 0.2$, $H_s = 0.0$, $D_u = 0.1$, $S_r = 0.1$, $S_c = 1$, $\lambda = 0.1$.

for both CD channels using machine learning algorithms. Figure 7.15a illustrates the effect of D_u in convergent channels, increase the D_u the temperature, decreases in convergent channels but in divergent channels temperature also increase shows in Figure (7.16). The effect of H_s is illustrated in Figures 7.17a and 7.18a for CD channel using machine learning, increase in the H_s decrease temperature in both channels. Figures 7.19a–7.22a represented the concentration effect against parameters S_r and λ,when S_r increase then concentration profiles also increase in CD channels in figures. An inverse behavior is seen in both CD channels when λ is increased as shown in Figures 7.21a and 7.22a.

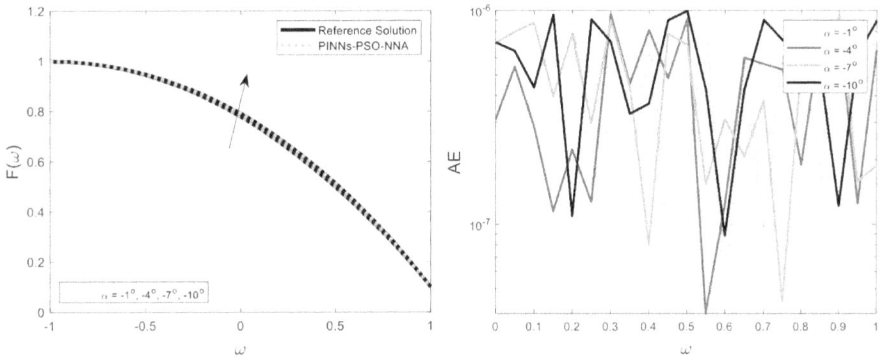

(a) (b)

FIGURE 7.11 (a) Velocity profile from PINNs-PSO-NNA for $\boldsymbol{Re = 5, C = 0.1, M = 1}$. (b) Absolute error between the reference solution and PINNs-PSO-NNA.

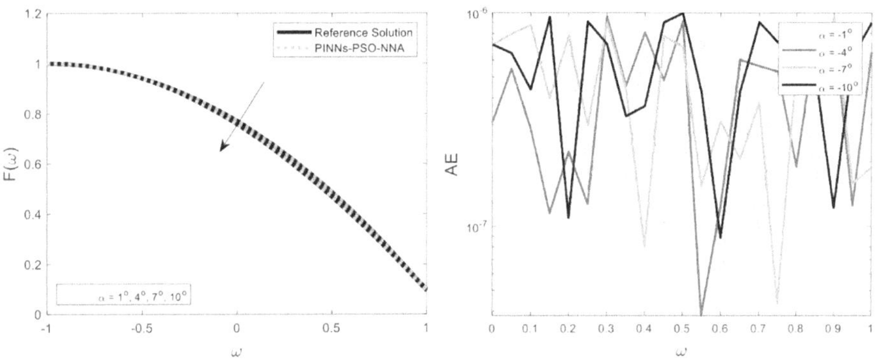

(a) (b)

FIGURE 7.12 (a) Velocity profile from **PINNs-PSO-NNA** for $Re = 5$, $C = 0.1$, $M = 1$. (b) Absolute error between the reference solution and PINNs-PSO-NNA.

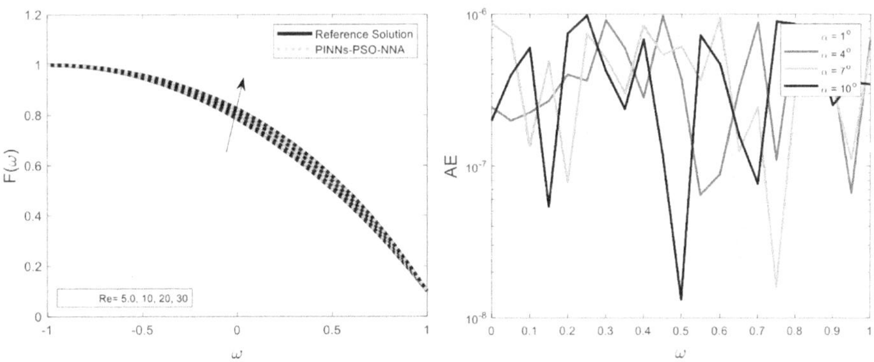

(a) (b)

FIGURE 7.13 (a) Velocity profile from **PINNs-PSO-NNA** for $Re = -5$, $C = 0.1$, $M = 3$. (b) Absolute error between the reference solution and PINNs-PSO-NNA.

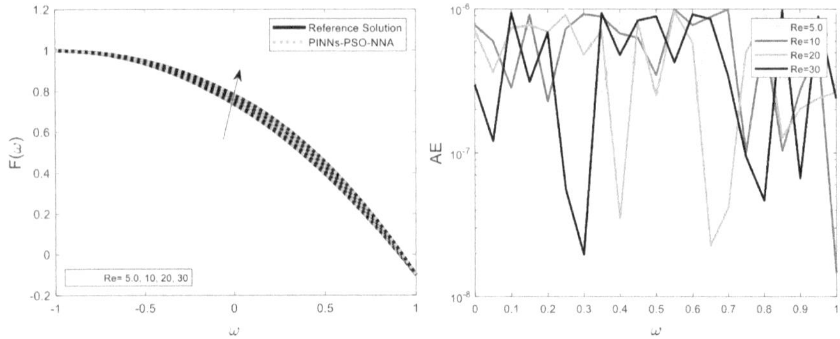

(a) (b)

FIGURE 7.14 (a). Velocity profile from **ANN-PSO-NNA** for $Re = -5$, $C = -0.1$, $M = 3$. (b) Absolute error between the reference solution and PINNs-PSO-NNA.

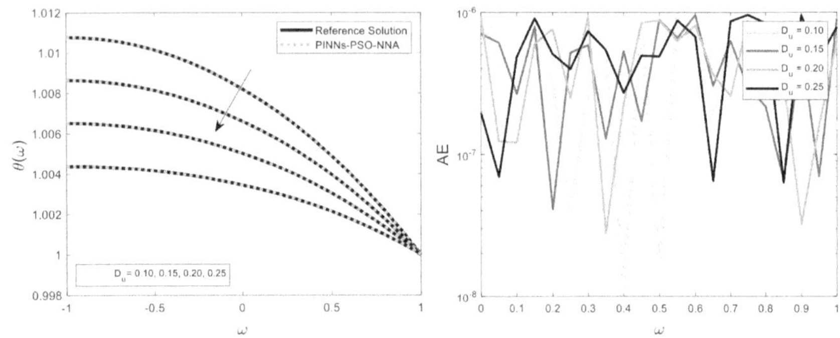

(a) (b)

FIGURE 7.15 (a) Temperature profile from PINNs-PSO-NNA for $\boldsymbol{Re = 5}$, $\boldsymbol{\alpha = -5°}$, $\boldsymbol{C = 0.1}$, $\boldsymbol{M = 3}$, $\boldsymbol{N_r = 0.3}$, $\boldsymbol{Pr = 0.1}$, $\boldsymbol{Ec = 0.2}$, $\boldsymbol{H_s = 0.0}$, $\boldsymbol{S_r = 0.1}$, $\boldsymbol{S_c = 1}$, $\boldsymbol{\lambda = 0.1}$. (b) Absolute error between the reference solution and PINNs-PSO-NNA.

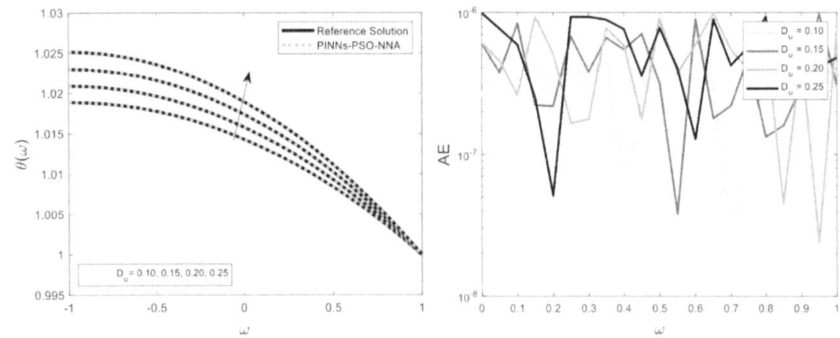

(a) (b)

FIGURE 7.16 (a) Temperature profile from PINNs-PSO-NNA for $\boldsymbol{Re = 5}$, $\boldsymbol{\alpha = 5°}$, $\boldsymbol{C = 0.1}$, $\boldsymbol{M = 3}$, $\boldsymbol{N_r = 0.3}$, $\boldsymbol{Pr = 0.1}$, $\boldsymbol{Ec = 0.2}$, $\boldsymbol{H_s = 0.0}$, $\boldsymbol{S_r = 0.1}$, $\boldsymbol{S_c = 1}$, $\boldsymbol{\lambda = 0.1}$. (b) Absolute error between the reference solution and PINNs-PSO-NNA.

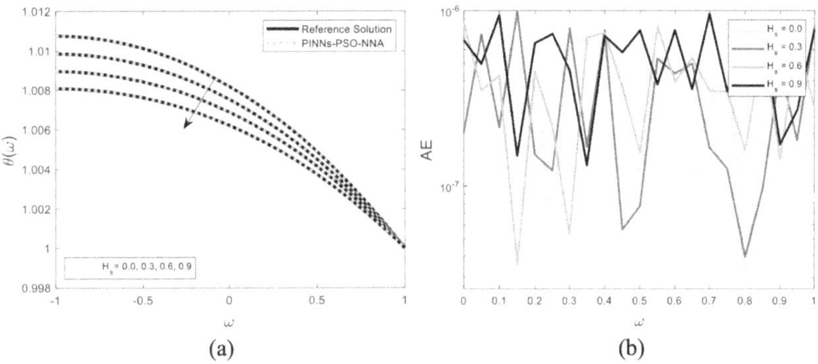

FIGURE 7.17 (a) Temperature profile from PINNs-PSO-NNA for $\boldsymbol{Re = 5}$, $\boldsymbol{\alpha = 5°}$, $\boldsymbol{C = 0.1}$, $\boldsymbol{M = 3}$, $\boldsymbol{N_r = 0.3}$, $\boldsymbol{Pr = 0.1}$, $\boldsymbol{Ec = 0.2}$, $\boldsymbol{D_u = 0.1}$, $\boldsymbol{S_r = 0.1}$, $\boldsymbol{S_c = 1}$, $\boldsymbol{\lambda = 0.1}$. (b) Absolute error between the reference solution and PINNs-PSO-NNA.

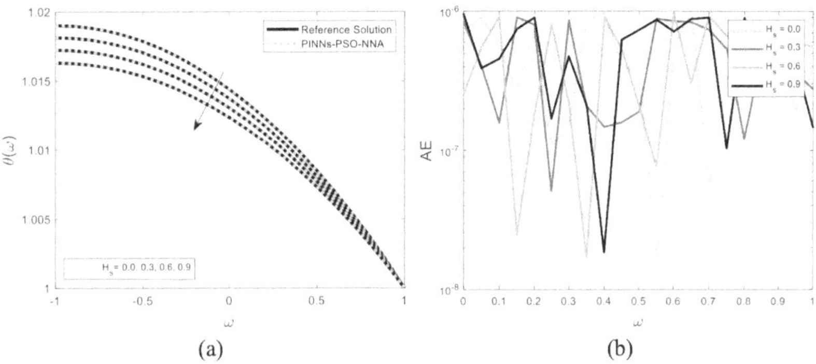

FIGURE 7.18 (a) Temperature profile from PINNs-PSO-NNA for $\boldsymbol{Re = 5}$, $\boldsymbol{\alpha = 5°}$, $\boldsymbol{C = 0.1}$, $\boldsymbol{M = 3}$, $\boldsymbol{N_r = 0.3}$, $\boldsymbol{Pr = 0.1}$, $\boldsymbol{Ec = 0.2}$, $\boldsymbol{D_u = 0.1}$, $\boldsymbol{S_r = 0.1}$, $\boldsymbol{S_c = 1}$, $\boldsymbol{\lambda = 0.1}$. (b) Absolute error between the reference solution and PINNs-PSO-NNA.

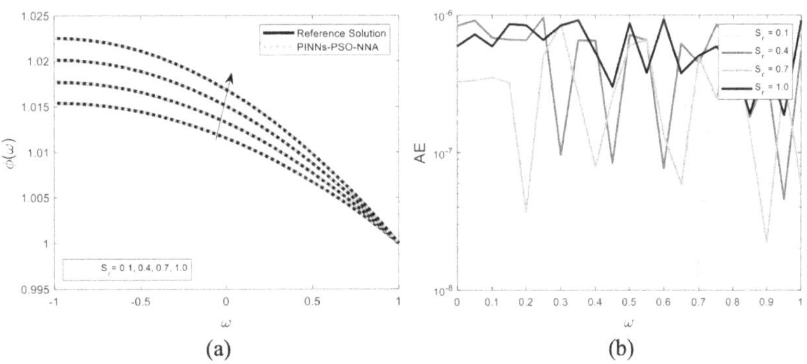

FIGURE 7.19 (a) Concentration profile from **PINNs-PSO-NNA** for $\boldsymbol{Re = 30}$, $\boldsymbol{\alpha = -5°}$, $\boldsymbol{C = 0.1}$, $\boldsymbol{M = 3}$, $\boldsymbol{N_r = 0.3}$, $\boldsymbol{Pr = 0.1}$, $\boldsymbol{Ec = 0.2}$, $\boldsymbol{H_s = 0.0}$, $\boldsymbol{D_u = 0.4}$, $\boldsymbol{S_c = 1}$, $\boldsymbol{\lambda = 0.2}$. (b) Absolute error between the reference solution and PINNs-PSO-NNA.

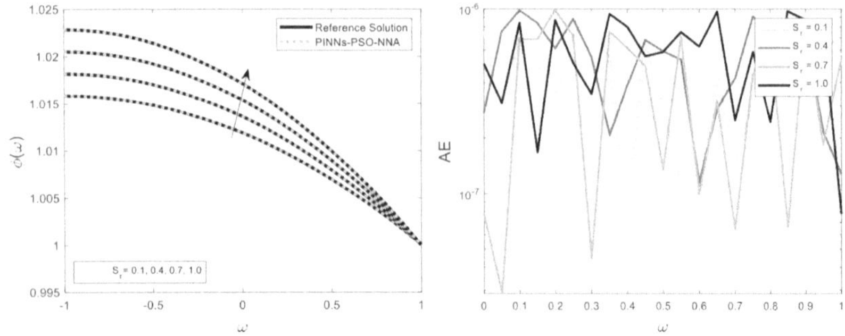

(a) (b)

FIGURE 7.20 (a) Concentration profile from PINNs-PSO-NNA for $\boldsymbol{Re = 0.10}$, $\boldsymbol{\alpha = 5°}$, $\boldsymbol{C = 0.1}$, $\boldsymbol{M = 3}$, $\boldsymbol{N_r = 0.3}$, $\boldsymbol{Pr = 0.1}$, $\boldsymbol{Ec = 0.2}$, $\boldsymbol{H_s = 0.0}$, $\boldsymbol{D_u = 0.4}$, $\boldsymbol{S_c = 0.5}$, $\boldsymbol{\lambda = 0.2}$. (b) Absolute error between the reference solution and PINNs-PSO-NNA.

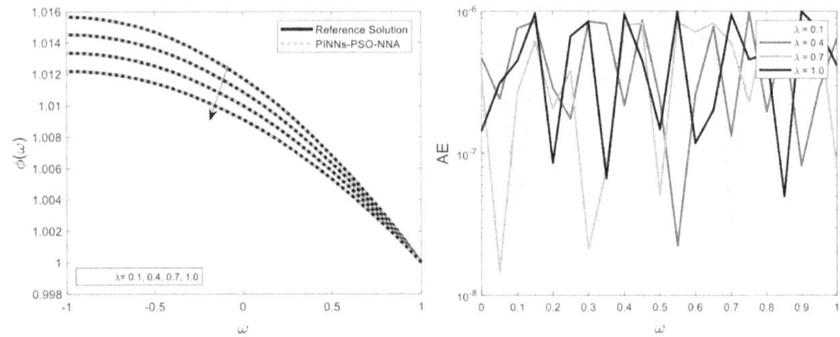

(a) (b)

FIGURE 7.21 (a) Concentration profile from PINNs-PSO-NNA for $\boldsymbol{Re = 5}$, $\boldsymbol{\alpha = 5°}$, $\boldsymbol{C = 0.1}$, $\boldsymbol{M = 3}$, $\boldsymbol{N_r = 0.3}$, $\boldsymbol{Pr = 0.1}$, $\boldsymbol{Ec = 0.2}$, $\boldsymbol{H_s = 0.0}$, $\boldsymbol{D_u = 0.4}$, $\boldsymbol{S_r = 0.1}$ $\boldsymbol{S_c = 1}$. (b) Absolute error between the reference solution and PINNs-PSO-NNA.

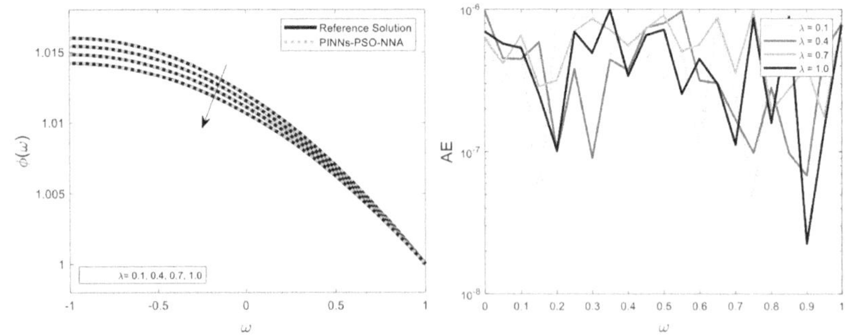

(a) (b)

FIGURE 7.22 (a) Concentration profile from **PINNs-PSO-NNA** for $Re = 0.10$, $\alpha = 5°$, $C = -0.1$, $M = 3$, $N_r = 0.3$, $Pr = 0.1$, $Ec = 0.2$, $H_s = 0.0$, $D_u = 0.4$, $S_r = 0.1$, $S_c = 0.1$. (b) Absolute error between the reference solution and PINNs-PSO-NNA.

7.5 CONCLUSION

In conclusion, this chapter explored the hydro-magnetic, incompressible JH fluid flow within stretchable-walled convergent/divergent (CD) channels, focusing on aspects that have received limited attention in the existing literature – namely, mass and heat transfer analyses using machine learning algorithm. The investigation uniquely contributes by examining the influences of Soret/Dufour, Joule heating, and chemical reaction effects on JH flow through the transformation of the governing coupled system of PDEs into ODEs using appropriate transformations.

- These nonlinear system ODEs were converted into the fitness (objective) function using PINNs.
- The optimization of the ANN-based fitness function used the PSO hybrid with NNA for ANN optimized weights and biases.
- Used the ANN optimized weights and biased in sigmoid function to predict the solution of velocity, temperature and concentration profiles.
- ANN-based results were compared with NDSolve for the validation of the propped algorithm.
- To find the efficiency of the propped algorithm, 200 independent runs were conducted.
- Further, ANN-BLMT approach used for velocity, concentration, and temperature profiles for dataset obtained from NDSolve.
- The comparisons are presented for both machine learning techniques with ANN-BLMT and proposed PINNs-PSO-NNA with well-established NDSolve methods for validation.
- The proposed PINNs-PSO-NNA is more accurate as compared to ANN-BLMT.
- Complex nonlinear DEs-based JH flow does not use traditional analytical and numerical approaches, therefore ANN-BLMT is not utilized. Whereas PINNs-PSO-NNA does not require target dataset therefore it is more useful and efficient algorithm as compared to supervised learning.

The ANN-based computed results underscore the impact of the Lorentz force generated by the magnetic field, revealing its dampening effect on fluid motion in both CD channels. Furthermore, the incorporation of heat source and Dufour effects leads to a sustained enhancement in temperature distribution, while the Soret number exhibits contrasting effects on concentration. The numerical outcomes highlight significant variations in the Nusselt number, elucidating the dependence on magnetic and radiation parameters, as well as the Soret number. Overall, this study enhances our understanding of the intricate dynamics involved in JH flow within CD channels and opens avenues for further exploration in this field. In future we will use quantum-based optimization algorithm to solve JH problems and minimize the computation cost.

7.6 NOMENCLATURE

CD	convergent/divergent channels
P	pressure
T	temperature

ρ	density
B_o	magnetic field
D	mass diffusivity coefficient
Tm	fluid mean temperature
K	chemical reaction coefficient
KT	thermal diffusion ratio
q	heat flux
cp	specific heat
K^*	mean absorption coefficient
C	stretching and shrinking parameter
Pr	Prandtl number
Re	Reynold number
Du	Dufour number
Ec	Eckert number
λ	chemical reaction parameter
Nr	radiation variable
Sr	Soret number
Sc	Schmidt number
ANN-BLMT	backpropagated Levenberg–Marquardt technique
ANN	artificial neural networks
PSO	particle swarm optimization
NNA	neural network algorithm
PSO-NNA	hybrid particle swarm optimization and neural network algorithm
ANN-PSO-NNA	artificial neural network integrated with particle swarm optimization hybrid with neural network algorithm
PINNs	physics-informed neural networks

Conflict of Interest: The authors of this chapter have no conflict of interest.
Data Availability: The author confirms that the data supporting the findings of this study are available within the manuscript.

REFERENCES

1. G. B. Jeffery, "L. The two-dimensional steady motion of a viscous fluid," Lond. Edinb. Dublin Philos. Mag. J. Sci, vol. 29, no. 172, pp. 455–465, 1915.
2. G. Hamel, "Spiralförmige Bewegungen zäher Flüssigkeiten," Jahresbericht Der Deutschen Mathematiker-Vereinigung, vol. 25, pp. 34–60, 1917.
3. J. L. Fitzjohn and W. L. Holstein, "Divergent flow in chemical vapor deposition reactors," J. Electrochem. Soc, vol. 137, no. 2, pp. 699–703, 1990.
4. A. A. Belevtsev, E. K. Isakaev, A. V. Markin, V. A. Khaimin, and V. F. Chinnov, "Analysis of metrological potentialities of a high-current arc in plasma generators with a diverging channel," High Temp, vol. 38, no. 5, pp. 667–674, 2000.
5. J. Deshler, "Experimental investigations of Jeffery-Hamel flows," APS, vol. 56, p. MQ.009, 2003.
6. J. Prins. "On the structure of compressor gas leakage flows. In: International compressor engineering conference". Purdue University, Purdue e-Pubs; 2006.
7. J. Peixinho, "Flow in a Slowly Divergent Pipe Section," in Seventh IUTAM Symposium on Laminar-Turbulent Transition, Dordrecht: Springer Netherlands, 2010, pp. 307–312.

8. M. G. Sobamowo, A. A. Yinusa, A. T. Akinshilo, and S. T. Aladenusi, "Homotopy analysis method to MHD-slip flow of an upper-convected Maxwell viscoelastic nanofluid in a permeable channel embedded in a porous medium," Int. J. Petrochem. Sci. Eng, vol. 5, no. 1, pp. 11–20, 2020.
9. G. Sucharitha, K. Vajravelu, and P. Lakshminarayana, "Magnetohydrodynamic nanofluid flow in a non-uniform aligned channel with joule heating," J. Nanofluids, vol. 8, no. 7, pp. 1373–1384, 2019.
10. G. Sucharitha, K. Vajravelu, and P. Lakshminarayana, "Effect of heat and mass transfer on the peristaltic flow of a Jeffrey nanofluid in a tapered flexible channel in the presence of aligned magnetic field," Eur. Phys. J. Spec. Top, vol. 228, no. 12, pp. 2713–2728, 2019.
11. M. V. Reddy and P. Lakshminarayana, "Mhd radiative flow of Williamson nanofluid with Cattaneo-Christov model over a stretching sheet through a porous medium in the presence of chemical reaction and suction/injection," J. Porous Media, vol. 25, no. 12, pp. 1–15, 2022.
12. I. Ullah, "Activation energy with exothermic/endothermic reaction and Coriolis force effects on magnetized nanomaterials flow through Darcy–Forchheimer porous space with variable features," Waves Random Complex Media, pp. 1–14, 2022.
13. M. Vinodkumar Reddy, G. Sucharitha, K. Vajravelu, and P. Lakshminarayana, "Convective flow of MHD non-Newtonian nanofluids on a chemically reacting porous sheet with Cattaneo-Christov double diffusion," Waves Random Complex Media, pp. 1–20, 2022.
14. Gerdroodbary, M. B., Takami, M. R., & Ganji, D. D., "Investigation of thermal radiation on traditional Jeffery–Hamel flow to stretchable convergent/divergent channels," Case Studies in Thermal Engineering, vol. 6, pp. 28–39, 2015.
15. Yarmand, H., Gharehkhani, S., Shirazi, S. F. S., Amiri, A., Montazer, E., Arzani, H. K., … & Kazi, S. N., "Nanofluid based on activated hybrid of biomass carbon/graphene oxide: Synthesis, thermo-physical and electrical properties," Int. Comm. Heat Mass Transfer, vol. 72, pp. 10–15, 2016.
16. Makinde, O. D., "Entropy-generation analysis for variable-viscosity channel flow with non-uniform wall temperature," Appl. Energy, vol. 85, no. 5, pp. 384–393, 2008.
17. Makinde, O. D., & Bég, O. A, "On inherent irreversibility in a reactive hydromagnetic channel flow," J. Ther. Sci., vol. 19, pp. 72–79, 2010.
18. L. Rosenhead, "The steady two-dimensional radial flow of viscous fluid between two inclined plane walls," Proc. R. Soc. Lond., vol. 175, no. 963, pp. 436–467, 1940.
19. K. Millsaps, K. Pohlhausen, "Thermal distribution in Jeffrey–Hamel flows between nonparallel plane walls," J. Aeronaut. Sci., vol. 20, pp. 187–9, 1953.
20. N. Riley, "Heat transfer in Jeffery-Hamel flow," Quart. J. Mech. Appl. Math, vol. 42, no. 2, pp. 203–211, 1989.
21. E. L. Fraenkel, "Laminar flow in symmetrical channels with slightly curved walls, i. On the Jeffery-Hamel solutions for flow between plane walls," Proc. R. Soc. Lond, vol. 267, no. 1328, pp. 119–138, 1962.
22. R. M. Terrill, "Slow laminar flow in a converging or diverging channel with suction at one wall and blowing at the other wall," Z. Angew. Math. Phys., vol. 16, no. 2, pp. 306–308, 1965.
23. H. Schlichting and K. Gersten, Boundary-Layer Theory, Berlin, Heidelberg: Springer Berlin Heidelberg, 2017.
24. J. S. Roy and P. Nayak, "Steady two dimensional incompressible laminar visco-elastic flow in a converging or diverging channel with suction and injection," Acta Mech, vol. 43, no. 1–2, pp. 129–136, 1982.
25. S. R. Stow, P. W. Duck, and R. E. Hewitt, "Three-dimensional extensions to Jeffery–Hamel flow," Fluid Dyn. Res, vol. 29, no. 1, pp. 25–46, 2001.
26. S. M. Moghimi, G. Domairry, S. Soleimani, E. Ghasemi, and H. Bararnia, "Application of homotopy analysis method to solve MHD Jeffery–Hamel flows in non-parallel walls," Adv. Eng. Softw., vol. 42, no. 3, pp. 108–113, 2011.

27. M. Hatami, M. Sheikholeslami, M. Hosseini, and D. D. Ganji, "Analytical investigation of MHD nanofluid flow in non-parallel walls," J. Mol. Liq., vol. 194, pp. 251–259, 2014.
28. N. Ahmed, A. Abbasi, U. Khan, and S. T. Mohyud-Din, "Thermal radiation effects on flow of Jeffery fluid in converging and diverging stretchable channels," Neural Comput. Appl., vol. 30, no. 8, pp. 2371–2379, 2018.
29. A. Mishra, A. K. Pandey, A. J. Chamkha, and M. Kumar, "Roles of nanoparticles and heat generation/absorption on MHD flow of Ag–H2O nanofluid via porous stretching/shrinking convergent/divergent channel," J. Egyptian Math. Soc., vol. 28, no. 1, 2020.
30. H. Berrehal and G. Sowmya, "Heat transfer analysis of nanofluid flow in a channel with non-parallel walls," J. Mech. Sci. Technol., vol. 35, no. 1, pp. 171–177, 2021.
31. F. Habiyaremye, M. Wainaina, and M. Kimathi, "The effect of heat and mass transfer on unsteady MHD nanofluid flow through convergent-divergent channel," Int. J. Fluid Mech. Therm. Sci., vol. 8, no. 1, p. 10, 2022.
32. S. Ullah, I. Ullah, A. Ali, K. Shah, and T. Abdeljawad, "Investigation of cross-diffusion effect on radiative Jeffery-Hamel flow in convergent/divergent stretchable channel with Lorentz force and Joule heating Alex. Eng. J., vol. 86, pp. 289–297, 2024.
33. I. U. Rahman, M. Sulaiman, F. K. Alarfaj, P. Kumam, and G. Laouini, "Investigation of non-linear MHD Jeffery–Hamel blood flow model using a hybrid metaheuristic approach," IEEE Access, vol. 9, pp. 163214–163232, 2021.
34. M. A. Z. Raja and R. Samar, "Numerical treatment for non-linear MHD Jeffery–Hamel problem using neural networks optimized with interior point algorithm," Neurocomputing, vol. 124, pp. 178–193, 2014.
35. P. Chandra and R. Das, "A hybrid RSA-IPA optimizer for designing an artificial neural network to study the Jeffery-Hamel blood flow with copper nanoparticles: Application to stenotic tapering artery Results Eng., vol. 20, no. 101542, p. 101542, 2023.
36. M. N. Aslam, N. Shaukat, A. Riaz, I. Khan, and S. Niazai, "Machine learning intelligent based hydromagnetic thermal transport under Soret and Dufour effects in convergent/divergent channels: A hybrid evolutionary numerical algorithm," Sci. Rep., vol. 13, no. 1, pp. 1–20, 2023.
37. Abu-Arqub, O., Abo-Hammour, Z., & Momani, S, "Application of continuous genetic algorithm for non-linear system of second-order boundary value problems," Appl. Math. Inf. Sci., vol. 8, no. 1, p. 235, 2014.
38. Ahmad, I., Raja, M. A. Z., Bilal, M., & Ashraf, F, "Bio-inspired computational heuristics to study Lane–Emden systems arising in astrophysics model," SpringerPlus, vol. 5, pp. 1–23, 2016.
39. Raja, M. A. Z., Shah, F. H., Tariq, M., Ahmad, I., & Ahmad, S. U. I, "Design of artficial neural network models optimized with sequential quadratic programming to study the dynamics of non-linear Troesch's problem arising in plasma physics," Neural Comput. Application, vol. 29, pp. 83–109, 2018.
40. Ahmad, I., & Mukhtar, A, "Stochastic approach for the solution of multi-pantograph differential equation arising in cell-growth model," Appl. Math. Comput., vol. 261, pp. 360–372, 2015.
41. Ahmad, I., & Bilal, M, "Numerical solution of Blasius equation through neural networks algorithm," Am. J. Comput. Math., vol. 4, no. 03, pp. 223–232, 2014.
42. Ahmad, I., Ahmad, S. U. I., Bilal, M., & Anwar, N, "Stochastic numerical treatment for solving Falkner–Skan equations using feedforward neural networks," Neural Comput. Application, vol. 28, pp. 1131–1144, 2017.
43. M. N. Aslam et al., "An ANN-PSO approach for mixed convection flow in an inclined tube with ciliary motion of Jeffrey six constant fluid," Case Stud. Therm. Eng., no. 103740, p. 103740, 2023.
44. Xu, B., Jie, D., Li, J., Yang, Y., Wen, F., & Song, H, "Integrated scheduling optimization of u-shaped automated container terminal under loading and unloading mode," Computers & Industrial Engineering, vol. 162, p. 107695, 2021.

45. M. N. Aslam, A. Riaz, N. Shaukat, M. W. Aslam, and G. Alhamzi, "Machine learning analysis of heat transfer and electroosmotic effects on multiphase wavy flow: A numerical approach," Int. J. Numer. Methods Heat Fluid Flow, 2023.
46. Huang, W., Wang, H., & Chen, Q, "Neural network predictions can be misleading evidence from predicting crude oil futures prices," E3S Web of Conferences, vol. 253, p. 02015, 2021). EDP Sciences.
47. Wang, H., Cui, Q., & Du, H. (2022). Modeling and optimization of water distribution in mineral processing considering water cost and recycled water. Computational Intelligence and Neuroscience, *2022*.
48. Zhang, M., Li, H., Pan, S., Lyu, J., Ling, S., & Su, S, "Convolutional neural networks-based lung nodule classification: A surrogate-assisted evolutionary algorithm for hyperparameter optimization," IEEE Trans. Evolution. Comput., vol. 25, no. 5, pp. 869–882, 2021.
49. Narang, N., Sharma, E., & Dhillon, J. S., "Combined heat and power economic dispatch using integrated civilized swarm optimization and Powell's pattern search method," Applied Soft Computing, vol. 52, pp. 190–202, 2017.
50. S. Panda and N. P. Padhy, "Comparison of particle swarm optimization and genetic algorithm for FACTS-based controller design," Appl. Soft Comput. J., vol. 8, no. 4, pp. 1418–1427, 2008, doi: 10.1016/j.asoc.2007.10.009.
51. M. O. Okwu and L. K. Tartibu, "Particle swarm optimisation," Stud. Comput. Intell., vol. 927, pp. 5–13, 2021, doi: 10.1007/978-3-030-61111-8_2.
52. Y. Yu and S. Yin, "A Comparison between Generic Algorithm and Particle Swarm Optimization," ACM Int. Conf. Proceeding Ser., pp. 137–139, 2020, doi: 10.1145/3429889.3430294.
53. A. Stacey, M. Jancic, and I. Grundy, "Particle swarm optimization with mutation," 2003 Congr. Evol. Comput. CEC 2003 - Proc., vol. 2, pp. 1425–1430, 2003, doi: 10.1109/CEC.2003.1299838.
54. R. Eberhart and J. Kennedy, "New optimizer using particle swarm theory," Proc. Int. Symp. Micro Mach. Hum. Sci., pp. 39–43, 1995, doi: 10.1109/mhs.1995.494215.
55. R. Toushmalani, "Gravity inversion of a fault by particle swarm optimization (PSO)," Springerplus, vol. 2, no. 1, pp. 1–7, 2013, doi: 10.1186/2193-1801-2-315.
56. Bassi, Mishra, and Omizegba, "Automatic tuning of proportional-integral-derivative (PID) controller using particle swarm optimization (PSO) algorithm," Int. J. Artif. Intell. Appl., vol. 2, no. 4, pp. 25–34, 2011, doi: 10.5121/ijaia.2011.2403.
57. A. A. A. Esmin, R. A. Coelho, and S. Matwin, "A review on particle swarm optimization algorithm and its variants to clustering high-dimensional data," Artif. Intell. Rev., vol. 44, no. 1, pp. 23–45, 2015, doi: 10.1007/s10462-013-9400-4.
58. S. Rana, S. Jasola, and R. Kumar, "A review on particle swarm optimization algorithms and their applications to data clustering," Artif. Intell. Rev., vol. 35, no. 3, pp. 211–222, 2011, doi: 10.1007/s10462-010-9191-9.
59. A. M. Ibrahim and M. A. Tawhid, "A hybridization of cuckoo search and particle swarm optimization for solving non-linear systems," Evol. Intell., vol. 12, no. 4, pp. 541–561, 2019, doi: 10.1007/s12065-019-00255-0.
60. Q. He, L. Wang, and B. Liu, "Parameter estimation for chaotic systems by particle swarm optimization," Chaos, Solitons and Fractals, vol. 34, no. 2, pp. 654–661, 2007, doi: 10.1016/j.chaos.2006.03.079.
61. A. Sadollah, H. Sayyaadi, and A. Yadav, "A dynamic metaheuristic optimization model inspired by biological nervous systems: Neural network algorithm," Appl. Soft Comput, vol. 71, pp. 747–782, 2018.

8 Machine Learning Process on Double Diffusive Convection in a Parallelogram-Shaped Cavity

Bengisen Pekmen and Hakan F. Oztop

8.1 INTRODUCTION

Double diffusive convection (DDC) occurs in existence of thermal and solutal density differences. DDC has attracted the interest of scientists in numerous environmental and industrial applications, including nuclear waste storage, magma chambers, solar ponds, metal solidification, and star mixing etc.

In literature, there are many studies on DDC. Morega and Nishimura [1] numerically solved the problem of DDC utilizing Chebyshev collocation method in a square cavity. They analyzed the effect of buoyancy ratio on the flow structure for rectangular closed space. Nishimura et al. [2] studied the numerical solution on oscillatory DDC with the secondary cell flow structure occurring for a certain range of buoyancy ratio. They discussed the bifurcation structures of the oscillatory flow in the considered system. Mamou et al. [3] presented an analytical and numerical study on DDC in a rectangular enclosure. Chamkha [4] investigated the unsteady DDC in a porous cavity by finite difference method (FDM). Results reveal that average Nusselt (Nu) and Sherwood (Sh) numbers are reduced by heat generation. Bairi [5] used air-filled parallelogram-shaped enclosures with discrete isothermal heat sources to investigate the effects of natural convection. He made transient analysis and found that the temporal evolution of the average Nu number at each band of the hot wall is determined for all treated configurations. Entropy generation due to DDC in a closed space with partial vertical heating and salting sources is analyzed by Oueslati et al. [6] using the finite volume method (FVM) and a full multigrid technique. Total entropy generation and Bejan number were observed to oscillate with the same frequency. Qin et al. [7] made a numerical investigation of DDC in a rectangular closed space with horizontal temperature and concentration gradients. They used compact difference method with full fourth-order accuracy and high resolution in space. The influence of Prandtl and Lewis numbers on flow structure, the temperature and concentration distribution are studied. Geometry is important to analyze the double-diffusive natural convection and therefore, trapezoidal-shaped enclosure is tested by

DOI: 10.1201/9781032688121-8

Gholizadeh et al. [8] with a partially heated active right sidewall by using the FDM with adaptation of partial differential equation (PDE) technique to generate regular grid distribution. They observed that increment on the Lewis number causes improving the mass transfer performance. DDC in a rectangular cavity is examined by Aly et al. [9] utilizing the incompressible smoothed particle hydrodynamic (ISPH) method considering Soret and Dufour effects. The increase in Soret with a decrease in Dufour number results in an increment in average Nu and a decrement in average Sh number. Effects of magnetic field on DDC in a square closed space are studied numerically by Reddy and Murugesan [10] using Galerkin's weighted residual finite element method (FEM) using velocity-vorticity formulation. They found that changing of Ha number plays important role on both Nu number and Sh number up to 78%. Stajnko et al. [11] applied boundary element method to simulate 3D DDC in a cubic porous encloure. Their results report that the rise in absolute value of buoyancy coefficient enhances the overall heat and mass transfer. Another study on effects of magnetic field on DDC is investigated by Mahapatra et al. [12] in which a trapezoidal cavity filled with nanofluids and different aspect ratios are chosen. They also tested the inclination angle of the cavity and they found that the mass transfer ratio in case of nanofluids is more effective than that of base fluid. ISPH method is embraced by Raizah et al. [13] for solving magnetohydrodynamic DDC in a nanofluid filled square cavity having either an open straight pipe or V-pipe. Their results uncover that more convective flow and intense fluid are obtained in V-shaped pipe. Chakkingal et al. [14] performed OpenFoam to simulate 3D DDC in a fluid filled enclosure with and without obstacles. Their results delineate that heat transfer is suppressed by obstacles, and mass transfer is enhanced in the modification of flow and the distribution of concentration by cylinders. Hussain et al. [15] applied FEM to solve DDC in a staggered cavity filled with Casson fluid considering the effect of an inclined uniform magnetic field. They found that the Hartmann number adversely affects the entropy generation. Eshaghi et al. [16] examined DDC in a H-shaped cavity filled with a hybrid nanofluid in the presence of an inclined baffle and with or without corrugations attached to the upper rib. Baffle angle 60 and -60 degree without corrugations result in the highest average Nu and Sh numbers. Numerical work of DDC in rectangular cavities with different aspect ratios is tested by Yang and Zhao [17]. Yan et al. [18] tested the numerical methods on 2D DDC in rectangular enclosures based on a high resolution upwind compact stream function model. Alomari et al. [19] performed FEM in order to simulate DDC in a curvilinear cavity involving fins and nanoparticles. Their results reveal that the optimal mass transfer occurs if the buoyancy ratio is at –4.

As indicated above that the geometry is the most important parameter on DD natural convection. In this context, one of the oldest study is found in reference [20] in which natural convection flow in a differentially heated parallelogram. Their results demonstrate that the average Nu number is found maximum when the aspect ratio of the parallogram is close to zero. The effects of oblique angle on natural convection is also studied by Baytas and Pop [21]. Analysis of natural convection via entropy generation approach in porous rhombic enclosures for various thermal aspect ratios has been performed by Anandalakshmi and Basak for heatline visualization [22] and entropy generation for the same geometry [23]. A numerical simulation of natural convection in a parallelogram shaped enclosure containing volumetric heat source with non-uniformly heated side walls has been performed by Hussain et al. [24]. Dutta et al. [25, 26] made a study on effects of rhombic cavity which is filled with nanofluids and under magnetic field. They showed that both Hartmann number and volume fraction of nanofluids are effective parameters on natural convection as well as parameters of rhombic geometry. Dutta et al. [27] performed a numerical study to analyze thermofluids and entropy generation for a 2D rhombic closed space having a wavy non-uniformly heated top wall of two different number of undulations. It is shown that the values of average Nu number at the top wall are less for undulated top wall, and higher concentration of nanoparticles always provide higher heat transfer. Thermal entropy generation is the dominating factor for lower Rayleigh number. In Ghani et al. [28], Cu–water nanofluid flow of natural convection is investigated in a parallelogram having an opening on the right wall and a heat source on the left wall by employing FVM.

In the last year, artificial intelligence (AI) studies is rapidly integrated into heat transfer problems. Adaptive network based fuzzy interface system is used by Selimefendigil and Oztop [29] for average Nu number modeling in MHD forced convection flow problem of CuO–water nanofluid over circular cylinders in a channel. Using COMSOL results, an artificial neural network (ANN) modeling for prediction of electric potential, temperature and power generation of thermoelectric generator device is performed in [30]. Alizadeh et al. [31] implemented multilayer perceptron in Python for prediction of effective heat transfer using ANSYS results simulating Cu–water nanofluid under uniform magnetic field. Shah et al. [32] modeled local Nu number and entropy generation by ANN considering radiation coefficient, enclosure angle and Rayleigh number as inputs and utilizing data from LBM numerical results. Colak et al. [33] conducted multilayer perceptron NN modeling based on Levenberg–Marquardt (LM) algorithm in a shell and helically coiled tube heat exchanger problem. Hussain et al. [34] concentrated on tree algorithms to model the whole fluid flow behavior in a porous media involving NEPCM, bacteria, centered L-shaped adiabatic baffle simulating the problem by FEM. Shilpa and Leela [35] modeled heat and mass transfer characteristics by ANN based on LM back-propagation algorithm using FDM numerical results on magnetohydrodynamics mixed convective flow. Elshehabey [36] employed ANN modeling on average Nu and Bejan numbers using FEM numerical data in solution of nanofluid natural convection flow in an inclined cavity involving H-shaped obstacle. Shang et al. [37] investigated the multilayer NN modeling of thermal conductivity in a nanofluid flow problem. Shilpa et al. [38] applied FEM for solving mixed convection flow in a vertical annulus in the presence of induced magnetic field, and they modeled the average Nu number by ANN based on LM using the numerical results. Tikadar and Kumar [39] presented random forest, extreme gradient boosting, support vector regressor and ANN for getting the best surrogate model in multiobjective optimization framework of a pin fin-metal foam problem. Rachedi et al. [40] also used ANN for modeling average Nu number from numerical results obtained from FVM solution of natural convection flow in a cavity with a constant heat flux along the left vertical wall.

In this study, double diffusive natural convection flow in a Fe_3O_4-water filled paralellogram is numerically and statistically investigated. In the numerical part, global radial basis function (RBF) is implemented with cubic polyharmonic spline RBF and Gauss–Chebyshev–Lobatto (GCL) grid points. The outcomes in some parameter variations are presented as well as quantitative results. In statistical part, a dataset is created in different dimensionless parameter variations from the numerical runs. The average Nu and the average Sh numbers are modeled by neural networks. To the best of Authors' knowledge, statistical modeling is firstly considered on a DDC problem in a parallelogram-shaped cavity.

8.2 PROBLEM DEFINITION

The two-dimensional, time-independent double diffusive natural convection flow of Fe_3O_4-water nanofluid is considered in a parallelogram-shaped cavity depicted in Figure 8.1 in which $x = 1$ bottom length is maintained in inclined walls. The nanofluid is Newtonian and the flow is laminar and incompressible. The base fluid and nanoparticles are in thermal equilibrium. The left and the right walls are the adiabatic walls. The bottom wall is the heated wall and the top wall is the cold wall. The parallelogram has an inclination angle θ.

Buoyancy effect is included by Boussinessq approximation due to the temperature and concentration differences. The viscous dissipation and the radiation impact are neglected.

Fe_3O_4 and water's thermophysical characteristics, are tabulated in Table 8.1.

Based on the single-phase model, physical characteristics of Fe_3O_4–water nanofluid may be expressed as:

$$\rho_{nf} = (1-\phi)\rho_f + \phi\rho_s, \tag{8.1}$$

$$(\rho c_p)_{nf} = (1-\phi)(\rho c_p)_f + \phi(\rho c_p)_s, \tag{8.2}$$

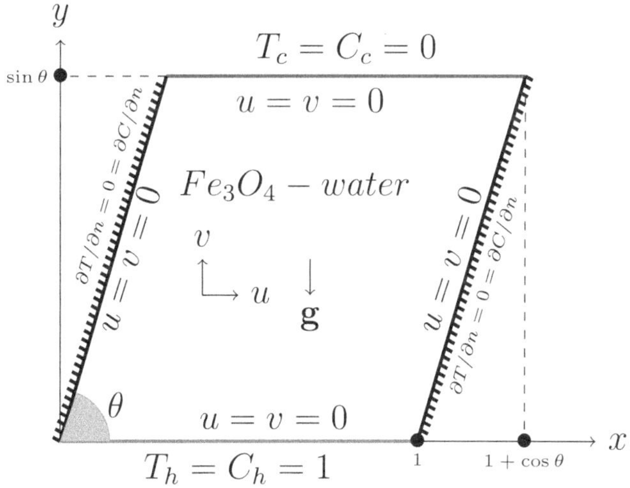

FIGURE 8.1 Problem configuration.

TABLE 8.1
Thermophysical Properties of Water and Fe_3O_4 [41]

Property	Water	Fe_3O_4
$\rho\ (kg/m^3)$	997.1	5200
$c_p[J/(kgK)]$	4179	670
$k[W/(mK)]$	0.613	6
$\beta\times10^{-5}(1/K)$	21	1.3

$$(\rho\beta)_{nf}=(1-\phi)(\rho\beta)_f+\phi(\rho\beta)_s, \tag{8.3}$$

$$\alpha_{nf}=\frac{k_{nf}}{(\rho c_p)_{nf}}, \tag{8.4}$$

$$\mu_{nf}=\mu_f(1-\phi)^{-2.5}, \tag{8.5}$$

$$\frac{k_{nf}}{k_f}=\frac{k_s+2k_f-2\phi(k_f-k_s)}{k_s+2k_f+\phi(k_f-k_s)}, \tag{8.6}$$

where nf,f,s denote the nanofluid, the fluid and the solid, respectively, ρ is the density, c_p is the specific heat at constant pressure, β is the thermal expansion coefficient, α is the thermal diffusivity, μ is the dynamic viscosity (model by Brinkman's model [42] for a nanofluid), k is the thermal conductivity (model by Maxwell's model [43] for a nanofluid), and ϕ is the concentration of nanoparticles.

The dimensional physical equations as a combination of continuity, momentum, energy and concentration equations are

$$\nabla\cdot\mathbf{u}=0, \tag{8.7a}$$

$$v_{nf}\nabla^2u=\frac{1}{\rho_{nf}}\frac{\partial p}{\partial x}+u\frac{\partial u}{\partial x}+v\frac{\partial u}{\partial y}, \tag{8.7b}$$

$$v_{nf}\nabla^2v=\frac{1}{\rho_{nf}}\frac{\partial p}{\partial y}+u\frac{\partial v}{\partial x}+v\frac{\partial v}{\partial y}-\frac{g}{\rho_{nf}}\left(\left(\rho\beta_T\right)_{nf}(T-T_c)+\left(\rho\beta_S\right)_{nf}(C-C_c)\right), \tag{8.7c}$$

$$\frac{\alpha_{nf}}{\alpha_f}\nabla^2T=(\mathbf{u}\cdot\nabla)T, \tag{8.7d}$$

$$D_c\nabla^2C=(\mathbf{u}\cdot\nabla)C, \tag{8.7e}$$

where $\mathbf{u}=\langle u,v\rangle$ is the velocity vector, T is the temperature, p is the pressure, C is the concentration, g is the gravitational acceleration and D_c is the diffusivity coefficient of concentration.

In order to get dimensionless equations, the non-dimensional variables are defined by:

$$(x', y') = \frac{(x, y)}{L}, (u', v') = \frac{(u, v)L}{\alpha}, p' = \frac{pL^2}{\rho_{nf}\alpha_f^2}, T' = \frac{T - T_c}{T_h - T_c}, C' = \frac{C - C_c}{C_h - C_c}. \quad (8.8)$$

These variables are replaced in dimensional equations and then the prime notations are removed. Once the non-dimensional form of $u - v - p - T - C$ equations are obtained, by definition of velocity components in terms of stream function ψ as $u = \frac{\partial \psi}{\partial y}, v = -\frac{\partial \psi}{\partial x}$ and by applying the definition of vorticity $\omega = \nabla \times \mathbf{u}$ to momentum equations, stream function-vorticity form of non-dimensional equations are deduced as follows:

$$\nabla^2 \psi = -\omega, \quad (8.9a)$$

$$\frac{\alpha_{nf}}{\alpha_f} \nabla^2 T = u \frac{\partial T}{\partial x} + v \frac{\partial T}{\partial y}, \quad (8.9b)$$

$$\frac{1}{Le} \nabla^2 C = u \frac{\partial C}{\partial x} + v \frac{\partial C}{\partial y}, \quad (8.9c)$$

$$Pr \frac{\nu_{nf}}{\nu_f} \nabla^2 \omega = u \frac{\partial \omega}{\partial x} + v \frac{\partial \omega}{\partial y} - \frac{(\rho\beta)_{nf}}{\rho_{nf}\beta_f} RaPr \left(\frac{\partial T}{\partial x} - N_r \frac{\partial C}{\partial x} \right), \quad (8.9d)$$

where the parameters Pr, Ra, Le, N_r are defined by:

$$Pr = \frac{\nu_f}{\alpha_f}, \; Ra = \frac{g\beta_f (T_h - T_c)L^3}{\alpha_f \nu_f}, \; Le = \frac{\alpha_f}{D_c}, \; N_r = \frac{\beta_s (C_h - C_c)}{\beta_T (T_h - T_c)}. \quad (8.10)$$

Boundary conditions may be set down as follows:

$$u = v = 0 = \psi, \quad \frac{\partial T}{\partial n} = \frac{\partial C}{\partial n} = 0 \text{ on } 0 < x < \cos(\theta), 0 \le y \le \sin(\theta), \quad (8.11a)$$

$$u = v = 0 = \psi, \quad \frac{\partial T}{\partial n} = \frac{\partial C}{\partial n} = 0 \text{ on } 1 < x < 1 + \cos(\theta), 0 \le y \le \sin(\theta), \quad (8.11b)$$

$$u = v = 0 = \psi, \quad T_h = C_h = 1 \text{ on } y = 0, \quad (8.11c)$$

$$u = v = 0 = \psi, \quad T_c = C_c = 0 \text{ on } y = 1. \quad (8.11d)$$

In order to be able to calculate gradients along the inclined left and right walls, parametrizations along these walls are firstly done as,

$$\mathbf{r}(t) = \langle t\cot(\theta), t \rangle, 0 \le t \le \sin(\theta) \quad (8.12)$$

$$\mathbf{r}(t) = \langle 1 + t\cot(\theta), t \rangle, 0 \le t \le \sin(\theta), \quad (8.13)$$

and then the tangent vectors for both parametrized lines are calculated as

$$\mathbf{r}'(t) = \langle \cot(\theta), 1 \rangle. \tag{8.14}$$

Thus, rotating the tangent vector along the left wall 90° counter clockwise, and the tangent vector along the right wall 90° clockwise, normal vectors along the left and right inclined walls are found as

$$\mathbf{n} = \langle -1, \cot(\theta) \rangle, \quad \mathbf{n} = \langle 1, -\cot(\theta) \rangle. \tag{8.15}$$

8.3 NUMERICAL AND STATISTICAL APPROACHES

8.3.1 Numerical Method

In this part, a short explanation on the numerical approximation process is to be given. The radial basis function (RBF) collocation method on non-uniform GCL grid points is used.

RBFs are formulated based on the radial distance $r = \sqrt{(x - x_j)^2 + (y - y_j)^2}$, with (x_j, y_j) representing the nodes used in the collocation method, and (x, y) denoting the field point under consideration. Within a domain Ω, this leads to the creation of a distance matrix r_{ij} that measures the distances between points x_i and x_j.

An unknown function φ can be represented as follows

$$\varphi_i = \sum_{j=1}^{N_b+N_i} f_{ij}\bar{\alpha}_j, \quad i = 1, \ldots, N_b + N_i; \qquad \varphi = F\bar{\alpha}, \tag{8.16}$$

where N_b is the total number of boundary nodes, N_i is the number of interior nodes, F is the matrix constructed column-wise from r_{ij}^3, and with dimensions $(N_b + N_i)^2$, $\bar{\alpha}$ is a vector of unknown coefficients. The calculation of spatial derivatives using this framework is straightforward and can be expressed as

$$\frac{\partial \varphi}{\partial x} = \frac{\partial F}{\partial x} F^{-1}\varphi = D_x\varphi, \quad \frac{\partial \varphi}{\partial y} = \frac{\partial F}{\partial y} F^{-1}\varphi = D_y\varphi, \tag{8.17}$$

$$\nabla^2\varphi = \frac{\partial^2 \varphi}{\partial x^2} + \frac{\partial^2 \varphi}{\partial y^2} = \left(\frac{\partial^2 F}{\partial x^2} + \frac{\partial^2 F}{\partial y^2} \right) F^{-1}\varphi = D_2\varphi. \tag{8.18}$$

By inserting numerical approximations into Eqn. (8.9), an iterative system is obtained as follows:

$$D_2\psi^{m+1} = -\omega^m \tag{8.19a}$$

$$u^{m+1} = D_y\psi^{m+1}, \quad v^{m+1} = -D_x\psi^{m+1},$$

$$\left(\frac{\alpha_{nf}}{\alpha_f} D_2 - M \right) T^{m+1} = 0, \tag{8.19b}$$

$$\left(\frac{1}{Le}D_2 - M\right)C^{m+1} = 0, \tag{8.19c}$$

$$\left(Pr\frac{\nu_{nf}}{\nu_f}D_2 - M\right)\omega^{m+1} = -\frac{(\rho\beta)_{nf}}{\rho_{nf}\beta_f}RaPrD_x\left(T^{m+1} - N_rC^{m+1}\right), \tag{8.19d}$$

where m is the iteration level, and M is the matrix $diag(u)D_x + diag(v)D_y$.

Once the stream function equation is solved, a convergence accelerator parameter τ $(0 < \tau < 1)$ is used as

$$\omega^{m+1} = \tau\omega^{m+1} + (1-\tau)\omega^m. \tag{8.20}$$

The termination criterion is set as

$$\frac{\left\|\psi^{n+1} - \psi^n\right\|_\infty}{\left\|\psi^{n+1}\right\|_\infty} + \frac{\left\|T^{n+1} - T^n\right\|_\infty}{\left\|T^n\right\|_\infty} + \frac{\left\|C^{n+1} - C^n\right\|_\infty}{\left\|C^n\right\|_\infty} + \frac{\left\|\omega^{n+1} - \omega^n\right\|_\infty}{\left\|\omega^n\right\|_\infty} < 10^{-5} \tag{8.21}$$

The used non-uniform GCL nodes are calculated by

$$\frac{b+a}{2} - \frac{b-a}{2}\cos\left(\frac{i\pi}{N-1}\right), \quad i = 0,\ldots,N-1, x, y \in [a,b]. \tag{8.22}$$

In the current problem, $x \in [0,1], y \in [0, \sin(\theta)]$.

The average Nu number and the average Sh numbers, respectively, along the bottom heated wall are calculated by the formula

$$\overline{Nu} = -\frac{k_{nf}}{k_f}\int_0^1 \frac{\partial T}{\partial y}dx, \quad \overline{Sh} = -\int_0^1 \frac{\partial C}{\partial y}dx. \tag{8.23}$$

In numerical approximation of $\overline{Nu}$ and $\overline{Sh}$ along the bottom heated wall, Simpson's rule on non-uniform grids [44] is used.

8.3.2 Machine Learning Modeling Techniques

In this section, two machine learning (ML) processes will be briefly explained. One of them is very popular one known as (artificial) neural networks, and the other one is the multivariate adaptive regression splines (Mars).

The metric to measure the goodness of fit of any model is determined by the mean squared error (MSE) defined by

$$MSE = \frac{\sum_{i=1}^{N_d}(y_i - \hat{y}_i)^2}{N_d}, \tag{8.24}$$

where N_d is the data size, y_i is the actual data and $\hat{y}_i$ is the predicted data.

8.3.2.1 Neural Networks

The information in a multilayer feedforward NN flows from left to right between the first (input) layer and the last (output) layer passing through the hidden layers.

An example demonstration in case of three layer neural network is displayed in Figure 8.2. In input layer, x_i denote the inputs in our problem, namely $x_1 = Ra, x_2 = Le, x_3 = N_r, x_4 = \phi, x_5 = \theta$. Hidden layers are presented by layer size 5 as an example on this configuration. In output layer, two outputs are illustrated in which $z_1^o = \overline{Nu}$ and $z_2^o = \overline{Sh}$.

The mathematical expressions between layers to get one output z_1^o using the inputs x_1, x_2, x_3, x_4, x_5 may be written as follows:

$$y_k^{(1)} = f\left(\sum_{i=1}^{5} W_{ki}^{(1)} x_i + b_k\right), \tag{8.25a}$$

$$y_k^{(2)} = f\left(\sum_{j=1}^{L_s} W_{kj}^{(2)} y_k^{(1)} + b_k\right), \tag{8.25b}$$

$$y_k^{(3)} = f\left(\sum_{j=1}^{L_s} W_{kj}^{(3)} y_k^{(2)} + b_k\right), \tag{8.25c}$$

$$z_1^{(o)} = f\left(\sum_{k=1}^{L_s} W_{1k}^{(4)} y_k^{(3)} + b\right), \tag{8.25d}$$

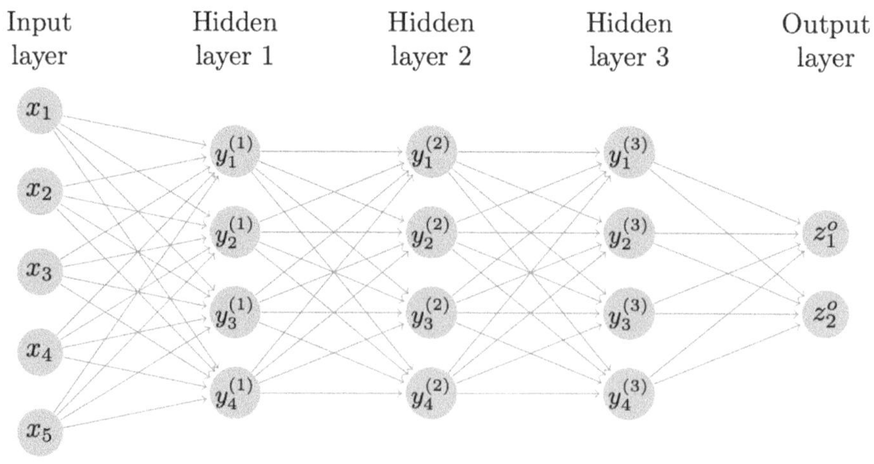

FIGURE 8.2 Trilayer neural network (TNN) example.

where $k = 1,\ldots, L_s$, L_s is the layer size. W is the weight matrix, b is bias vector and f is the activation function which is chosen as rectified linear unit $f(x) = \max(x, 0)$ in the current study. In the last equation, the bias b is a number (1×1). The matrix W and the bias vector are initialized randomly and updated based on an optimization algorithm minimizing the residual.

The dataset is obtained by performing numerical method in distinct combinations of x_i's, and the computed results for z_1^o and z_2^o build the dataset with inputs. NN model is created by training a train data, and tested on a test data. Therefore, the first big data at hand is separated into train and test data.

8.3.2.2 Multivariate Adaptive Regression Splines

Mars calculates a function f which is defined as:

$$f = \sum_{i=1}^{M_b} \tilde{C}_i B_i, \tag{8.26}$$

in which the basis functions are denoted by B_i, the number of basis functions by M_b, and the unknown coefficients by $\tilde{C}_i$.

The piecewise continuous functions that define the basis functions B_i are as follows

$$h[-(x - \tilde{z}_i)] = \begin{cases} \tilde{z}_i - x & if\ \tilde{z}_i > x \\ 0 & \text{otherwise} \end{cases},$$

$$h[+(x - \tilde{z}_i)] = \begin{cases} x - \tilde{z}_i & if\ x > \tilde{z}_i \\ 0 & \text{otherwise} \end{cases}, \tag{8.27}$$

$\tilde{z}_i$'s are called as knots from the dataset.

There are two stages to Mars' processing. The first stage is called the forward stage, during which an algorithm based on fast search creates an overfitted model and adds basis functions to it. The lowest generalized cross validation is obtained at this stage by determining a collection of basis functions. Until the predetermined maximum number of basis functions is achieved, activities are halted. The overfitted model developed in the forward stage is trimmed in the second stage, known as the backward stage. An ideal model is created at the conclusion of these phases.

This model function is configured as Friedman explains [45]

$$I + \sum_{i=1}^{N_c} \left(\tilde{C}_i \prod_{j=1}^{N_{in}} sign(h((Input(j) - \tilde{K}) \times \tilde{D}), \tilde{D}) \right), \tag{8.28}$$

where ± 1 indicates the right or left direction of the corresponding basis function, N_{in} denotes the number of inputs, N_c denotes the number of coefficients, K denotes the cuts (knots) which are the associated values in the jth input, and $I = \tilde{C}_0$, the intercept, is the first entry of $\tilde{C}$ with $\tilde{D} = 0$ and $\tilde{K} = 0$.

The functions h and $sign$ are identified by

$$h(x) = \max(x, 0), \qquad sign(x, \tilde{D}) = \begin{cases} 1 \ if \ \tilde{D} = 0 \\ x \ if \ \tilde{D} \neq 0 \end{cases}. \tag{8.29}$$

8.4 NUMERICAL AND STATISTICAL CALCULATIONS

In this section, numerical and statistical observations are dictated. In-house codes and all computations are carried in MacOS computers using MATLAB R2024a.

8.4.1 Validation and Grid Independency

The current numerical approach, global RBF method, is validated on a benchmark study presented by Quertatani et al. [46] in which FVM combined with multigrid technique is applied. Furthermore, 256^2 number of fine grids are used in that reference. In Table 8.2, $\overline{Nu}$ values along the heated wall are compared. As is noted, global RBF with GCL grids and $f = r^3$ capture the good agreement with Ref.'s results using smaller number of grid points.

In Table 8.3, the current numerical approach is also affirmed by a DDC flow problem presented by Qin et al. [7] in which a fourth-order accurate FDM is used. In that study, rectangular cavity has height 2. The present results using GCL nodes and $f = r^3$ are in good agreement with the results in Ref. [7].

In the present investigation, an example GCL node distribution is displayed in Figure 8.3 fixing the angle θ at 45°. Table 8.4 appraises the grid independence. 41×41 number of GCL grid points are performed at all calculations.

TABLE 8.2
$\overline{Nu}$ Comparison when $Ha = 0$, $\gamma = 0$

Ra	Present (N, τ)	Ref. [46]
10^3	0.9999 (31, 0.01)	1.0004
10^4	2.1546 (37, 0.0075)	2.1581
10^5	3.9020 (37, 0.0075)	3.9103

TABLE 8.3
$\overline{Nu}$ and $\overline{Sh}$ Comparisons Keeping $Pe = 0.71$, $N_r = 0.8$, $Ra = 10^5$

	Present		Ref. [7]	
Le	$\overline{Nu}$	$\overline{Sh}$	$\overline{Nu}$	$\overline{Sh}$
0.5	2.1638	1.5358	2.204	1.564
1	2.8277	2.8277	2.834	2.834
2	3.3145	4.2821	3.310	4.244
4	3.5140	5.8022	3.511	5.696

TABLE 8.4
Grid Independence Analysis with $Ra = 10^5$, $Le = 1$, $N_r = 0.5$, $\theta = 45°$, $\phi = 0.04$

N	$\|\psi\|_{max}$	$\overline{Nu}$	$\overline{Sh}$
29	10.59	2.7650	2.6305
33	10.59	2.8253	2.6872
37	10.59	2.8705	2.7296
41	10.59	2.9054	2.7623
45	10.59	2.9331	2.7883

8.4.2 Some Numerical Results

In numerical simulations, Prandtl number is fixed at 6.8. Nanoparticle concentration is also hold at $\phi = 0.04$ unless otherwise declared. The other parameters range in $10^2 \leq Ra \leq 10^5$, $-5 \leq N_r \leq 5$, $1 \leq Le \leq 10$, $30° \leq \theta \leq 90°$. The visualized contours are displayed by streamlines ψ, isotherms T, isoconcentration lines C and vorticity contours ω.

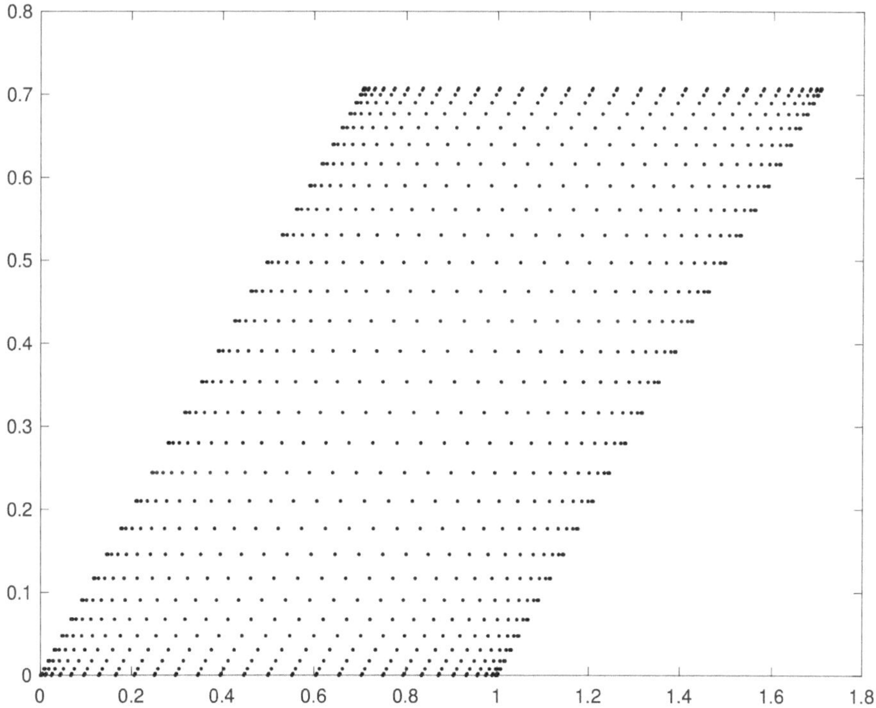

FIGURE 8.3 A demonstration of GCL grid distribution at $\theta = 45°$.

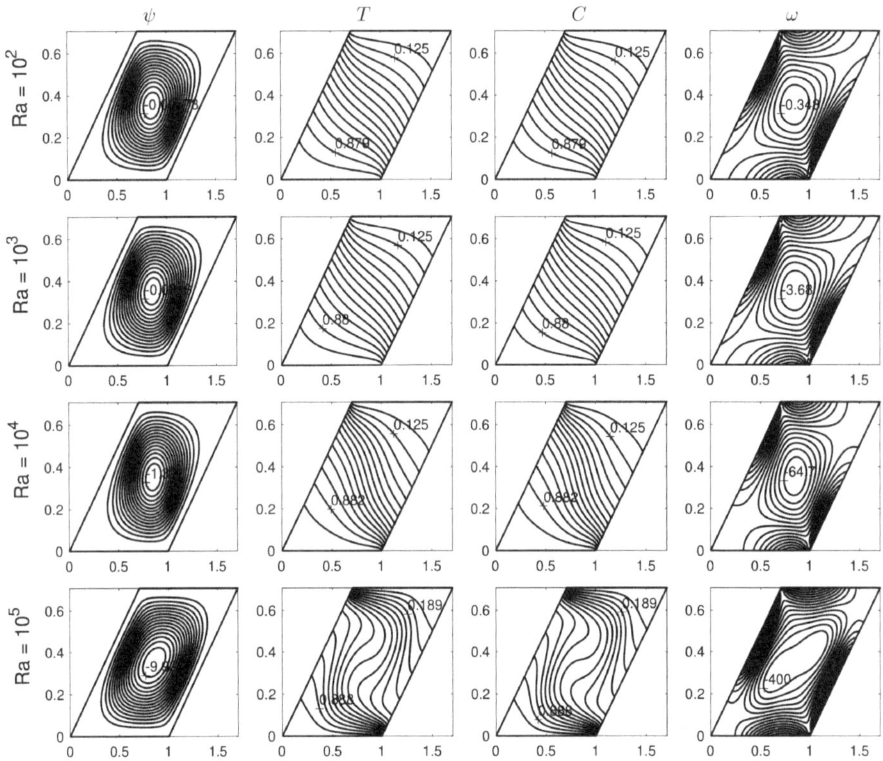

FIGURE 8.4 Rayleigh number variation when $N_r = 0.5$, $Le = 1$, $\theta = 45°$.

In Figure 8.4, the variation in *Ra* number is examined keeping $N_r = 0.5$, $Le = 1$, $\theta = 45°$. In streamlines, fluid flows faster as *Ra* rises due to the rise in buoyancy force. Thermal and solutal boundary layers along the heated bottom wall are pronounced at $Ra = 10^5$ due to the rise in buoyancy force. Table 8.5 verifies these outcomes. As *Ra* increases, absolute stream function rises, $\overline{Nu}$ and $\overline{Sh}$ grow at 186.7% and 197.7%, respectively.

In Figure 8.5, the alteration in buoyancy ratio parameter is inspected holding $Ra = 10^5$, $Le = 1$, $\theta = 45°$. The central values at primary cells in streamlines point to

TABLE 8.5
$\overline{Nu}$ and $\overline{Sh}$ Behavior in Case of *Ra* Variation

Ra	$\lvert \psi \rvert_{max}$	$\overline{Nu}$	$\overline{Sh}$
10^2	0.0083	1.0135	0.9279
10^3	0.0870	1.0123	0.9267
10^4	1.4231	1.1251	1.0519
10^5	10.5870	2.9054	2.7623

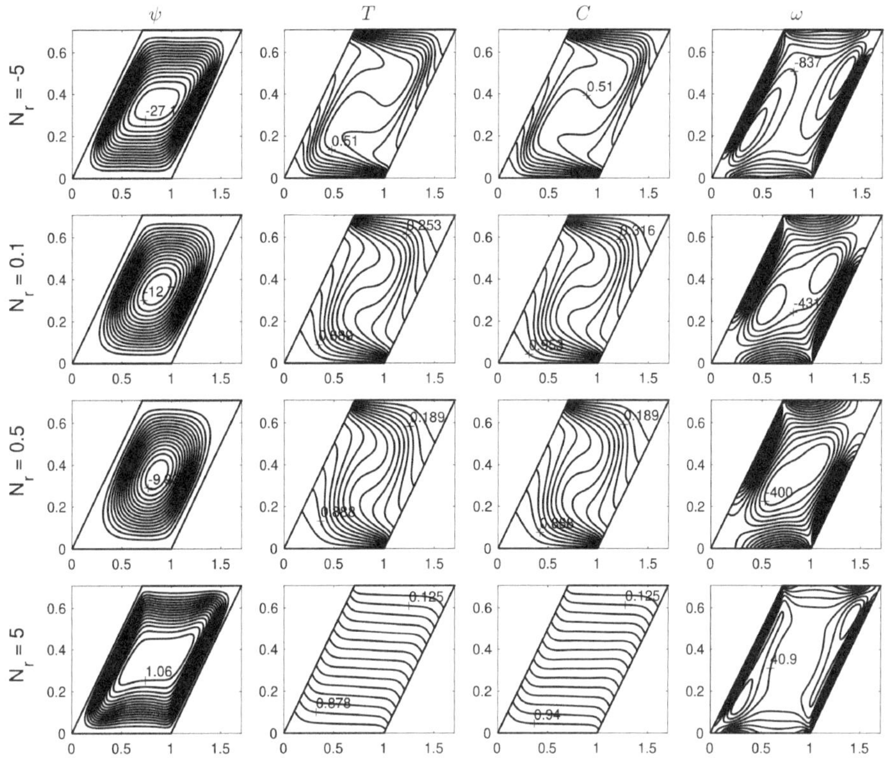

FIGURE 8.5 N_r variation when $Ra = 10^5$, $Le = 1$, $\theta = 45°$.

the decrease in fluid flow with the increase in N_r from −5 to 5. Convective behavior in isotherms and isoconcentration lines at $N_r = -5$ turns into conductive behavior at $N_r = 5$. Table 8.6 demonstrates the crucial drop in fluid velocity at a larger N_r value. $\overline{Nu}$ and $\overline{Sh}$ decreases 76.8% and 76.8%, respectively, as N_r increments.

In Figure 8.6, the change in Lewis number is tested carrying $Ra = 10^5$, $N_r = 0.5$, $\theta = 45°$. As expected Le variation affects significantly isoconcentration lines. Strong circulation in isoconcentration lines occurs. Table 8.7 upholds the significant rise in convective mass transfer if Le is increased

TABLE 8.6
$\overline{Nu}$ and $\overline{Sh}$ Behavior in Case of N_r Variation

N_r	$\lvert\psi\rvert_{max}$	$\overline{Nu}$	$\overline{Sh}$
−5	28.9345	5.4359	5.1549
0.1	13.5480	3.4272	3.2541
0.5	10.5870	2.9054	2.7623
5	1.1267	1.2630	1.1953

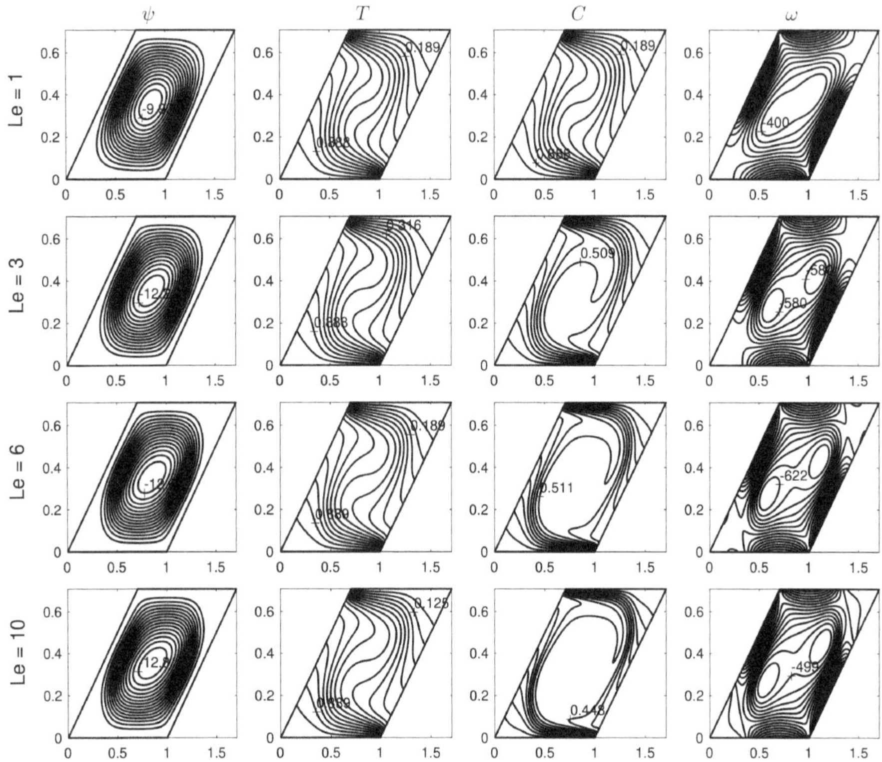

FIGURE 8.6 *Le* variation when $Ra = 10^5$, $N_r = 0.5$, $\theta = 45°$.

from 1 to 10 which increases 150.8%. The convective heat transfer rises 12.9% with the augmentation in *Le*.

In Figure 8.7, the distinct angles are controlled fixing $Ra = 10^5$, $N_r = 0.5$, $Le = 1$. Due to increasing of volume of the closed space, domination of convection is also increased and wavy shaped isotherms are observed. Concentration and isotherms are very similar to each other. Cat eye shaped vorticity is found for the highest value of inclination angle of the closed space. Flow strength is increased with increasing of inclination angle, too. Obviously, fluid flows very fast at $\theta = 90°$. The thermal

TABLE 8.7
$\overline{Nu}$ and $\overline{Sh}$ Behavior in Case of *Le* Variation

Le	$\vert\psi\vert_{max}$	$\overline{Nu}$	$\overline{Sh}$
1	10.5870	2.9054	2.7623
3	13.0276	3.1263	4.3872
6	13.4198	3.2234	5.7215
10	13.6154	3.2802	6.9290

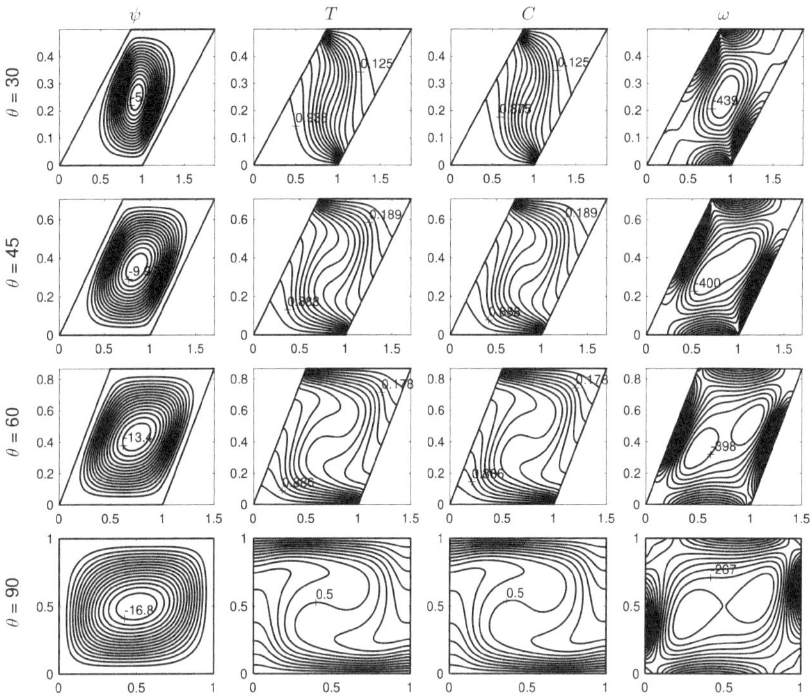

FIGURE 8.7 θ variation when $Ra = 10^5$, $N_r = 0.5$, $Le = 1$.

boundary layer as well as solutal boundary layer along the bottom wall is pronounced as θ varies from $\theta = 30°$ to $\theta = 90°$. Table 8.8 conducts the effect of the rise in θ as quantitative results. $\overline{Nu}$ and $\overline{Sh}$ surge 63.2% and 60.8%, respectively.

8.5 ML MODELING ON $\overline{Nu}$ AND $\overline{Sh}$

The data is collected from numerical calculations. The distinct parameter combinations are set firstly considering Ra(100, 500, 1000, 2500, 5000, 7500, 10^4, 25000, 50000, 75000, 10^5), Le(1, 5, 10), N_r(−10, −5, −1, −0.5, −0.1, 0.1, 0.5, 1, 5, 10), ϕ(0, 0.02, 0.04) and θ(30°, 45°, 60°, 90°) as inputs. With these inputs, the

TABLE 8.8
$\overline{Nu}$ and $\overline{Sh}$ Behavior in Case of θ Variation

θ	$\lvert \psi \rvert_{max}$	$\overline{Nu}$	$\overline{Sh}$
30°	5.9926	2.1041	2.0195
45°	10.5870	2.9054	2.7623
60°	14.3151	3.1644	3.0004
90°	17.9162	3.4337	3.2481

numerical computations are utilized for getting $\overline{Nu}$ and $\overline{Sh}$ numbers along the heated bottom wall. The *convergent* numerical outcomes construct the dataset. A mixed data of size 3353 is created. The random division of this dataset into train (80%) and test (20%) sets is done by syntax

```
cvpartition(length(dataset), 'Holdout',0.2),
```

which gives the train data of size 2683×7, and the test data of size 670×7.

8.5.1 NN Modeling

Since the division is random, considering 10 different random divisions, model resulting with smallest MSE results for getting output z_1^o is saved, and the corresponding divided data is used for training the second model for z_2^o. The syntax for neural network modeling is as follows:

```
models = fitrnet(trainn(:,1:5), trainn(:,colno), 'LayerSizes',Ls, ...
         'Activations', 'relu', 'Lambda', 0, ...
         'IterationLimit', 1000, 'Standardize', true),
```

where the true standardization means the normalization is carried out in input data, and *colno* stands for the column number of output to be modeled.

Table 8.9 conducts MSE results on train and test data in different layer sizes. The layer size involving two elements and three elements is referred to bilayer and TNN, respectively. As is noted that from one layer to three layers, MSE results on train data get smaller and smaller. MSE on test data is also found smaller in TNN than the other ones. The good fit in case of TNN with $L_s = 100,100,100$ is also depicted in Figure 8.8. The black dots lying around perfect prediction and small residual values confirm the good fit.

TABLE 8.9
Train and Test MSE Results

		MSE	
Layer Size	**Output**	**Train**	**Test**
50	$\overline{Nu}$	0.0122	0.0146
	$\overline{Sh}$	0.0652	0.0844
50,50	$\overline{Nu}$	0.0024	0.0025
	$\overline{Sh}$	0.0049	0.0112
50,50,50	$\overline{Nu}$	0.0015	0.0022
	$\overline{Sh}$	0.0048	0.0084
100,100,100	$\overline{Nu}$	0.0010	0.0021
	$\overline{Sh}$	0.0039	0.0097

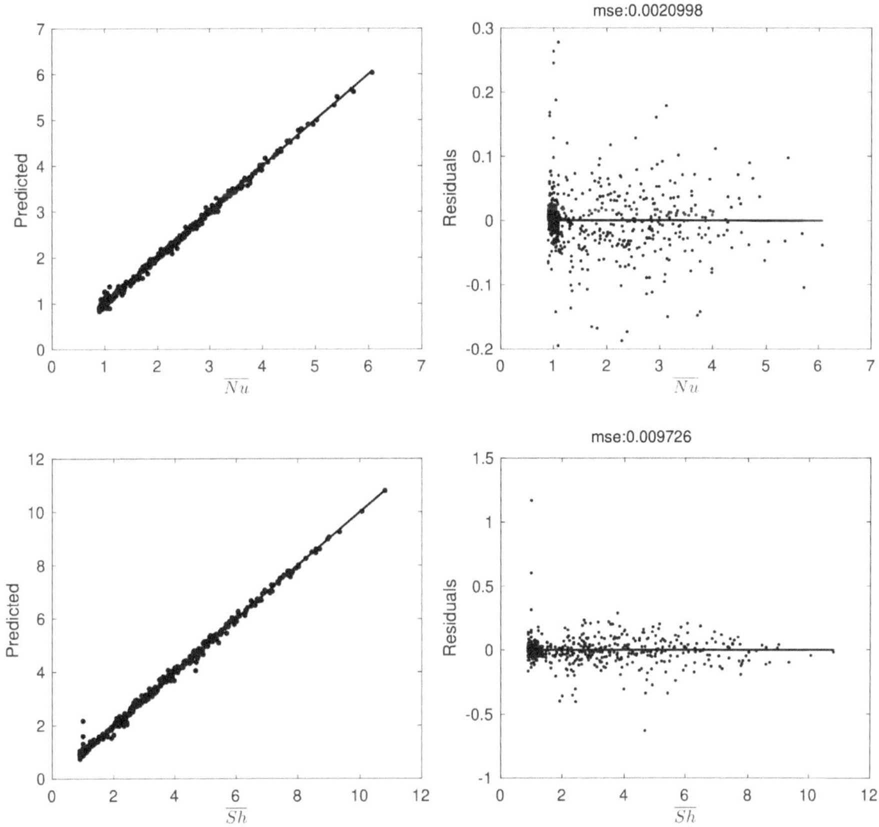

FIGURE 8.8 TNN modeling on test data with $L_s = 100,100,100$.

8.5.2 Mars Modeling

In this modeling, the dataset used in TNN modeling with layer size 100 is employed. The model is figured out in R-Project using the module earth, and the syntax is:

```
model=earth(trainData[,1:5],trainData[,colout],nk=1000,degree=5,
nprune=1000,thresh=1e-11,fast.k=100,penalty=-1,trace=2).
```

In this syntax, *colout* corresponds to the output column, *nk* is the optimal number of model terms created by forward pass, *nprune* is the optimal number of terms in the pruned model, *degree* is the maximum number of interacting input variables, *thresh* is the threshold used in forward stepping. *fast.k* is the optimal number of parent terms used in the forward step. *penalty=-1* calculates the generalized cross validation as MSE. *trace=2* enables us to follow the iterations. It is observed that smaller *nk*, *nprune*, *fast.k* gives fast results while the larger ones result in slower iterations. However, particularly, the larger *nk*, *nprune* make the MSE results better.

TABLE 8.10
MSE Results on Train and Test Data in Mars Modeling

Output	Train	Test
$\overline{Nu}$	0.00062	0.0044
$\overline{Sh}$	0.00559	0.0168

In Table 8.10, MSE metric results on train and test data are addressed. As noted the trained model with Mars is as good as TNN, but is not good at on test data as TNN. This means that more data is necessary for getting better predictions on unseen data. However, the model function is always the same whenever the model is rerun on the same data.

8.6 CONCLUSION

In this chapter, double diffusive natural convection flow of Fe_3O_4–water nanofluid is numerically and statistically investigated inside a parallelogram-shaped enclosure. The differentially heated and concentrated cavity has insulated adiabatic inclined walls, the hot bottom wall, and the cold top wall. The numerical process is achieved by global radial basis function method based on cubic spline polyharmonic splines carrying GCL grids. A dataset from the numerical calculations in distinct problem parameter variations is also created for analyzing the machine learning modeling on average Nu and Sh numbers. $\overline{Nu}$ and $\overline{Sh}$ are increasing functions of Ra, Le, θ, but are decreasing with the rise in N_r. For predicting $\overline{Nu}$, the trained model results in smaller MSE on test data in both TNN and Mars modeling. TNN gives better predictions than Mars on test data in view of smaller MSE metric results obtained for both $\overline{Nu}$ and $\overline{Sh}$.

ACKNOWLEDGMENTS

The authors declare that they have no known competing financial interests or personal relationships that could have appeared to influence the work reported in this chapter.

REFERENCES

1. A. M. Morega and T. Nishimura. Double diffusive convection by a Chebyshev collocation method. *Technology Report Yamaguchi University*, 5(5):259–276, 1996.
2. T. Nishimura, M. Wakamatsu, and A. Morega. Oscillatory double-diffusive convection in a rectangular enclosure with combined horizontal temperature and concentration gradients. *International Journal of Heat and Mass Transfer*, 41(11):1601–1611, 1998.
3. M. Mamou, P. Vasseur, and E. Bilgen. Analytical and numerical study of double diffusive convection in a vertical enclosure. *Heat and Mass Transfer*, 32(1–2):115–125, 1996.

4. A. J. Chamkha. Double-diffusive convection in a porous enclosure with cooperating temperature and concentration gradients and heat generation or absorption effects. *Numerical Heat Transfer: Part A: Applications*, 41(1):65–87, 2002.
5. A. Bairi. Transient free convection in passive buildings using 2D air-filled parallelogram-shaped enclosures with discrete isothermal heat sources. *Energy and Buildings*, 43(2–3):366–373, 2011.
6. F. Oueslati, B. Ben-Beya, and T. Lili. Double-diffusive natural convection and entropy generation in an enclosure of aspect ratio 4 with partial vertical heating and salting sources. *Alexandria Engineering Journal*, 52(4):605–625, 2013.
7. Q. Qin, Z. A. Xia, and Z. F. Tian. High accuracy numerical investigation of double-diffusive convection in a rectangular enclosure with horizontal temperature and concentration gradients. *International Journal of Heat and Mass Transfer*, 71:405–423, 2014.
8. M. M. Gholizadeh, R. Nikbakhti, J. Khodakhah, and A. Ghasemi. Numerical study of double diffusive buoyancy forces induced natural convection in a trapezoidal enclosure partially heated from the right sidewall. *Alexandria Engineering Journal*, 55(2): 779–795, 2016.
9. A. M. Aly and Z. A. S. Raizah. Double-diffusive natural convection in an enclosure filled with nanofluid using ISPH method. *Alexandria Engineering Journal*, 55(4): 3037–3052, 2016.
10. N. Reddy and K. Murugesan. Magnetic field influence on double-diffusive natural convection in a square cavity–A numerical study. *Numerical Heat Transfer, Part A: Applications*, 71(4):448–475, 2017.
11. J. K. Stajnko, J. Ravnik, and R. Jecl. Numerical simulation of three-dimensional double-diffusive natural convection in porous media by boundary element method. *Engineering Analysis With Boundary Elements*, 76:69–79, 2017.
12. T. R. Mahapatra, B. C. Saha, and D. Pal. Magnetohydrodynamic double-diffusive natural convection for nanofluid within a trapezoidal enclosure. *Computational and Applied Mathematics*, 37:6132–6151, 2018.
13. Z. A. S. Raizah and A. M. Aly. ISPH method for MHD double-diffusive natural convection of a nanofluid within cavity containing open pipes. *International Journal of Numerical Methods for Heat & Fluid Flow*, 30(7):3607–3634, 2020.
14. M. Chakkingal, R. Voigt, C. R. Kleijn, and S. Kenjereš. Effect of double-diffusive convection with cross gradients on heat and mass transfer in a cubical enclosure with adiabatic cylindrical obstacles. *International Journal of Heat and Fluid Flow*, 83:108574, 2020.
15. S. Hussain, S. Shoeibi, and T. Armaghani. Impact of magnetic field and entropy generation of Casson fluid on double diffusive natural convection in staggered cavity. *International Communications in Heat and Mass Transfer*, 127:105520, 2021.
16. S. Eshaghi, F. Izadpanah, A. S. Dogonchi, A. J. Chamkha, M. B. B. Hamida, and H. Alhumade. The optimum double diffusive natural convection heat transfer in h-shaped cavity with a baffle inside and a corrugated wall. *Case Studies in Thermal Engineering*, 28:101541, 2021.
17. J.-Q. Yang and B.-X. Zhao. Numerical investigation of double-diffusive convection in rectangular cavities with different aspect ratio i: High-accuracy numerical method. *Computers & Mathematics with Applications*, 94:155–169, 2021.
18. Y. Yan, S. Wu, K. Tian, and Z. Tian. Numerical simulation for 2D double-diffusive convection (DDC) in rectangular enclosures based on a high resolution upwind compact streamfunction model i: Numerical method and code validation. *Applied Mathematics and Mechanics*, 43(9):1431–1448, 2022.

19. M. A. Alomari, K. Al-Farhany, Q. H. Al-Salami, F. Q. A. Alyousuf, I. R. Ali, and N. Biswas. Numerical analysis of double-diffusive free convection in a curvilinear cavity filled with nanofluid and triple fins attached to the hot walls. *The European Physical Journal Plus*, 139(2):149, 2024.
20. J. M. Hyun and B. S. Choi. Transient natural convection in a parallelogram-shaped enclosure. *International Journal of Heat and Fluid Flow*, 11(2):129–134, 1990.
21. A. C. Baytas and I. Pop. Free convection in oblique enclosures filled with a porous medium. *International Journal of Heat and Mass Transfer*, 42(6):1047–1057, 1999.
22. R. Anandalakshmi and T. Basak. Heat flow visualization for natural convection in rhombic enclosures due to isothermal and non-isothermal heating at the bottom wall. *International Journal of Heat and Mass Transfer*, 55(4):1325–1342, 2012.
23. R. Anandalakshmi and T. Basak. Analysis of natural convection via entropy generation approach in porous rhombic enclosures for various thermal aspect ratios. *International Journal of Heat and Mass Transfer*, 64:224–244, 2013.
24. S. H. Hussain, A. K. Hussein, and K. A. Kadim. Numerical simulation of natural convection in a parallelogrammic enclosure containing volumetric heat source with non-uniformly heated left sidewall. *Heat Transfer–Asian Research*, 43(6):542–560, 2014.
25. S. Dutta, N. Goswami, A.K. Biswas, and S. Pati. Numerical investigation of magnetohydrodynamic natural convection heat transfer and entropy generation in a rhombic enclosure filled with cu-water nanofluid. *International Journal of Heat and Mass Transfer*, 136:777–798, 2019.
26. S. Dutta and T. Elnaqeeb. Numerical simulation of non-uniform heating due to magnetohydrodynamic natural convection in a nanofluid filled rhombic enclosure. *The European Physical Journal Special Topics*, 231(13):2879–2900, 2022.
27. S. Dutta, A. K. Biswas, and S. Pati. Analysis of natural convection in a rhombic enclosure with undulations of the top wall–a numerical study. *International Journal of Ambient Energy*, 43(1):87–97, 2022.
28. I. A. Ghani, F. A. Khammas, A. W. Mustafa, and A. K. Hussein. Mhd natural convection in a fully opened parallelogrammic enclosure filled with copper–water nanofluid and partially heated from its left sidewall. *Journal of Advanced Research in Fluid Mechanics and Thermal Sciences*, 74(2):120–145, 2020.
29. F. Selimefendigil and H. F. Oztop. Magnetic field effects on the forced convection of CuO-water nanofluid flow in a channel with circular cylinders and thermal predictions using ANFIS. *International Journal of Mechanical Sciences*, 146:9–24, 2018.
30. F. Selimefendigil and H. F. Oztop. Thermoelectric generation in bifurcating channels and efficient modeling by using hybrid CFD and artificial neural networks. *Renewable Energy*, 172:582–598, 2021.
31. R. Alizadeh, M. Mesgarpour, A. Ameri, J. M. N. Abad, and S. Wongwises. Artificial intelligence prediction of natural convection of heat in an oscillating cavity filled by CuO nanofluid. *Journal of the Taiwan Institute of Chemical Engineers*, 124:75–90, 2021.
32. M. A. Shah, K. Pan, M. Ibrahim, and T. Saeed. Use of neural network and machine learning in optimizing heat transfer and entropy generated in a cavity filled with nanofluid under the influence of magnetic field: A numerical study. *Engineering Analysis with Boundary Elements*, 139:113–131, 2022.
33. A. B. Colak, D. Akgul, H. Mercan, A. S. Dalkılıç, and S. Wongwises. Estimation of heat transfer parameters of shell and helically coiled tube heat exchangers by machine learning. *Case Studies in Thermal Engineering*, 42:102713, 2023.
34. S. Hussain, F. Ertam, M. B. B. Hamida, H. F. Oztop, and N. H. Abu-Hamdeh. Passive control of, energy storage of NePCM, heat and mass transfer with gamma-shaped baffle in a thermo-bioconvection system using CFD and artificial intelligence. *International Communications in Heat and Mass Transfer*, 144:106764, 2023.

35. B. Shilpa and V. Leela. An artificial intelligence model for heat and mass transfer in an inclined cylindrical annulus with heat generation/absorption and chemical reaction. *International Communications in Heat and Mass Transfer*, 147:106956, 2023.
36. H. M. Elshehabey. Predicting heat transfer rate and system entropy based on combining artificial neural network with numerical simulation. *International Journal of Numerical Methods for Heat & Fluid Flow*, 34(6):2480–2512, 2024.
37. Y. Shang, K. A. Hammoodi, A. ad Alizadeh, K. Sharma, H. Rajab, M. Ahmed, M. Kassim, H. Maleki, and S. Salahshour. Artificial neural network hyperparameters optimization for predicting the thermal conductivity of MXene/graphene nanofluids. *Journal of the Taiwan Institute of Chemical Engineers*, 164:105673, 2024.
38. B. Shilpa, V. Leela, and M. D. Alsulami. Artificial neural network approach for mhd mixed convection and entropy generation in a vertical annulus with time periodic thermal boundary conditions in the presence of radial and induced magnetic field. *Numerical Heat Transfer, Part B: Fundamentals, Pages* 1–27, 2023.
39. A. Tikadar and S. Kumar. Machine learning approach to predict heat transfer and fluid flow characteristics of integrated pin fin-metal foam heat sink. *Numerical Heat Transfer, Part B: Fundamentals*, pages 1–26, 2023.
40. K. Rachedi, H. Ragueb, D. Behnous, A. Tahiri, B. Manser, and M. A. Ait Chikh. Analysis of natural convection heat transfer in a rectangular cavity with discrete heat flux: Implications for building thermal management using artificial neural networks. *Numerical Heat Transfer, Part B: Fundamentals*, pages 1–23, 2024.
41. M Sheikholeslami, T Hayat, and AJIJOH Alsaedi. Numerical study for external magnetic source influence on water based nanofluid convective heat transfer. *International Journal of Heat and Mass Transfer*, 106:745–755, 2017.
42. H. C. Brinkman. The viscosity of concentrated suspensions and solutions. *Journal of Chemical Physics*, 3:571–581, 1952.
43. J. C. Maxwell-Garnett. Colors in metal glasses and in metallic films. *Phil. Trans. R. Soc. A*, 203:385–420, 1904.
44. R. Khan, M. R. Hossain, and S. Parvin. Numerical integration schemes for unequal data spacing. *American Journal of Applied Mathematics*, 5(2):48–56, 2017.
45. J.H. Friedman. Multivariate adaptive regression splines. *The Annals of Statistics*, 19:11–67, (1991).
46. Nasreddine Ouertatani, Nader Ben Cheikh, Brahim Ben Beya, and Taieb Lili. Numerical simulation of two-dimensional Rayleigh–Bénard convection in an enclosure. *Comptes Rendus Mécanique*, 336(5):464–470, 2008.

Index

A

Activation function, 2, 6, 8, 146, 147, 167, 169, 176, 193, 231
Adam optimizer, 4, 12, 18, 169, 173
Adaptive moment estimation. *See* Adam optimizer
Adaptive network, 72, 78, 101, 224
Adaptive network-based fuzzy inference system (ANFIS), 72, 77, 101, 224
Air preheater (APH), 157–163, 167, 172, 174, 179, 180
Allen–Cahn equation, 12, 17
ANSYS, 126, 224
Ant colony optimization (ACO), 101, 195
Artificial intelligence (AI), 106, 108, 155, 224
Artificial neural networks (ANNs), 38, 42, 60, 101, 108, 137, 146, 155, 188, 218, 224
  back-propagation, 53, 109, 110, 114, 118, 127
  biological neural networks, 146
  convolutional neural networks (CNN), 71, 101, 155
  grouped method of data handling (GMDH), 38, 45
  Levenberg–Marquardt technique (ANN-BLMT), 192, 206–208, 217, 218
  multilayer perceptron neural networks (MLPNNs), 72, 73, 77, 78, 101, 112, 114, 118
  perceptron feed-forward ANN. *See* multilayer perceptron neural networks (MLPNNs)
  physics-informed neural networks (PINNS), 1–12, 17, 71, 72, 101, 156, 157, 159, 160, 164–181, 193, 198, 217, 218

B

Bayesian optimization, 29
Bias operator, 197
Boiling heat transfer, 68, 69, 72, 74, 75
Boundary
  boundary conditions (BCs), 55, 79, 141, 156, 162, 164–166, 168–171, 191, 194, 227
  boundary layer, 145, 234, 237
  turbulence boundary layer, 75, 101
Boussinesq approximation, 225
  natural convection, 222–225
Brinkman model, 226
Brownian motion, 111, 121–123
Broyden–Fletcher–Goldfarb–Shanno (L-BFGS), 4
Buongiorno model, 120, 121
Buoyancy, 123, 151, 222, 234
Burgers' equation, 12

C

CFD. *See* Computational fluid dynamics
Colburn *j*-factor, 42, 52, 59
Computational fluid dynamics (CFD), 22, 60, 101, 155
  for heat exchangers, 52–53, 180
  for metal foams, 78
COMSOL, 56, 224
Conjugate Heat Transfer
  for metal foams, 78–81
Convection
  convection term, 12
  double diffusive convection (DDC), 222–225
  natural convection. *See* Buoyancy
Convergence, 36, 57, 207
  accelerator parameter, 229
Crossover, 35–37

D

Darcy, 64
  for metal foams, 77
Darcy–Forchheimer, 120, 122, 123
Decision Tree Regression (DTR), 73, 75, 101
Deep Galerkin method (DGM), 9, 16
Deep learning (DL), 155; *See also* Physics-informed deep learning
Deep reinforcement learning (DRL), 71, 101
Differentiable program architecture (DPA), 178
Differential transform method (DTM), 188
Diffusivity
  mass diffusivity coefficient, 191
  thermal diffusivity, 226
Double diffusive convection (DDC). *See* Convection
Dufour number, 192, 218

E

Eckert number, 192, 218
Elitism, 36
Evolutionary algorithm (meta-heuristic), 195
Extreme learning machines (ELM), 72, 101

F

Feed-forward NN (FNN), 4
Fin heat sinks, 37–39
Finite elements method (FEM), 3, 4, 50
Finite volume method (FVM), 63, 9, 80, 222
Form coefficient, 78, 79, 85, 100
Full factorial design (FFD), 42, 172
Fuzzy logic-based algorithms, 72, 101

G

Galerkin, 137
Galerkin method (DGM), 9, 120, 137
Gauss–Chebyshev–Lobatto (GCL), 225
Gaussian process regression (GPR), 74, 101
Gene-based algorithms, 75
Genetic algorithms (GAs), 21, 29, 32, 60, 73, 75, 101, 189
  initial population, 34
Genetic programming (GP), 75, 101, 113, 176
Geothermal heat exchanger, 126
Gradient boosting (GB), 73, 101, 224
Graphene
  in fluids, 112–117, 122, 124
Gravitational acceleration, 59, 226
Grouped method of data handling (GMDH). *See* Artificial neural networks (ANNs)

H

Halton, 6
Hammersley sequence, 6
Hartmann number, 144, 223, 224
Heat exchanger (HX), 23, 24, 38, 51–53, 60
  air preheater (APH), 157–163, 167, 172, 174, 179, 180
  axial finned-tube, 74
  in studies, 67
  nanofluid applications, 124–126
  optimization, 32, 33
  plate, 81
Heat recovery and ventilation (HRV), 49, 60
Heat sink, 30, 37–45, 60
  in metal foams, 77, 78
  in studies, 67–72, 77
Heat transfer coefficient, 53, 59, 82, 85, 90, 123–125
  boiling, 69, 72
  in studies, 72–75
Hidden layers, 5, 8, 72, 92, 146, 169, 173, 177, 230
Homotopy analysis method (HAM), 188
Hydro-magnetic thermal transport, 192
Hypernetworks, 156, 160, 166, 171, 175, 180, 181

I

Interface Conditions, 162, 164, 168

J

Jacobian elliptic functions, 188
Jeffery-Hamel (JH), 187–189, 207
  governing equations, 194
Jet impingement, 42, 69, 74
Joule heating, 121, 187, 189, 190, 192, 217

K

Kriging model (KRI), 38, 60

L

Lasso regularization regression (LRR), 77
Latent heat, 48
Latent heat thermal energy storage (LHTES), 45
Latin hypercube sampling (LHS), 6, 34
Layer-wise locally adaptive activation functions (L-LAAF), 8, 16
Least square method (LSM), 188
Levenberg–Marquardt. *See* Artificial neural networks (ANNs)
Lewis number, 222, 227, 230
Lid-driven cavity, 13, 15
LM algorithm (LMA), 137, 224
Lorentz force, 144, 188, 217

M

Machine learning (ML), 2, 64, 101, 137, 151, 155, 187, 188, 192, 193
  modeling techniques, 229
Magnetic field (MF), 135, 189, 191, 218, 223, 224
Magnetic number (M), 198
Magnetohydrodynamic (MHD), 120, 121, 127, 135, 187, 223, 224
Mass transfer, 137, 138
Mathematica, 189
MATLAB, 50, 56, 115, 206, 232
Maxwell's model, 226
Mean squared error (MSE), 4, 70, 72, 101, 116, 130, 151, 165, 168, 169, 181, 207, 238
Metal foam (MF), 39, 42, 43, 56, 63, 64, 77–81, 85, 90, 101
Micro-channels, 21, 22
Microfluidic, 22
Monte Carlo sampling methods, 34
Multi-objective genetic algorithm (MOGA), 40, 50, 52, 53
Multi-objective optimization (MOO), 30, 31, 47

Multiple linear regression (MLR), 69, 74, 101
Multivariate adaptive regression splines (Mars), 229, 231
  modeling, 239, 240
Mutation, 36, 37

N

Nanofluid (NF), 108, 187, 223, 224
  flow characteristics, 119–123
  heat exchanger systems, 123–126
  magnetohydrodynamic (MHD), 135
  relative thermal conductivity, 111–113
  renewable energy systems, 127–128
  specific heat, 118, 119
  thermal conductivity, 109–113, 226
  thermophysical properties, 109
  viscosity, 114–117, 226
Nanoparticles, 108, 110–119, 226
Nanotubes, 117, 125, 131
Natural convection. *See* Buoyancy
Navier–Stokes equations, 15
NDSolve, 188, 189, 198, 207, 208, 217
Nelder–Mead algorithm, 29
Neural network (NN). *See* Artificial neural networks (ANNs)
Neural networks algorithm (NNA), 189, 196, 218
Neural operators, 157
NeuralODE, 2
NSGA-II algorithm, 42, 45, 53, 60
Number of transfer units (NTU), 124, 160–162
Nusselt number, 41, 42, 59, 100
  flow over a circular cylinder, 83

O

Objective function, 23, 29, 37, 56
OpenFoam, 80, 223
Optimization
  case studies, 38
  challenges, 27, 28
  derivative-free optimization (DFO), 29
  gradient-free, 29
  heat exchangers, 51–54
  metal foam, 42
  multi-objective, 21–23, 30
  particle swarm (PSO), 29, 195, 217, 218
  porous channels, 54, 57
  process, 23–25
  sensible heat, 47
Ordinary differential equations (ODE), 2

P

Pareto front, 30, 31, 41, 42, 57
Partial differential equations (PDEs), 1–5, 69, 121, 156, 160, 162, 164, 177, 192
Particle swarm optimization (PSO). *See* Optimization
Peclet number (Pe), 162
Periodic boundary conditions, 12, 79
Permeability, 63, 64, 79, 80, 85, 100
Phase change material (PCM), 42–49, 60
Phase Coupled SIMPLE, 125
Physics-encoded NN (PENN), 2
Physics-encoded recurrent convolutional NN (PERCNN), 2
Physics-guided NN (PGNN), 2
Physics-informed deep learning (PIDL), 156, 157, 164, 180
Physics-informed neural networks (PINNs). *See* Artificial neural networks (ANNs)
Physics-informed neural operators (PINOs), 157
Physics-informed symbolic networks (PISNs), 157, 176
Porosity, 55, 56, 59, 79, 101
Porous
  cavity, 222
  channels, 38, 54, 55
  media, 55, 68, 70, 72, 135
  metal foams, 63
  wavy porous fins, 67, 68, 70, 72
Prandtl number (Pr), 79, 100, 192, 218, 222
Pressure drop
  experimental validation, 81, 82
  in metal foam, 85
  reducing, 31, 39, 41, 42, 56
  in studies, 67–75
Pruned differentiable programs, 177
PSO-NNA, 195, 217, 218
Python, 224

R

Random forest (RF), 73, 101
Random search, 29
Rayleigh number (Ra), 224, 227, 234
Rectified linear unit (ReLU), 8, 231
Recurrent neural networks (RNNs), 159
Ree–Eyring fluid, 122
Response surface models (RSM), 38, 47, 48, 60, 74
Reynolds number, 42, 56, 101, 192, 198, 211, 218
Ridge regularization regression (RRR), 77, 78
Root mean square error (RMSE), 78, 101
Roseland approximation, 140
Runge-Kutta, 13, 120–123

S

Sampling, 6, 7
Scaled exponential linear units (SELU), 8
Schmidt number, 145, 192, 218
Sensible heat, 47

Shape factor, 111, 139, 142
Sherwood number, 120–123, 142, 222, 229
Simplex, 29
Simulated annealing, 25, 29, 195
Sisko fluid model, 122
Skin friction, 67, 70, 120–123, 142
Sobol sequence, 6, 12
Solar radiation heat flux, 128
Solutal, 142, 222, 234, 237
  boundary layer, 234, 237
Soret and Dufour effects, 120, 123, 135, 136, 189, 190, 223
Soret number, 192, 218
Support vector machine (SVM), 38, 47, 60, 69, 74, 75, 77, 78, 101, 111
Swarm optimization. *See* Optimization
Symbolic regression (SR), 157, 160, 175–178

T

Thermal radiation, 120, 127
Thermophoresis, 121–123, 135
Thermophoretic particle deposition (TPD), 135
TOPSIS, 45
Tournament selection, 35
Transfer learning, 10–12, 70, 101, 156, 160, 166, 170, 171
Transfer operator, 197
Tree-based algorithms, 73, 74, 75, 176
Trilayer neural network (TNN), 230, 238, 240
Turbulence boundary layer (TBL), 69, 75, 101

V

VIKOR, 45
Volume fraction
  in nanofluids, 111, 116, 119, 123, 127, 128

W

Wall Heat Flux, 90
Weighted-residual method (WRM), 188

X

Xavier normal weight initialization, 12

Z

Zukauskus correlation, 83